Operational Modes of Bioreactors

BIOTOL

BOOKS IN THE BIOTOL SERIES

The Molecular Fabric of Cells
Infrastructure and Activities of Cells

Techniques used in Bioproduct Analysis
Analysis of Amino Acids, Proteins and Nucleic Acids
Analysis of Carbohydrates and Lipids

Principles of Cell Energetics
Energy Sources for Cells
Biosynthesis and the Integration of Cell Metabolism

Genome Management in Prokaryotes
Genome Management in Eukaryotes

Crop Physiology
Crop Productivity

Functional Physiology
Cellular Interactions and Immunobiology
Defence Mechanisms

Bioprocess Technology: Modelling and Transport Phenomena
Operational Modes of Bioreactors

In vitro Cultivation of Micro-organisms
In vitro Cultivation of Plant Cells
In vitro Cultivation of Animal Cells

Bioreactor Design and Product Yield
Product Recovery in Bioprocess Technology

Techniques for Engineering Genes
Strategies for Engineering Organisms

Principles of Enzymology for Technological Applications
Technological Applications of Biocatalysts
Technological Applications of Immunochemicals

Biotechnological Innovations in Health Care

Biotechnological Innovations in Crop Improvement
Biotechnological Innovations in Animal Productivity

Biotechnological Innovations in Energy and Environmental Management

Biotechnological Innovations in Chemical Synthesis

Biotechnological Innovations in Food Processing

Biotechnology Source Book: Safety, Good Practice and Regulatory Affairs

BIOTOL BIOTECHNOLOGY BY OPEN LEARNING

Operational Modes of Bioreactors

PUBLISHED ON BEHALF OF :

Open universiteit and **Thames Polytechnic**

Valkenburgerweg 167
6401 DL Heerlen
Nederland

Avery Hill Road
Eltham, London SE9 2HB
United Kingdom

BUTTERWORTH
HEINEMANN

Butterworth-Heinemann Ltd
Linacre House, Jordan Hill, Oxford OX2 8DP

PART OF REED INTERNATIONAL BOOKS

OXFORD LONDON BOSTON
MUNICH NEW DELHI SINGAPORE SYDNEY
TOKYO TORONTO WELLINGTON

First published 1992

British Library Cataloguing in Publication Data
A catalogue record for this book is
available from the British Library

Library of Congress Cataloguing in Publication Data
A catalogue record for this book is
available from the Library of Congress

ISBN 0 7506 1508 7

Composition by Thames Polytechnic
Printed and Bound in Great Britain by
Thomson Litho, East Kilbride, Scotland

The Biotol Project

The BIOTOL team

OPEN UNIVERSITEIT, THE NETHERLANDS
Prof. M. C. E. van Dam-Mieras
Prof. W. H. de Jeu
Prof. J. de Vries

THAMES POLYTECHNIC, UK
Prof. B. R. Currell
Dr J. W. James
Dr C. K. Leach
Mr R. A. Patmore

This series of books has been developed through a collaboration between the Open universiteit of the Netherlands and Thames Polytechnic to provide a whole library of advanced level flexible learning materials including books, computer and video programmes. The series will be of particular value to those working in the chemical, pharmaceutical, health care, food and drinks, agriculture, and environmental, manufacturing and service industries. These industries will be increasingly faced with training problems as the use of biologically based techniques replaces or enhances chemical ones or indeed allows the development of products previously impossible.

The BIOTOL books may be studied privately, but specifically they provide a cost-effective major resource for in-house company training and are the basis for a wider range of courses (open, distance or traditional) from universities which, with practical and tutorial support, lead to recognised qualifications. There is a developing network of institutions throughout Europe to offer tutorial and practical support and courses based on BIOTOL both for those newly entering the field of biotechnology and for graduates looking for more advanced training. BIOTOL is for any one wishing to know about and use the principles and techniques of modern biotechnology whether they are technicians needing further education, new graduates wishing to extend their knowledge, mature staff faced with changing work or a new career, managers unfamiliar with the new technology or those returning to work after a career break.

Our learning texts, written in an informal and friendly style, embody the best characteristics of both open and distance learning to provide a flexible resource for individuals, training organisations, polytechnics and universities, and professional bodies. The content of each book has been carefully worked out between teachers and industry to lead students through a programme of work so that they may achieve clearly stated learning objectives. There are activities and exercises throughout the books, and self assessment questions that allow students to check their own progress and receive any necessary remedial help.

The books, within the series, are modular allowing students to select their own entry point depending on their knowledge and previous experience. These texts therefore remove the necessity for students to attend institution based lectures at specific times and places, bringing a new freedom to study their chosen subject at the time they need and a pace and place to suit them. This same freedom is highly beneficial to industry since staff can receive training without spending significant periods away from the workplace attending lectures and courses, and without altering work patterns.

Contributors

AUTHORS

J. Bes, Hogeschool Rotterdam en Omstreken, Delft, The Netherlands

Ir B. Geraats, Gist-brocades nv, Delft, The Netherlands

Dr P. G. M. Hesselink, TNO-Environment and Energy, Delft, The Netherlands

Dr C. K. Leach, Leicester Polytechnic, Leicester, UK

Dr G. Mijnbeek, Bird Engineering bv, Schiedam, The Netherlands

TECHNOLOGY AND EDITORIAL ADVISORS

Dr G.M. Hall, Loughborough University, Loughborough, UK

Dr G. Mijnbeek, Bird Engineering BV, Schiedam, The Netherlands

Dr M. C. E. van Dam-Mieras, Open universiteit, Heerlen, The Netherlands

Dr W. J. P. White, The Old School, Croft, Leicester, UK

SENIOR TECHNOLOGY ADVISOR

Professor Ir K. Ch. A. M. Luyben, Technical University Delft, The Netherlands

SCIENTIFIC AND COURSE ADVISORS

Professor M. C. E. van Dam-Mieras, Open universiteit, Heerlen, The Netherlands

Dr C. K. Leach, Leicester Polytechnic, Leicester, UK

ACKNOWLEDGEMENTS

Grateful thanks are extended, not only to all authors and advisors, but to all those who have contributed to the development of this book. They include Mrs A. Allwright, Miss K. Brown, Dr R. O. Jenkins, Dr M. de Kok, Miss J. Skelton, Professor R. Spier and Dr G. Subramanian. Our particular thanks go to Dr Ir N. M. G. Oosterhuis of Bird Engineering bv, Schiedam, the Netherlands.

The development of the BIOTOL project is funded by **COMETT, The European Community Action Programme for Education and Training for Technology**. Additional support is provided by the Open universiteit of The Netherlands and by Thames Polytechnic.

Project Manager Dr J.W. James

Contents

How to use an open learning text

An open learning text presents to you a very carefully thought out programme of study to achieve stated learning objectives, just as a lecturer does. Rather than just listening to a lecture once, and trying to make notes at the same time, you can with a BIOTOL text study it at your own pace, go back over bits you are unsure about and study wherever you choose. Of great importance are the self assessment questions (SAQs) which challenge your understanding and progress and the responses which provide some help if you have had difficulty. These SAQs are carefully thought out to check that you are indeed achieving the set objectives and therefore are a very important part of your study. Every so often in the text you will find the symbol Π, our open door to learning, which indicates an activity for you to do. You will probably find that this participation is a great help to learning so it is important not to skip it.

Whilst you can, as a open learner, study where and when you want, do try to find a place where you can work without disturbance. Most students aim to study a certain number of hours each day or each weekend. If you decide to study for several hours at once, take short breaks of five to ten minutes regularly as it helps to maintain a higher level of overall concentration.

Before you begin a detailed reading of the text, familiarise yourself with the general layout of the material. Have a look at the contents of the various chapters and flip through the pages to get a general impression of the way the subject is dealt with. Forget the old taboo of not writing in books. There is room for your comments, notes and answers; use it and make the book your own personal study record for future revision and reference.

At intervals you will find a summary and list of objectives. The summary will emphasise the important points covered by the material that you have read and the objectives will give you a check list of the things you should then be able to achieve. There are notes in the left hand margin, to help orientate you and emphasise new and important messages.

BIOTOL will be used by universities, polytechnics and colleges as well as industrial training organisations and professional bodies. The texts will form a basis for flexible courses of all types leading to certificates, diplomas and degrees often through credit accumulation and transfer arrangements. In future there will be additional resources available including videos and computer based training programmes.

Preface

The potential of biotechnology to contribute to a great variety of industrial and environmental processes is widely acknowledged. This potential extends across many sectors of human activity but especially focuses on improvements in agriculture, food manufacture, health care and environmental management. Biotechnology is seen as a clean technology and an attractive alternative to the more conventional technologies. It is also regarded as offering opportunities to develop a vast array of new and valuable products.

The successful development of biotechnology depends upon a concerted multi-disciplinary effort in which advances in our knowledge in many biological disciplines is integrated with the application of sound engineering principles and practices. Thus, although the basic biological and biochemical research disciplines are the primary driving force of innovation for new products and processes, it is the application of the engineering sciences that are decisive in bringing these products and processes to fruition.

BIOTOL texts are designed to provide a learning resource for those who seek to contribute to the development of biotechnological enterprises. Within the spectrum of BIOTOL materials, four texts have been specifically designed to enable readers to develop engineering knowledge and expertise essential to the design and operation of bioprocesses. These texts are: 'Bioprocess Technology: Modelling and Transport Phenomena', Operational Modes of Bioreactors', 'Bioreactor Design and Product Yield' and 'Product Recovery in Bioprocess Technology'.

This text is the second in this series. It builds upon the knowledge of the underpinning principles of mass, heat and momentum transport and modelling of bioprocess to develop an understanding of the key processes which enable the design and operation of bioreactors. The text begins with a brief chapter which enable readers to orientate themselves and to fully appreciate the role this text plays in developing relevant expertise and knowledge. This is followed by a chapter on oxygen transfer, a key issue in bioreactors. Subsequent chapters consider transfer processes in reactors of various design and configuration but especially focused on Continuous Stirred Tank reactors, bubble columns and airlift reactors. A major section of the text examines the principles underpinning the performance of reactors in which the biocatalyst has been immobilised. The final chapter explains the process of sterilisation, an essential ingredient in bioprocesses involving pure cultures and an important contributor to product quality control.

This text acts as a pre-requisite to developing strategies for bioreactor design and scale up (discussed in the BIOTOL text 'Bioreactor Design and Product Yield') and the procedures adopted for the downstream processing of reactor products (described in the BIOTOL text 'Product Recovery in Bioprocess Technology').

It is impossible to break with the tradition of including an expression of gratitude to all who have contributed to the development and production of this text. Like other BIOTOL texts on process technology, this text has been prepared by an experience team of technologists and teachers. Their distinction is reflected in the care they have shown in the selection of material, the depth of treatment and the logical manner of its development and presentation enabling the production of a high quality learning material.

Our thanks recognises all who have contributed.

We encourage readers to make full use of the opportunities provided by the authors. Attempt the in-text activities specifically designed to facilitate your learning and use the self assessed questions to check your progress. We are confident that this text will enable you to develop your knowledge and understanding of an important aspect of biotechnology thereby enhancing your ability to contribute to these exciting and important endeavours.

Scientific and Course Advisors: Professor M. C. E. van Dam-Mieras
Dr C. K. Leach

Introduction

Introduction

1.1 The commercial context of bioreactor use

Fermentation technology, which involves the cultivation of bacteria, fungi and yeasts, is an ancient art. Beer, wine and bread are produced by the action of yeasts, while cheese and yogurt are produced by the action of bacteria. The use of bioreactors to produce fermentation products is rather a modern development. This particularly developed along with the need for aseptic and controlled conditions for antibiotic production in the 1940's. Nowadays, bioreactors are applied in a large number of processes which not only include the products obtained from micro-organisms, but also those obtained by the cultivation of plant and animal cells. In the last few years products such as insulin and interferon, which are produced by genetically engineered micro-organisms, have been added to the list of fermentation products. Furthermore, bacteria are extensively used in waste water treatment systems. Table 1.1 gives an insight of the scale of the bioreactors used in various biotechnological processes.

Π Examine Table 1.1 carefully. Which process uses the largest volume reactors and which the smallest? What factors do you think influence the choice of the size of the reactors used? You will see if your reasoning is correct in the next paragraphs.

Since the 1940's, conventional stirred tank and bubble column reactors have been the most commonly used reactors in industrial biotechnological processes. However, in the last two decades new bioreactors have been developed and successfully applied. A prominent example is the airlift loop reactor developed in the seventies by ICI for the production of single cell protein (SCP). In the field of waste water treatment large scale bioreactors have been developed. As biotechnology expands, numerous developments are expected depending upon the price and the scale of the annual production needed.

three categories of products

In Figure 1.1, the relation between price and world annual production is shown. It is evident from the figure that generally as the production quantity increases the cost of the product decreases. Production can be increased by applying large scale systems. In general, fermentation products can be divided into three categories A, B and C shown as regions in Figure 1.1. The division is mainly on the basis of price per kg. Region A is characterized by very high price/low volume and contains performance (bio) chemicals such as factor VIII, interferon, human growth factor, etc. Region B represents commodities containing products such as industrial enzymes and penicillin which have high prices and are produced in relatively large volume. Bulk chemicals such as bakers' yeast, ethanol and citric acid which have low prices and are produced in large volumes, make up region C.

organisms/cells	product groups	examples	bioreactor volume
bacteria/yeasts/fungi (change of consistency, pH, nutritional value or aroma/conservation)	beer wine cheese bread sojasauce sauerkraut		1000 m^3 200 m^3
bacteria/yeasts/fungi (products)	alcohols	ethanol sorbitol	$50m^3$
	amino acids	sodium glutamate methionine lysine phenylalanine asparagine	100 m^3
	organic acids	citric acid acetic acid lactic acid gluconic acid	100 m^3
	vitamins	vit.C vit.B2 vit.E	10-50 m^3
	enzymes	amylases glucose-isomerase lactase pectinase protease	10-50 m^3
	polysaccharides	xanthan cyclodextrins pullulan	20-50 m^3
	sweetening agents	aspartame	200 m^3
	antibiotics	penicillin cephalosporin	
genetically-engineered bacteria/yeasts (products)	pharmaceuticals	insuline interferon human growth factor	10 m^3
bacteria/yeasts (biomass)	bakers/brewers' yeast		200 m^3
	single cell protein	pruteen mycoprotein	1500 m^3 (continu)
bacteria	waste water treatment		27000 m^3 (aerob) 5500 m^3 (anaerob)
animal cells	pharmaceuticals vaccines	interferon	0.1-1 m^3
hybridoma cells	monoclonal antibodies		
plant cells	pharmaceuticals	shikonine	

Table 1.1 Scale of bioreactors used in biotechnological processes (adapted from Hanemaaijer J H and Timmer J M K (1987), VMT 12 23-27).

for high priced products the cost of reactor not so important

For the products of category A, the accent is not on the cost of the bioreactor. Here, the demand is to obtain a product of very-high quality rather than high productivity. In this product group quality control contributes approximately 50% of the product cost. It becomes important to understand the physiology and the interaction of the micro-organism with its environment in the bioreactor.

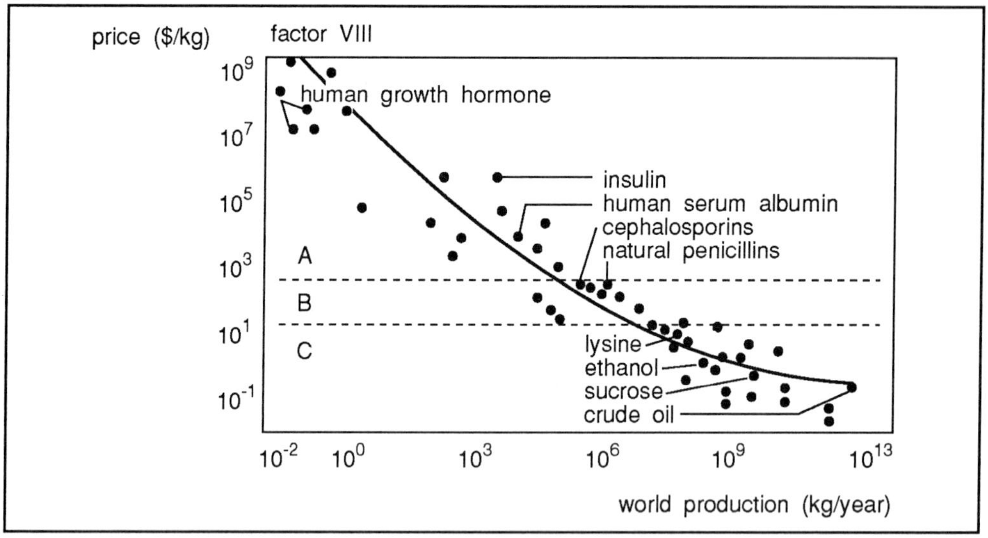

Figure 1.1 Relationship between the market price in kg^{-1} against volume of world production in kg $year^{-1}$ (1987 prices). Divisions A, B and C are explained in the text. (Data from van Suijdam, J.C. (1987), Chemical Magazine. June/July, 385-388).

Some authors argue that with very high priced products there is no need to switch over from relatively low efficiency, conventional bioreactors to more efficient new generation bioreactors. This view is, however, in no way universally held. It must be realised that often with high priced products there is a need for more instrumentation and computers are required to regulate quality control. Due to the high price of the product, there is a strong tendency to use small bioreactors in order to prevent the contamination of the whole volume of the culture which might occur when using a large bioreactor. Furthermore, good quality containment demands may also exist. As the scale of the bioreactor increases, it becomes more difficult to control the containment of the micro-organisms. Providing the cost of running the bioreactor is only a small part of the total cost, then the balance is to use the tried and tested methods.

Π Examine Figure 1.1 carefully and see if you can decide whether or not there is pressure to use new, more efficient bioreactors for products in categories B and C.

In region B, in contrast to region A, there is a tendency to develop new bioreactors. Genetically engineered micro-organisms may be used and fermentation conditions may be improved to obtain a higher productivity.

pressure for new bioreactors for large volume products

Category C is the group where the annual production is very high and bioreactors with very large volumes are used. It is in this category that most of the developments in bioreactors have taken place and research is still continuing. Clearly with products in category C, a more efficient bioreactor can substantially reduce the total overall cost.

It should, however, be self-evident that any commercial enterprise needs to consider the cost/benefits ratio of using particular bioreactor types. This involves both a knowledge of what types of bioreactors are available, and what advantages and limitations each type offers. This text is designed to, not only provide such knowledge, but also to explain the operation and application of the various types of bioreactors.

1.2 Pre-requisite knowledge and layout of the text

Π Just sit back for a moment and make a list of as many items as you can that need to be considered before choosing or designing a bioreactor.

We do not intend to try to guess all of the items you will have included on your list. We would anticipate, however, that you will have included such items as:

- how big should the vessel be;
- how should nutrients be supplied;
- how much nutrient will be needed;
- how much product(s) (waste or valued) will be made;
- how will the product(s) be collected;
- what factor(s) will limit the production capacity of the vessel;
- how will physical parameters (eg temperature, pH etc) influence the performance of the system and how will these parameters be monitored and controlled;
- can the system be operated aseptically;
- will the bio-system withstand the shear forces set up in the vessel?

There are indeed a lot of technical factors that need to be considered. We can also add economic (capital and operational etc) and safety factors.

transfer processes important in bioreactors

Many of the parameters that need to be considered involve transfer processes. We can particular identify mass and energy transfers. Thus nutrients have to be transferred efficiently from the medium into the cells. Likewise gases such as oxygen may have to be transferred from gas phase (air), through the medium, to the cells. It is not surprising, therefore, that much effort is given to considering transfer processes in the design and operation of bioreactors. This text has been written on the assumption that the reader is familiar with the fundamentals of mass and energy transfer. There are, however, many helpful 'reminders' built into the text. For those unfamiliar with these aspects of process technology, the supporting BIOTOL text 'Bioprocess Technology: Modelling and Transport Phenomena' is recommended.

We begin this text with a description of oxygen and gas-liquid mass transfer. Our main reason for doing so is that oxygen availability often governs the productivity of a bioreactor. Oxygen solubility in media is low. If it is realised that 1g of bacterial cells (eg *Azotobacter*) is capable of consuming all of the oxygen in 1 litre of air-saturated medium in about 7-10 seconds, then it should be clear that oxygen availability is often a problem. Many of the advances that have been made in bioreactor design relate to this issue.

Armed with this knowledge, we will then launch into the discussion of bioreactor design and operation. To do this we have adopted a two phased development of the discussion. In the first phase (Chapters 3 and 4), we provide an overview of the major types of bioreactors. In Chapter 3, we discuss one of the oldest and still most widely used type of reactors, the Conventional Stirred Tank Reactor (CSTR). In the subsequent chapter, we move on to discuss bubble column and air lift reactors as well as the newer development in reactor design.

In the second phase specially attention is paid to the particular problems and attributes of immobilised systems. In the final chapter, we examine the problems of sterilisation and the need for aseptic operation. This discussion of bioreactors is further extended in the BIOTOL text 'Bioreactor Design and Product Yield' and the isolation of the products from the reactor medium (down-stream processing) is dealt with in the BIOTOL text 'Product Recovery in Bioprocess Technology'.

Summary and objectives

This brief introduction provided a broad commercial framework in which bioreactors are used and explained what pre-knowledge is assumed amongst readers.

Now that you have completed this chapter you should be able to:

- identify low volume/high price and high volume/low price biotechnology products and explain in broad terms the pressures (or otherwise) for improvements in bioreactor design;
- understand that mass and energy transport phenomena are of great importance in bioreactor operation and identify that, in aerobic systems, oxygen limitation is a major factor to consider.

Oxygen and gas-liquid mass transfer

Oxygen and gas-liquid mass transfer

2.1 Introduction

2.1.1 Layout of the chapter

In many biotechnological processes gases are essential reactants. This chapter examines the transfer of gaseous substrates to biocatalysts in liquid fermentation media and some important factors affecting this process. Mathematical relations are derived and explained for common situations. The theory and practical implementation of methods for measuring and analysing the process of gas-liquid mass transfer in bioreactors will also be dealt with.

There are many applications of the issues covered by this chapter and the problems of gas-liquid mass transfer are encountered in a wide spectrum of biotechnological processes. Understanding gas-liquid mass transfer is of course, vitally important in bioreactor operation, scaling up and process control.

In this chapter, we will first deal with some of the general aspects of gases as reactants. We will examine gas solubility in liquid before discussing gas-liquid mass transfer and the methods available to, experimentally, analyse gas solubility and mass transfer.

2.1.2 Supply and removal of gaseous reactants

homogeneous and heterogeneous systems

All chemical and biochemical reactions using catalysts involve transportation of reactants to and from the catalyst. In homogeneous systems consisting of one phase only, mass transfer processes are relatively straightforward with respect to kinetics and mathematical description. Most biotransformations however, are carried out in heterogeneous multi-phase systems. This implies that several mass transfer steps of reactants over distinct phases are involved in the bioconversion process.

Consider a bioreactor containing the enzyme polyphenol oxidase immobilised in a solid support. This enzyme can be used to catalyse the conversion of the amino acid tyrosine to the pharmaceutically useful dihydroxyphenylalanine (L-DOPA).

We can represent the reaction by:

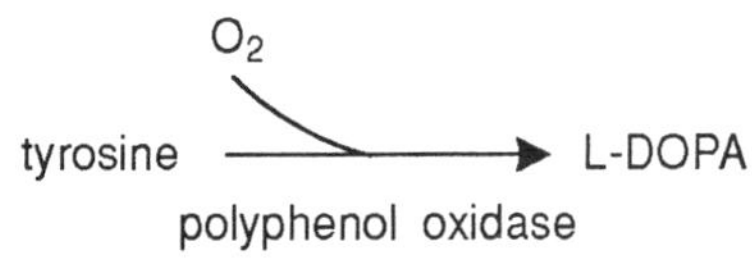

Such a system is multi-phase system.

In this instance, an oxygen molecule has to dissolve from the gas phase into the liquid medium, reach the immobilised polyphenol oxidase in the solid phase before it is incorporated into tyrosine to yield L-DOPA.

sterols as a multiphase system

Another example of a multiphase system is steroid fermentation where a sterol molecule has to be solubilised from its crystaline state into the fermentation liquid, transported to the *Mycobacterium* solid state biocatalyst which selectively degrades the C-17 sterol side chain to yield the hormone androstenedione (AD).

Π Write down a reason why the sterol has to be solubilised from its crystalline state.

The solubility of sterols in aqueous media is very low (in the order of ng-mg ml^{-1}). When sterols are added to the *Mycobacterium* system, this has to be done with significant amounts of sterol in order to make commercially acceptable quantities. The levels of sterols used are well above their levels of solubility in the reactor milieu. We can represent this transfer by the sequence.

sterol (crystalline) ⟶ sterol (soluble) $\xrightarrow[\textit{Mycobacterium}]{\text{transported to}}$ product

Although micro-organisms, plant and animal cells, organelles and enzymes are able to use a wide variety of gases as substrates, only dissolved gas molecules can be used as reactants in liquid media. For oxygen, this implies that an oxygen molecule has to escape from its gas bubble and dissolve into the fermentation fluid before it can be used by the biocatalyst. More generally, mass transport from a gas phase via a liquid phase to a solid phase and vice versa usually occurs in bioreactions involving gaseous reactants. Often, this mass transfer is the rate limiting step in the total bioconversion process.

damage caused by oxygen depletion

We should also remember the importance of adequate oxygen supply to cells or micro-organisms in aerobic processes. This importance is reflected by the severe problems caused by even a short period of oxygen shortage. Frequently irreversible damage, such as cell death, is the result. We must also remember that adequate removal of gaseous waste products, such as CO_2 in penicillin production by *Penicillium* species, is required to prevent inhibition of the process and inactivation of the biocatalyst.

oxygen consumption in relation to oxygen solubility

The rate of oxygen utilisation and the need for replacement by transfer will be demonstrated by the following example. Suppose an actively respiring culture of 1 g l^{-1} cell dry weight is growing at a growth rate of $\mu = 0.67$ h^{-1}; then, the amount of biomass will double in 1 h. For this purpose, ca. 2 gl^{-1} of glucose will be consumed for biomass and energy generation. As about 2.5 moles of oxygen are required by growing cells for metabolising 1 mole of glucose, in the order of 0.89 g l^{-1} of dissolved oxygen will be consumed in that time. The maximum solubility of oxygen in aqueous media is ca. 9 mg at 20°C and at 1 atmosphere of air pressure. This means that dissolved oxygen is consumed at a rate of about 100 times its maximum solubility per hour, even at a modest biomass concentration of 1 g l^{-1}. These figures are fairly typical. Some micro-organisms (eg *Azotobacter spp.*) have particularly high rates of oxygen uptake. Experiments have demonstrated that 1 g l^{-1} of these cells can consume all the oxygen in 1 litre of air saturated media in about 7 seconds at about 30°C! Think how quickly a biomass density of 20 g l^{-1} would consume all of the oxygen in air saturated media. Clearly, continuous oxygen supply, dissolution and transfer to the cells is required to keep the biomass growing and viable.

Gases are usually introduced in a bioreactor by sparging. In addition to supply (and removal) of gaseous reactants, sparging generates convection mixing and this promotes mixing of the fermentation liquid. The contribution of gas bubble induced convection to mixing can be substantial. For instance, the rising of gas bubbles is the only mixing mechanism in air-lift reactors, which are usually very well mixed and have high oxygen supply rates.

2.1.3 Henry's Law

Π Henry's law states that the mass of a slightly soluble gas that dissolves in a definite mass of a liquid at a given temperature is very nearly directly proportional to the partial pressures of that gas. This statement holds for gases which do not react chemically with the solvent. Write down a mathematical form of Henry's law.

Apolar gases like oxygen are only sparingly soluble in aqueous media which consist mainly of highly polar H_2O molecules. A general correlation between the partial gas pressure and the solubility of sparingly soluble gases in liquids is provided by Henry's Law.

$$C_g = H_g * p_g \tag{E - 2.1}$$

in which: C_g = concentration of dissolved gas 'g' in solvent 's'; p_g = partial pressure of 'g' in the gas phase; H_g = Henry coefficient for 'g' in 's' at a specific temperature.

The units of C_g are usually mmol l^{-1}, mg l^{-1} or dimensionless mole fractions; p_g is expressed in Pascal (Pa) or atmosphere (atm) and H_g corresponds to the applied C_g units. In some texts a dimensionless Henry (He_g), distribution coefficient is used. He_g is correlated to H_g via the gas constant and the absolute temperature by:

$$H_g = He_g/RT \tag{E - 2.2}$$

in which: R = gas constant (8.31 J $mol^{-1}.K^{-1}$); T = temperature (K).

The linear Henry relation is only valid at modest concentrations and pressures but despite this limitation it has been successfully used. Table 2.1 shows the maximum solubilities (C*), also called saturation concentrations, as well as the corresponding Henry coefficients for some gases in water.

Π Air contains about 21% oxygen. Using the data supplied in Table 2.1, calculate the C* for oxygen in water at 20°C and 1 Pa of air pressure.

You should have calculated C* to be about 9 mg l^{-1} (or 0.294 mmol l^{-1})

Since $C_g = H_g \times p_g$ and $p_g = 0.21$ Pa

Then $C_g = 1.4 \times 0.21$ mmol l^{-1} = 0.294 mmol l^{-1}

= 0.294 x 32 mg l^{-1} = approx. 9 mg l^{-1}

(since the molar weight of O_2 is 32g mol^{-1})

gas	C* (mmol l^{-1})	C* (mg l^{-1})	H_g (mmol $l^{-1}Pa^{-1}$)
oxygen	1.4	44.8	1.4
nitrogen	0.76	21.3	0.76
carbon dioxide	40.0	1600.0	40.0
methane	1.5	24.0	1.5
ethane	2.1	63.0	2.1
propane	2.9	127.6	2.9
butane	6.7	388.6	6.7

Table 2.1 Maximum solubilities of several gases in pure water at 20°C and 1 Pa of partial gas pressure.

As mentioned before, water is a poor solvent for most apolar gases. In many organic solvents saturation concentrations of such gases are much higher (Table 2.2). This is one of the reasons why several biotransformations proceed faster in these solvents or in two liquid phase systems consisting of an aqueous and an organic liquid phase.

Ostwald coefficient

Bunsen coefficient

In some literature, the gas solubility is expressed by the dimensionless Bunsen or Ostwald coefficients. The Ostwald coefficient is defined as the ratio of the volume of a specific gas 'g' absorbed per unit volume of solvent. The Bunsen coefficient corresponds to the volume of 'g' reduced to 0°C and 1 Pa absorbed per unit volume of solvent at p_g = 1 Pa. As these coefficients are not widely used in biotechnology, they will not be discussed further in this chapter. You should however be aware that they exist.

liquid	O_2		CO_2	
	C* (g/l)	H_g (mmol l^{-1} Pa^{-1})	C* (g/l)	H_g (mmol l^{-1} Pa^{-1})
acetone	300.8	9400	15.3	347
benzene	354.6	11 080	41.6	946
l-butanol	364.5	11 390	51.9	1179
butylacetate	299.5	9360	22.0	499
tetra[a]	256.6	8020	43.6	909
cyclohexane	275.8	8620	62.2	1414
DCL[b]	490.9	15 340	28.6	650
ethanol	331.5	10 360	35.3	802
ethylbenzene	375.1	11 160	49.5	1124
methanol	335.7	10 490	26.4	601
l-tetradecene	343.0	10 720	68.2	1551
toluene	341.4	10 670	42.5	965
m-xylene	348.5	10 890	46.9	1067

Table 2.2 Maximum solubilities and Henry-coefficients for oxygen and carbon dioxide in several liquids at 20°C and 1 Pa of gas pressure (after Luhring and Schumpe, 1989, J Chem Eng. Data 34,250-252).
a = tetrachloromethane; b = 1,2-dichloroethane.

2.2 Factors influencing O_2-solubility in an aqueous-system

The values of C* and H_g shown in Table 2.1 are only valid for a restricted set of conditions. They are influenced by temperature and by solute present in the water. Both temperature and solutes influence the structure of water and therefore will influence C*. H_g and C* are also influenced by partial gas pressure. We will examine each of these influences.

Effects of partial gas pressure

At p_g > ca. 1 Pa, serious deviations from the linear Henry's law occur. This results in a non-linear relation of C_g with p_g leading to lower values for C* than would be expected from Equation 2.1.

Effects of temperature

The decrease in oxygen solubility in aqueous media with increasing temperature can be adequately represented by the empirical relation in Equation 2.3:

$$C^* = p_g/p_t\,(67.75 - 1.887T + 0.0369T^2 - 0.000309T^3) \qquad \text{(E - 2.3)}$$

in which: p_t = total gas pressure; T = temperature in °C.

(This equation was derived by Tresdale, Downing and Lowden, 1955, J Appl. Chem 5, 53-39)

Effects of solutes

effects of salt and glucose are additive

The concentration of solutes is another factor influencing the solubility of gases in aqueous systems. In most fermentation media, salts and nutrients like glucose are present. Several empirical formulas exist which describe the effect of their concentration on oxygen solubility. In one approach, the effect of salts and glucose is considered as additive. For example:

$$\log(C^*_w/C^*_e) = \log(C^*_w/C^*_{gl}) + \log(C^*_w\,C^*_s) \qquad \text{(E - 2.4)}$$

in which: C^*_e = maximum oxygen solubility in system used; C^*_w = maximum oxygen solubility in pure water; C^*_s = maximum oxygen solubility in salt solution; C^*_{gl} = maximum oxygen solubility in glucose solution.

A linear relationship between C^*_{gl} and the glucose concentration S_{gl} ($g\,l^{-1}$) exists:

$$C^*_{gl} = C^*_w\,(1 - 0.0012\,.\,S_{gl}) \qquad \text{(E - 2.5)}$$

The correlation of C* with the salt concentration is expressed in terms of ionic strength as measured by the conductivity λ in $\Omega\,.\,cm^{-1}$ of the salt solution:

$$\log(C^*_w/C^*_s) = a_0 + a_1\,\lambda + a_2\,\lambda^2 \qquad \text{(E - 2.6)}$$

The coefficients (a) in Equation 2.6 are independent of the glucose concentration.

SAQ 2.1

Calculate the decrease in C* for oxygen when 30 g of glucose are added to 1 litre water at 20°C and 1 Pa of air pressure.

Note air contains 21% oxygen and Hg can be obtained from Table 2.1.

2.3 Gas-liquid mass transfer

2.3.1 Resistance barriers in gas-liquid mass transfer

In a bioreactor with an aqueous liquid medium and a solid phase biocatalyst like an enzyme, a microbial cell or an aggregate, transport of gaseous substrate and products is an ongoing process. In such a heterogeneous system, there are several mass transfer steps to be distinguished. The various stages of the mass transfer process for gaseous reactants are given in Figure 2.1. The scheme in Figure 2.1 is valid for other non-aqueous phase reactants like solids and sparingly water miscible liquids as well.

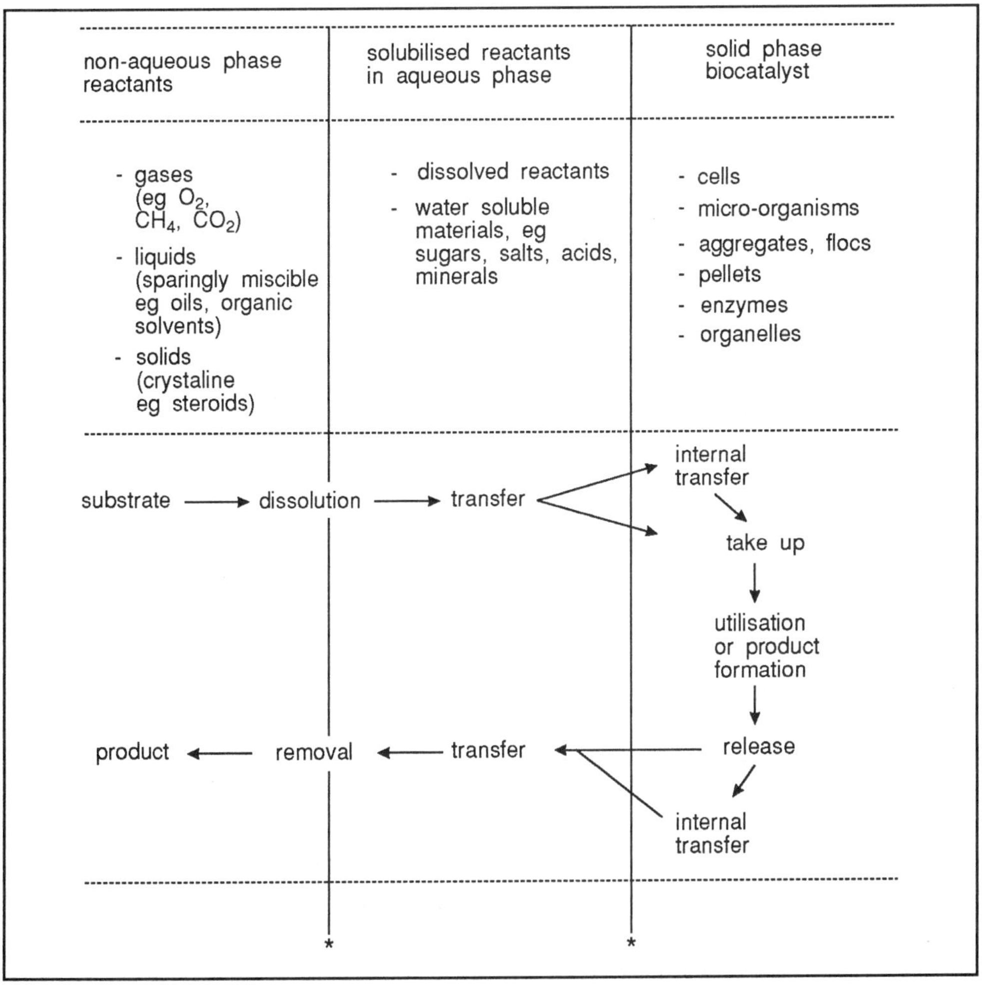

Figure 2.1 Generalised scheme of mass transfer of sparingly soluble reactants to and from biocatalysts. Phase boundaries are indicated by *.

resistance barriers in oxygen transfer

In the case of aerobic systems, oxygen molecules meet with several barriers. Figure 2.2 shows seven main mass transfer resistances and their effects on the oxygen concentration when oxygen molecules escape from an air bubble, become water solubilised and are transferred to the reaction site of a solid phase biocatalyst. Note that in principle the same scheme is valid for sparingly water miscible liquids or solids, the gas bubble being replaced by a second liquid phase droplet or solid particle, respectively.

According to the two film model depicted in Figure 2.2, an O_2 molecule in a gas bubble of homogeneous composition enters the gas film (barrier I) and crosses the gas-liquid interface (II) before solubilising and moving through the liquid film (III) around the gas bubble. After transportation through the bulk liquid (IV), the molecule enters the liquid film (V) around the solid particle, passes the liquid-solid interface (VI) and diffuses through the solid biocatalyst aggregate (VII) to the reaction site. Note that a solid biocatalyst aggregate usually has stagnant water within the aggregate; see 2.3.2. After these seven purely physical barriers, biochemical processes like substrate take up or reactions might constitute other rate determining steps.

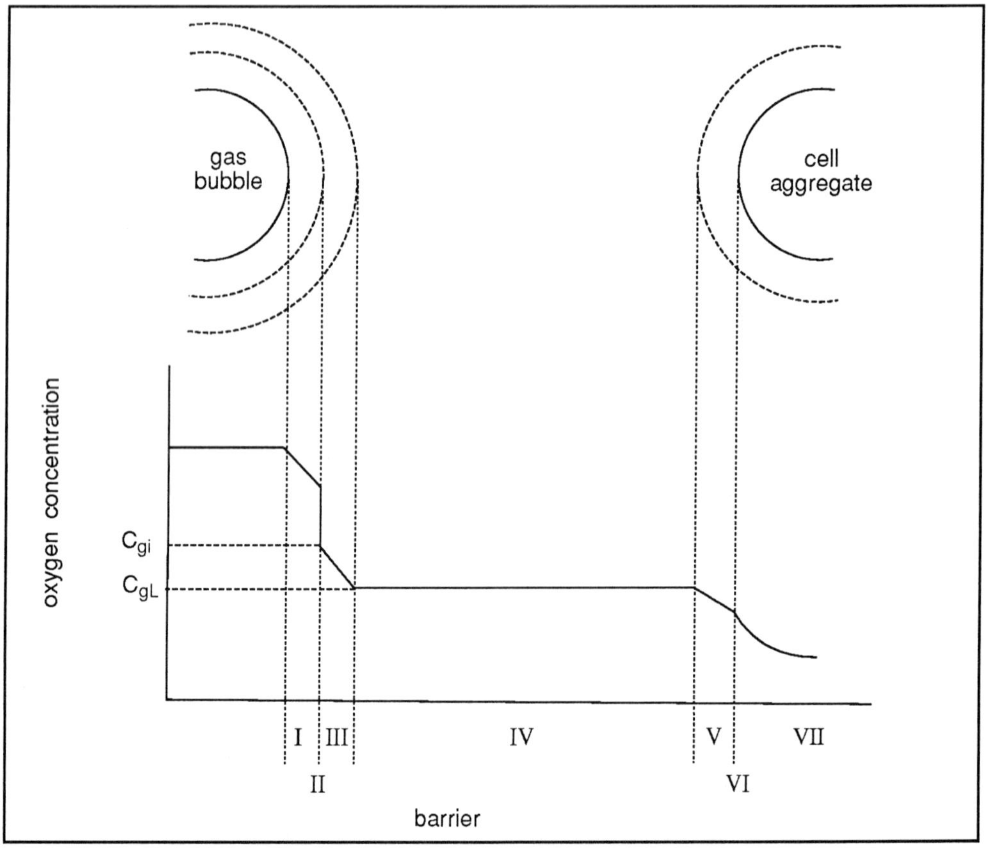

Figure 2.2 Mass transfer resistances and the resulting concentration profile for gas molecules moving from a gas bubble to a solid phase biocatalyst in a liquid medium (see text for details).

An extra barrier will arise within the gas bubble whenever it is of non-homogeneous composition due to inadequate mixing. Such a situation could be met with small bubbles rising slowly or when surface active agents influence the bubble to behave as though having a rigid surface. Then an extra diffusive process is required for transfer of O_2 molecules to the outer layer of the bubble. Another barrier will arise if the bulk liquid (IV) is very viscous and/or not well mixed. Then, variations in local O_2-concentrations will occur, mostly leading to a decrease in O_2-concentrations around the solid biocatalyst.

Π How could we depict the effects of a viscous liquid on the oxygen concentration profile shown in Figure 2.2?

One way to do this would be to draw another gradient for barrier IV as shown below.

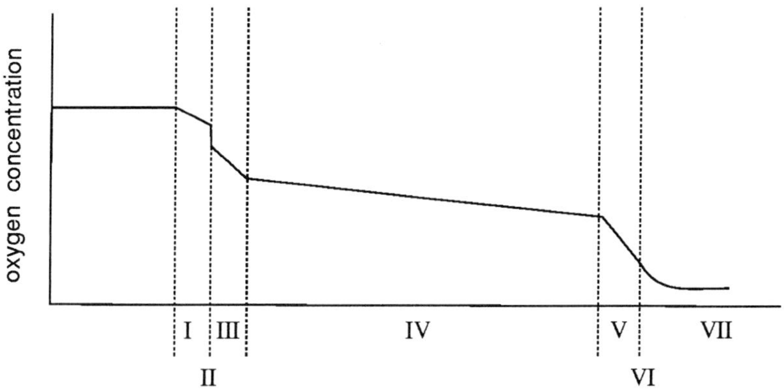

In practice, however, the slope of the line for barrier IV need not be constant. The shape of the line would depend on the extent of mixing.

2.3.2 Identification of rate limiting steps

All physical mass transfer steps in Figure 2.2 occur in series, therefore the slowest process will be rate controlling. Identification of this rate-controlling step allows a simplified mathematical description of gas-liquid mass transport in such systems. More important, in practice, it indicates which process step should be focused on first when improving mass transfer limited bioreactor performance.

rate limiting step likely to be in liquid phase

Mass transfer through films and aggregates (I, III, V and VII) is diffusive by nature and therefore influenced by the diffusion coefficient, D_g, of the solubilised gas molecule in the respective phases. In gases or gas films, diffusion coefficients of the relatively small gas molecules like oxygen are in the order of 10^{-4} m^2 s^{-1} in contrast to about 10^{-9} m^2 s^{-1} in liquid or liquid films of low viscosity such as we encounter with most aqueous solutions. This implies that one of the liquid film resistances (III $\rightarrow$ V) will dominate over the resistance in the gas film (I). Whenever present, microbial pellets and immobilised biosystems consist of partly 'co-immobilised' stagnant water. Therefore, such a barrier can be described as a pseudo-liquid resistance. Consequently, diffusion coefficients within such solid phases are in the order of 10^{-9} - 10^{-10} m^2 s^{-1}, and thus barrier VII might become important.

convection is faster than diffusion

If the bulk liquid is well mixed, barrier IV will not have a significant effect on mass transfer. Mass transport resulting from convection is much faster than diffusional mass transport in liquids. Therefore, in most agitated aqueous fermentation broths there is no appreciable bulk liquid mass transfer resistance. However, in the case of very viscous and/or badly mixed bulk liquid media, convectional mass transport is strongly decreased and a significant barrier will be met. This situation is regularly found in bioreactors in which extracellullar polysaccharides like xanthan or gellan gums are produced. In a xanthan production process, the viscosity of the fermentation broth might increase from 1-10 mPa.s to over 50 000 mPa.s, leading to severe mixing problems, stagnant zones and a non-homogeneous bulk liquid phase.

gas-liquid interface can be a significant barrier

At the gas-liquid interface (II) a decrease in concentration of solubilised gas molecules will occur due to the much lower solubility of gases in aqueous liquids than in the original gas phase of the bubble. This effect might constitute a serious barrier and cannot simply be overlooked (next paragraph). Barrier II will be less important when a gas like oxygen is transferred to organic solvents with much higher solubilities for such sparingly water soluble gases (Table 2.2).

The liquid-solid interface (VI) is usually relatively unimportant with separate cells, enzymes, microbial pellets or immobilised biocatalysts as the solid phase. Generally, there is not such a strong decrease in solubility of the gaseous molecules as is observed at the gas-liquid interface (II).

SAQ 2.2

Identify the mass transfer resistance in the supply of gases most probably being the rate-controlling barrier in the following bioprocesses. Use the nomenclature for the various barriers (ie I-VIII) used in Figure 2.1.

1) Penicillin production by pellets of *Penicillium* sp.
2) Growth of *E. coli* in a medium with 5% yeast extract.
3) Hydroxylation of tyrosine to L-DOPA by the enzyme polyphenoloxidase entrapped in alginate.
4) Production of xanthan gum by *Xanthomonas* sp. in the early and the final stage of a fermentation run.
5) Surface aerated culturing of shear sensitive animal cells.

SAQ 2.3

Represent the concentration profile of CO_2 as developed by a *Penicillium chrysogenum* pellet and transported to an air bubble schematically. Indicate mass transfer barriers as in Figure 2.2.

2.4 Mathematical description of gas-liquid mass transfer

The overall rate of gas transfer in a bioreactor as shown in Figure 2.2 is determined by the rate-controlling step. For most systems, this will be the transport across the liquid film around the gas bubble (barrier III). In such a stagnant film, mass transfer is a diffusional process. In general, this can be described by Fick's law for unidirectional steady state diffusion.

$\phi'' = -D_g \, . \, dC_g/dx$ (E - 2.7)

in which:

ϕ''- mass flow per area (mol s^{-1} m^{-2})

D_g = diffusion coefficient of gas 'g' in liquid medium used (m^2s^{-1})

x = length (m)

At steady state, ϕ'' is constant and therefore

$dC_g/dx = (C_{gi} - C_{gl})/\delta$ (E - 2.8)

where: C_{gi} = concentration of 'g' at gas-liquid interface; C_{gl} = concentration of 'g' in bulk liquid; δ = thickness of liquid film.

In our system, the absolute value of the total mass flow per unit time equals

$A_s \, \phi'' = A_s D_g \, \Delta C_g/\delta$ (E - 2.9)

where: A_s = total surface area of liquid film and $\Delta C_g = (C_{gi} - C_{gl})$

GTR

The mass transfer of dissolved gas molecules per unit time and volume is given by dC_g/dt and is also called the gas transfer rate (GTR). Then the total mass transfer in our system equals V_r. dC_g/dt where V_r = reaction volume. As in Equation 2.9: $A_s \, \phi'' = V_r \, dC_g/dt$ and we define $A_s/V_r = a$ and $k_L = D_g/\delta$, we obtain:

$GTR = dC_g/dt = k_La \, \Delta C_g$ (E - 2.10)

where: GTR = gas transfer rate (mol s^{-1} m^{-3}); V_r = reaction volume (m^3); a = specific surface area for exchange (m^2 m^{-3}); k_L = mass transfer coefficient (m s^{-1}).

Generally, it is assumed that the liquid at the gas-liquid interface is saturated with gas molecules, implying $C_{gi} = C^*$. Then, $\Delta C_g = (C^* - C_{gl})$, which is the maximum concentration difference that can be obtained. This concentration difference is the driving force for diffusion.

OTR

Equation 2.10 is one of the basic equations in describing gas-liquid mass transfer. It has proven to be of great practical value and is therefore widely applied in bioprocess engineering. In aerobic processes, the mass transfer of oxygen is of interest. In that case, the term oxygen transfer rate abbreviated to OTR is used instead of GTR and Equation 2.10 becomes:

$OTR = k_La \, (C^* - C_{gl})$ (E - 2.11)

volumetric mass transfer coefficient

The product k_La is called the volumetric mass transfer coefficient. This volumetric mass transfer coefficient determines the maximum OTR for a given situation with for example constant medium viscosity, gas flow and agitation rates. When oxygen transfer is limiting, the k_La therefore controls the overall process rate. Moreover, the k_La is considered as a critical parameter in scaling up bioreactors and bioconversion processes. When scaling up, a constant value for k_La on both scales is taken as a measure for reproducibility of reactor or process performance. In brief, the k_La is one of the most important factors in describing and modelling a bioreactor for aerobic processes.

SAQ 2.4

Estimate the maximum concentration of growing *Escherichia coli* that can be obtained in an Erlenmeyer with cotton plug and in an CSTR, having k_La values of 0.02 and 0.1 s^{-1}, respectively. Both reactors contain a glucose medium and are incubated at 30°C under 1 Pa of air pressure. Treat the medium as water and assume that 0.27 g of oxygen is consumed per hour by 1 g of *E. coli* dry weight during growth.

2.5 Empirical correlations for k_L and a

2.5.1 Mass transfer coefficient k_L

Many empirical correlations exist for the mass transfer coefficient k_L. These correlations refer to the practical situation and theory according to which they are derived. For obtaining equations 2.10 and 2.11 a steady state uni-directional diffusion through a stagnant film was assumed. Under these conditions:

$$k_L = D_g/\delta \qquad \text{(E - 2.12)}$$

In practical circumstances, there is usually not a real stagnant film with steady state mass flow. Some convective mass transport will occur and an unsteady state will be observed. Under these conditions, k_L is obtained from solving Fick's law for unsteady state diffusion. Depending on the assumptions made for solving that second order differential equation, there are two main solutions for k_L.

penetration theory

First, in Higbie's penetration theory, diffusion into a semi-infinite fluid film around a spherical particle (for example a bubble) is assumed. This renders:

$$k_L = 2\sqrt{(D_g/\pi t_e)} \qquad \text{(E - 2.13)}$$

in which t_e = exposure time (s)

The exposure time in Equation 2.13 is defined as

$$t_e = d_b/v_b \qquad \text{(E - 2.14)}$$

or the time required to travel one equivalent particle diameter d_b at particle rate v_b.

surface renewal theory

Alternatively, according to Danckwerts' surface renewal theory in which surface elements are exchanged at random rate:

$$k_L = \sqrt{(D_g\,.\,s)} \qquad \text{(E - 2.15)}$$

in which: s = surface renewal factor (s^{-1})

The surface renewal factor 's' is constant for a given degree of turbulence. The units of 's' are those of a frequency.

Note that both solutions (Equations 2.13 and 2.15) yield a correlation of $k_L \sim \sqrt{D_g}$

In the penetration and surface renewal theories, mobile bubble surfaces are assumed. Thus, rapid replacement of exported gas molecules will take place and the bubble will be of homogeneous composition. However, not all bubbles have mobile surfaces. For rigid particles with non-slip surfaces, it was shown by Moo-Young and Blanch in 1981 that:

$$k_L \sim (D_g)^{2/3} \qquad \text{(E - 2.16)}$$

Wesselingh showed in 1987 that bubbles with $d_b < 0.5$ mm will always be rigid, whereas at $d_b > 2.5$ mm surfaces will be mobile. In practice, bubbles with $d_b < 2$ mm are considered as having non-slip surfaces.

Stokes-Einstein equation

Summarising Equations 2.12, 2.13, 2.15 and 2.16, k_L is related to D_g as a main parameter to the power of the order of 0.5-1.0 depending on the hydrodynamic conditions. For media of low viscosity like most aqueous fermentation broths, D_g can be estimated by the Stokes-Einstein equation.

$$D_g = kT/6\,\pi\, r_s\, \eta \qquad \text{(E - 2.17)}$$

in which: k = Bolzmann constant ($1.38 * 10^{-23}$ K^{-1}); T = absolute temperature (K); r_s = solute sphere radius (m); η = viscosity of the solvent (Pa.s).

Equation 2.17 indicates that medium viscosity will be an important factor in practice for the value of D_g. Values for D_g are in the order of 0.5-2.5 * 10^{-9} m^2s^{-1} in pure water at 25°C. In media with high cell concentrations, D_g values decrease. However, D_g values are only 10% lower than in water at ca. 20 g l^{-1} of cell dry weight of totally dispersed micro-organisms. In immobilising matrices like alginate beads the D_g for oxygen is still about 2 * 10^{-9} m^2 s^{-1}.

k_La is usually determined

Unfortunately, direct calculation of k_L using Equations 2.12, 2.13, 2.15 or 2.16 is almost impossible as δ, t_e and s can hardly be quantified in most bioreactions. Therefore, the product k_La is usually determined instead of k_L. We will examine methods for determining k_La in a later section.

2.5.2 Specific surface area a

The specific surface area a for gas-liquid mass transfer was defined in Section 2.4 as:

$$a = A_s/V_r \qquad \text{(E - 2.18)}$$

For spherical gas bubbles of total volume V_g divided in n identical bubbles of radius r_b, the total exchanging area equals:

$$A_s = n\,.\,4\pi r_b^2 \qquad \text{(E - 2.19)}$$

and

$$V_g = n\,.\,4/3\,.\,\pi r_b^3 \qquad \text{(E - 2.20)}$$

By defining the gas hold-up ε_g, as being the partial volume of gas in the reactor liquid:

$$\varepsilon_g = V_g/V_r \qquad \text{(E - 2.21)}$$

and then combining Equations 2.20, 2.21 and 2.19, we get:

$$A_s = 3\varepsilon_g V_r/r_b \text{ or } a_s = 6\varepsilon V_r/d_b \qquad \text{(E - 2.22)}$$

with $2r_b = d_b$, the bubble diameter. Combining of Equation 2.22 with Equation 2.18 leads to:

$$a = 3\varepsilon_g/r_b \text{ or } a = 6\varepsilon_g/d_b \quad \text{(E - 2.23)}$$

The assumption that gas bubbles can be considered as spheres has proven to be a reasonable approximation in practice. However, gas bubbles are usually not spheres with identical diameters. Therefore, the Sauter mean diameter d'_b is used instead of d_b in Equation 2.23. The Sauter mean diameter is defined as:

$$d'_b = \Sigma d_b^3/\Sigma d_b^2 \quad \text{(E - 2.24)}$$

Equation 2.23 indicates that the bubble diameter will be one of the ruling parameters in gas-liquid mass transfer.

SAQ 2.5

5 l of a medium are placed in a bio-reactor. When air is pumped into the medium, the volume of the medium appears to increase to 6 l. Measurement of bubbles of air in the medium indicates that they have a diameter of 5 mm. Calculate the specific surface area (a) for the vessel. (Note: assume the bubbles all have the same diameter).

2.6 Factors influencing k_La

The volumetric mass transfer coefficient k_La is influenced by several factors including bioreactor geometry; gas flow (ϕ_g); superficial gas velocity (vs) ie the gas flow per reactor cross section; impeller type; impeller speed (N); power input for mixing per unit reaction volume (P/V_r); medium viscosity (η); surface active agents (C_{saa}) and anti-foam (C_{af}). As can be expected from this multitude of factors, no general formula correlating k_La to all these factors exists. Therefore, numerous empirical relations have been developed for specific practical situations. Generally, for Continuously Stirred Tank Reactors (CSTR) these relations look like:

$$k_La = \alpha\, \phi_g x\, N^Y\, (P/V_r)^z \quad \text{(E - 2.25)}$$

Values for α, x, y and z have been determined for a wide variety of reactor geometries, impeller types and additives. For reactor types like bubble columns and air-lift reactors, other correlations for k_La predominate:

$$k_La = \alpha''\, v_s x'' \text{ where } \alpha = \text{coefficient}, v_s = \text{superficial gas velocity} \quad \text{(E - 2.26)}$$

Note that especially the gas bubble diameter and the contribution of convection to mass transfer are influenced by the parameters in Equation 2.25 and 2.26. We will return to these relationships when we discuss the various types of bioreactors in later chapters.

We do not anticipate that you can derive these relationships. It may be helpful for you in the future to know where you might find values for the coefficients for the various reactor impeller types and additives. Table 2.3 provides some useful references.

Atkinson, B. and Mavituna, F. (1983), Gas-liquid mass transfer. In: Biochemical engineering and biotechnology handbook, MacMillan Publishers Ltd., Surrey, UK, PP. 727-801.

Moo-Young, M. and Blanch, H. (1981), Design of biochemical reactors; mass transfer criteria for simple and complex systems, Adv. Biochem. Eng. 19 1-69.

Schuegerl, K. (1981), Oxygen transfer into highly viscous media, Adv. Biochem. Eng. 19 72-174

Trilli, A. (1986), Scale-up of fermentations, in Manual of industrial microbiology and biotechnology, Demain, A.L and Solomon, N.A (eds.), American Society for Microbiology, Washington, D.C., pp. 277-307.

Table 2.3 Sources of informaion giving determined values for coeffecients relating k_La to reactor geometry, impeller type and chemical additions.

Table 2.4 gives an overview of the consequences of changes in some of the reaction conditions.

factor increased	effect on k_La	comments
ϕ_g	+	generation of convection: k_L >>
V_s	+	generation of convection: k_L >>
N	+	generation of convection: k_L >> and bubble break up: a >>
P/V_r	+	generation of convection: k_L >> and bubble break up: a >>
η	-	decrease in convection and diffusion: k_L << and bubble coalescence: a <<
C_{saa}	+	smaller bubbles: a >>
C_{af}	-	rigid gas-liquid interfaces: k_L << and bubble coalescence: a <<
biomass	0	if medium viscosity is relatively unaffected

Table 2.4 Qualitative effects of some factors on k_La. Symbols: +, - or 0 = positive, negative or no effect, respectively of k_L or a. ϕ_g = gas flow rate, V_s = superficial gas flow velocity, N = impeller speed, P/V = power volume rate, η = viscosity, C_{saa} = concentration of surface active agents, C_{af} = concentration of antifoam.

Π Examine this table carefully and note which of the factors increase k_La. In general, we need to maximise k_La in order to support more biomass in the reactor.

Figure 2.3 provides some experimentally determined graphs showing the effects described in Table 2.4. Surprisingly, the biomass concentration has no appreciable effect on k_La as long as medium viscosity is not significantly affected. Most totally dispersed cell cultures of eg *E. coli*, *Pseudomonas* and *Candida* with biomass concentrations up to 10-20 gl^{-1} cell dry weight do not influence medium viscosity substantially. On the other hand, *Penicillium* and *Aspergillus* cultures which contain mycelia and large cell pellets and increase medium viscosity during growth, resulting in lower k_La values.

Π Figure 2.3e has a different axis than that used in the other parts of Figure 2.3 (ie OTR, rather than k_La). What information would you need to convert this axis from OTR to k_La?

You should have come to the conclusion that you would need to know the saturation concentration of O_2 (ie C*) under the conditions under which the CSTR was operating. You would also need to know the actual concentration of oxygen in the reactor since:

$OTR = k_La (C^* - C_{gl})$ (see Equation 2.11)

Π Note the effect of the detergent Tween on k_La (Figure 2.3a).

typical k_La values

We can conclude that, it is very hard to predict k_La values for a given bioreactor set up. Too many factors have their influences. From practical experience however, the order of magnitude of some k_La values can be given. For a 1 litre Erlenmeyer flask with ca. 150 ml of aqueous medium which is incubated at ambient temperature at 150 rpm (reciprocal shaker), k_La is in the order of 0.02 s^{-1} with a cotton plug. Without a cotton plug, k_La is increased to about 0.05 s^{-1}. For a 15 l baffled CSTR with 10 l of aqueous medium stirred at 400 rpm and 10 l min^{-1} of air flow, k_La will be in the order of 0.1 s^{-1}. However, these values are only indicative and should be used with great care. We will now move on to the experimental analysis of parameters relating to oxygen mass transfer.

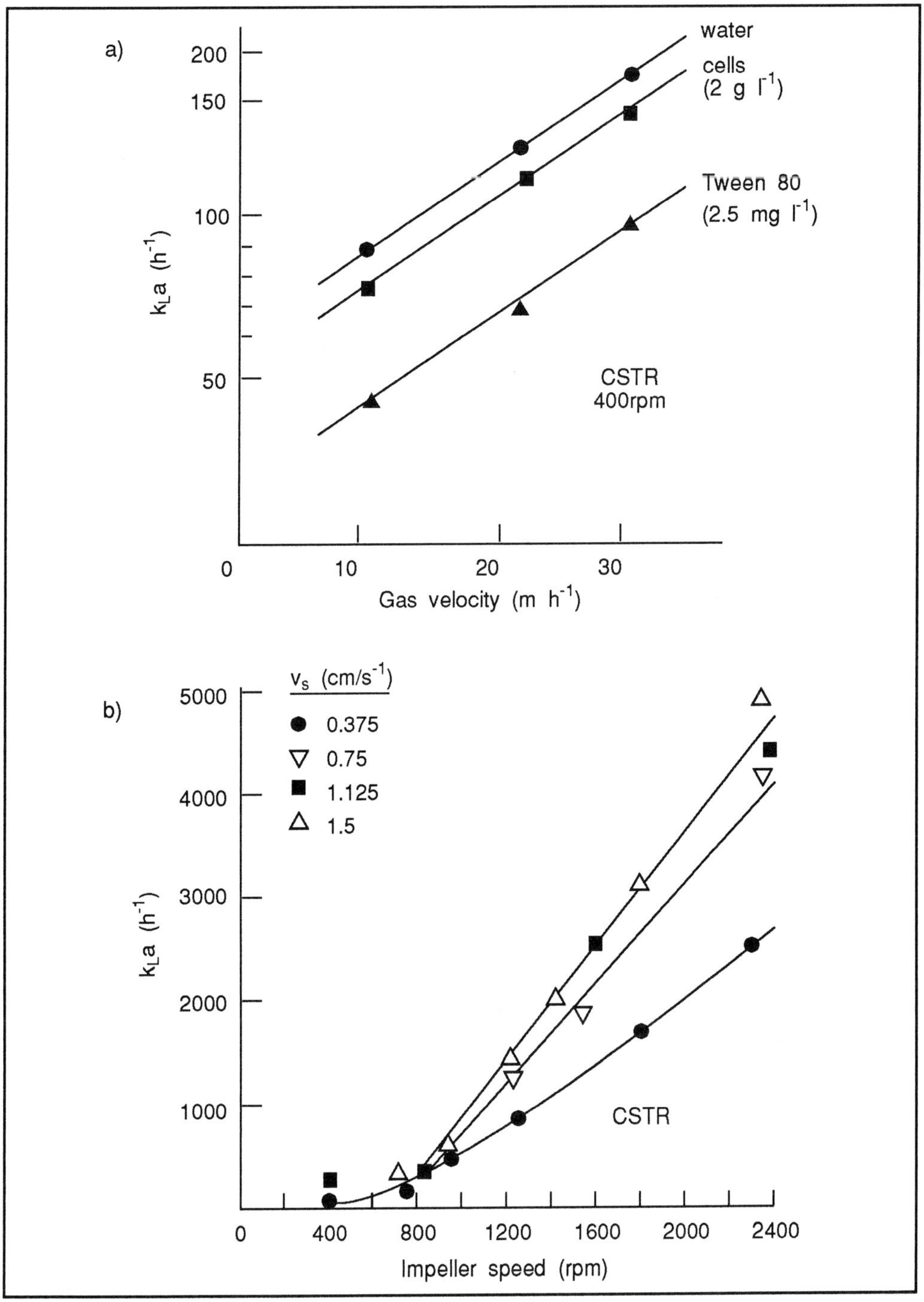

Figure 2.3 Influences of several factors on k_La (re-drawn and stylised from Atkinson and Mavituna 1983. Biochemical engineering and biotechnology handbook, MacMillan Publishers Ltd).
a) Influence of gas velocity on k_La in water and a culture in the presence and absence of the detergent Tween. b) Influence of impeller speed on k_La at a variety of gas velocities (v_s). c) Influence of dispersed biomass concentration on k_La for a CSTR at different stirrer speeds and gas velocities. d) Influence of mycelial biomass on k_La for a CSTR. e) Effects of antifoam concentration on OTR values for a CSTR.

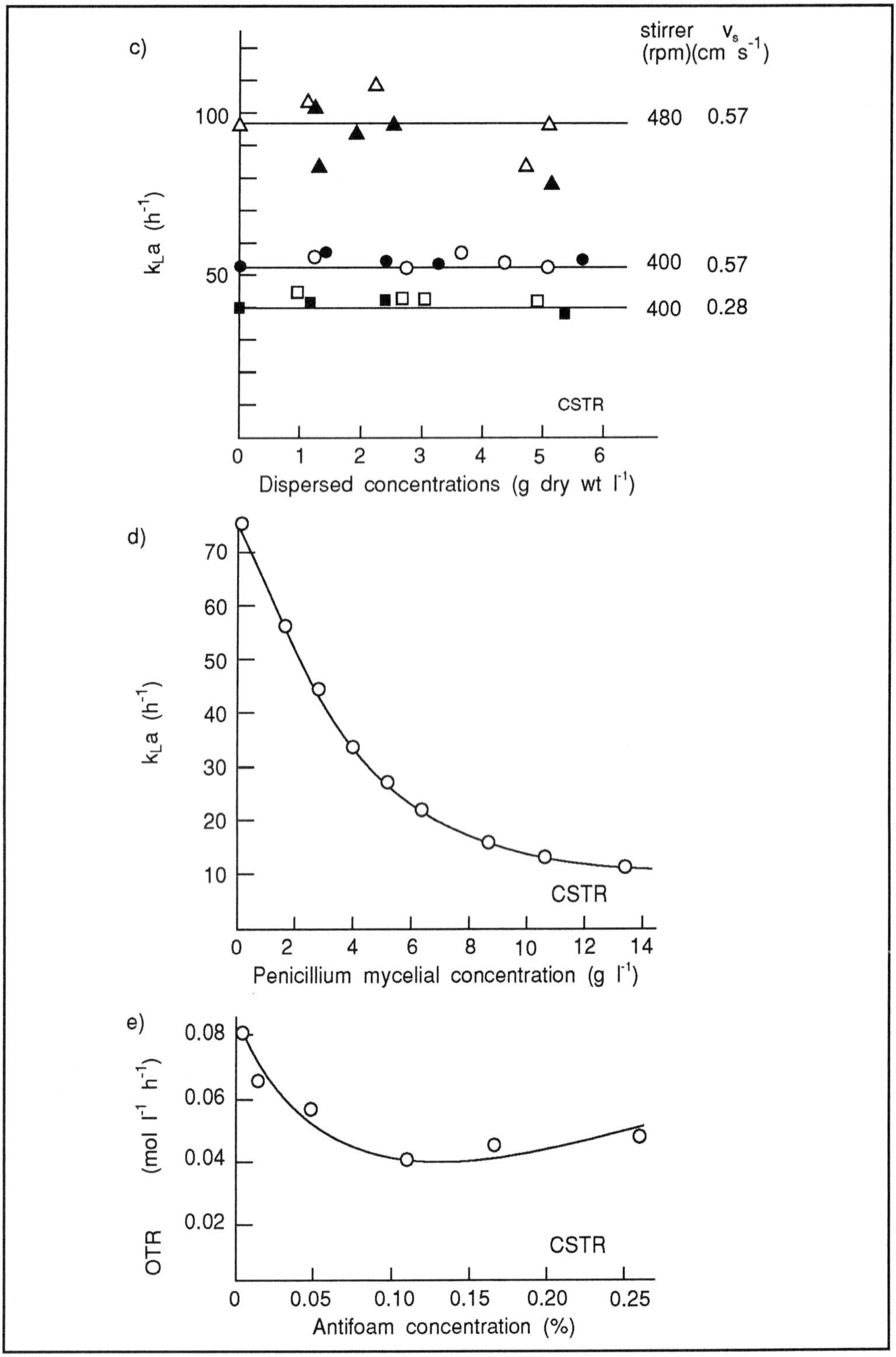

Figure 2.3 Continued

2.7 Measuring dissolved oxygen concentration (DOC) and dissolved oxygen tension (DOT)

As oxygen is a crucial substrate in aerobic biotransformation processes, the determination of the dissolved oxygen concentration (DOC) or the dissolved oxygen tension (DOT) is of great importance. The same is true for assessment of the OTR or more specifically, the k_La (see Section 2.8). Especially when scaling up bioreactors, a constant value of k_La is aimed at as a measure of reactor performance at larger scales.

The DOC and DOT can be determined by several methods. In principle, four types of methods can be identified: chemical, volumetric, tubing and, more recently, electrochemical methods using electrodes. For monitoring and control of bioreactors, the last method is usually applied. For the sake of completeness, the other methods will be treated briefly as well.

2.7.1 Chemical methods

Various chemical methods are available for titrimetric determination of the DOC. They are used for measuring the DOC in samples taken from eg fermenter fluid or from sewage water. Apart from off-line DOC determinations, chemical methods are employed for calibration of dissolved oxygen electrodes. The Winkler method is frequently used for this purpose.

The Winkler method

The Winkler method is one of the most widely used titrimetric methods. It involves the addition of an excess of a solution containing manganous ions and an iodometric procedure. Manganous ions are oxidised to permanganate by oxygen. The permanganate in turn forms iodine from iodide. Finally, the remaining iodide concentration can be determined by titrating with thiosulphate using starch as an indicator. Unfortunately, fermentation broths usually contain components which disturb this method by interfering with the redox reactions involved. Nevertheless, the method is useful in many cases, especially in DOC analysis in waste water for which it has become a standard method.

NADH oxidation

catalyst needed

Another important chemical method measures the oxidation of NADH to NAD^+ by oxygen. Because direct oxidation does not take place, NADH-oxidase, disrupted liver or heart mitochondria or phenazonium methosuphate (PMS) are added to catalyse the reaction. The conversion of NADH is followed spectrophotometrically at 340 nm.

Phenylhydrazine

This method involves the oxidation of phenylhydrazine. As this method is far less important than other methods, it will not be treated further.

disadvantages of chemical methods

The general drawbacks of all chemical methods are that they are rather laborious and usually not fast and sensitive enough for measurements in solutions with rapidly changing or limiting DOC's.

2.7.2 Physical methods

Volumetric methods

In the volumetric method, oxygen is removed from a liquid sample of known volume by formation of CO_2. The CO_2 is trapped and its volume is determined at known temperature and pressure. Application of the ideal gas law yields the number of moles of CO_2 and therefore of O_2. The original DOC in the sample is obtained by dividing by the sample volume. Although this method is less disturbed by interferences of medium components than for example the Winkler method, it is still not rapid and sensitive enough for monitoring DOC in fermentation broths.

Tubing method

extensive calibration required

In the tubing method, a thin-walled tube of semi-permeable material is immersed in the fermentation liquid. Usually, coiled and sterilisable PTFE (Teflon) or silicone tubes of about 10 m length, 1-3 mm diameter and a few hundreds of microns wall thickness are used. A slow metered stream of oxygen-free carrier gas like nitrogen is pumped through the tube end allowed to absorb the oxygen diffusing through the tube walls as described by Fick's law of diffusion. The oxygen content of the exit gas is monitored by a gas analyser or oxygen probe and this value can be related to the DOT in the fermentation liquid. Tube length, tube material, wall thickness, temperature and carrier gas velocity strongly influenced the oxygen diffusion process. Because of the need for extensive calibration, the use of this method is restricted.

2.7.3 Electrochemical methods

After their development by Leland Clark in the 1950's, membrane coated dissolved oxygen electrodes have become the most important instruments for monitoring DOT in bioconversions. The development of steam-sterilisable electrodes in the 1960's strongly supported their direct on-line application to bioreactors. Generally, the electrode membrane is selectively permeable to oxygen. Being impermeable to most ions like those of the electrolyte solution in the electrode, an oxygen probe measures without disturbing a process. These reasons as well as the accuracy and the user friendliness of dissolved oxygen electrodes account for their popularity amongst biotechnologists.

All oxygen electrodes are based on the reduction of dissolved oxygen at a noble metal cathode which is negatively polarised with respect to a reference anode. In polarographic electrodes, external voltage is required for negative polarisation and oxygen reduction, whereas in galvanic electrodes the required voltage results from the use of an electron supplying lead, zinc or cadmium anode.

A polarographic electrode usually contains a platinum or gold cathode, a silver/silverchloride anode and potassium chloride as electrolyte (Figure 2.4).

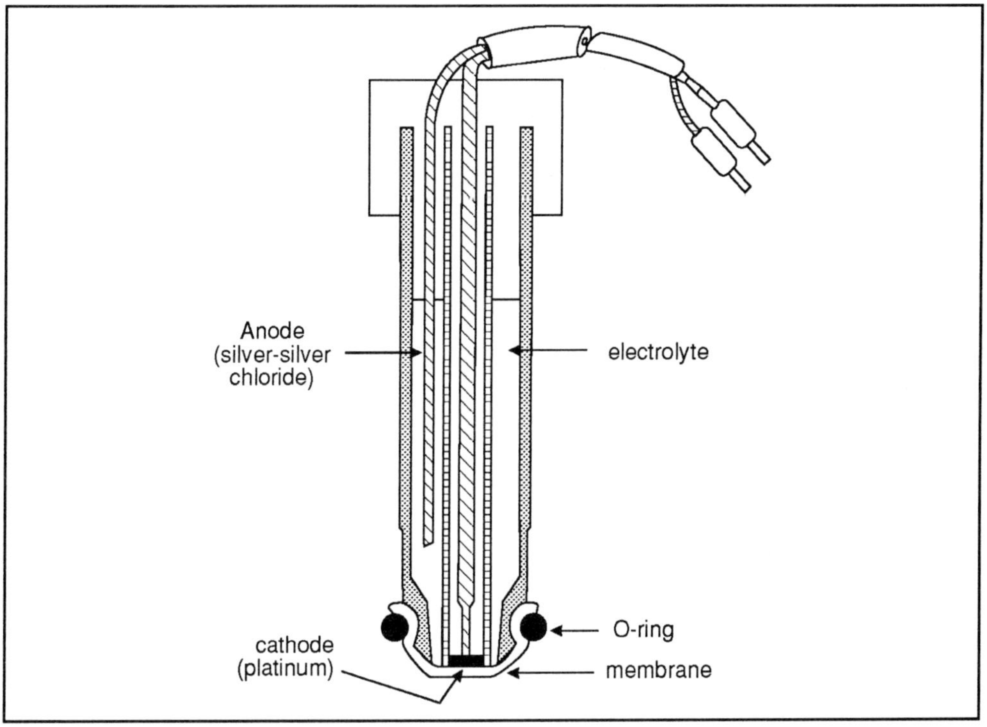

Figure 2.4 Polarographic dissolved oxygen electrode (see text for details).

When the anode of such an electrode is made 0.6-0.8 V negative, the following reactions take place:

cathode: $O_2 + 2H_2O + 2e^- \rightarrow H_2O_2 + 2OH^-$

$H_2O_2 + 2e^- \rightarrow 2OH^-$

anode: $Ag + Cl^- \rightarrow AgCl + e^-$

overall: $4Ag + O_2 + 2H_2O + 4Cl^- \rightarrow 4AgCl + 4OH^-$

The chloride ions are provided by the KCl electrolyte solution. For this reason and for removing the alkaline hydroxide ions, the solution has to be replaced from time to time.

The reduction of dissolved oxygen produces a voltage dependent current which can be depicted in a polarogram (current vs. voltage diagram) as in Figure 2.5. The reaction rate of oxygen reduction in the plateau region in Figure 2.5 is limited by the diffusion of dissolved oxygen through the electrode membrane into the electrode. This rate of diffusion is directly proportional to the oxygen activity or oxygen tension in the bioreactor liquid. Therefore, the current output is linear with the DOT in the plateau region of the polarogram. At higher voltages, the electrolysis of water producing hydrogen peroxide starts and disturbs the dissolved oxygen measurement by extra current generation.

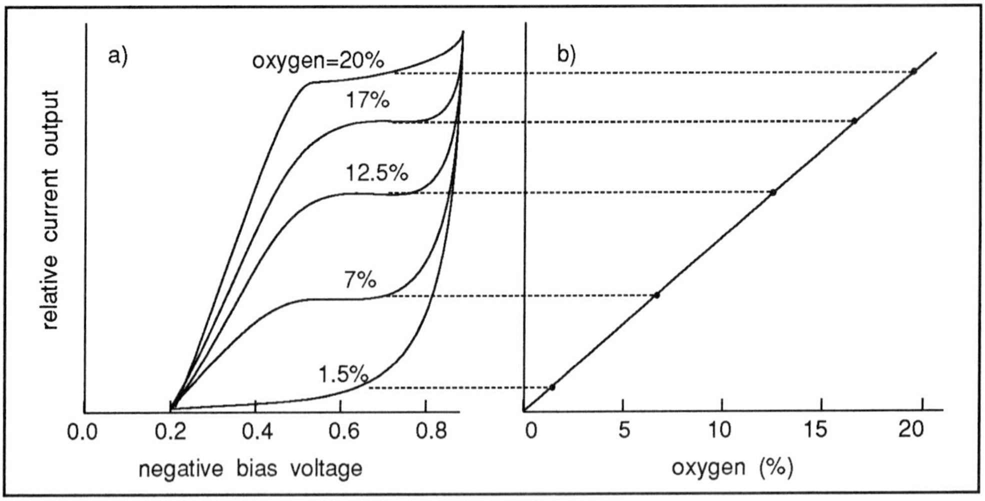

Figure 2.5 Polarogram of a dissolved oxygen electrode a) and the relation with the dissolved oxygen tension b).

In a galvanic electrode consisting of a silver cathode and a lead anode, the following reactions take place:

cathode: $O_2 + 2H_2O + 4e^- \rightarrow 4OH^-$

anode: $Pb \rightarrow Pb^{2+} + 2e^-$

overall: $2Pb + O_2 + 2H_2O \rightarrow 2Pb\,(OH)_2$

Some examples of galvanic electrodes are shown in Figure 2.6. By choosing the anode metal appropriately, the cathode will be charged 0.6-0.8 V negatively with respect to the anode. As with the polarographic electrode, the rate controlling step of oxygen reduction at the cathode is the rate of oxygen diffusion through the electrode membrane. Therefore, a linear relation between current output and DOT is observed with galvanic oxygen electrodes as well.

The response times of membrane coated oxygen electrodes is mainly dependent on membrane thickness and oxygen diffusivity in the membrane. In principle, it is insensitive to the nature of the surrounding fermenter liquid. However, one should expect an increase in response time and a decrease in current output if there is a substantial mass transfer resistance in the liquid film at the outside of the membrane. Typically electrodes have a response time of 1-10s.

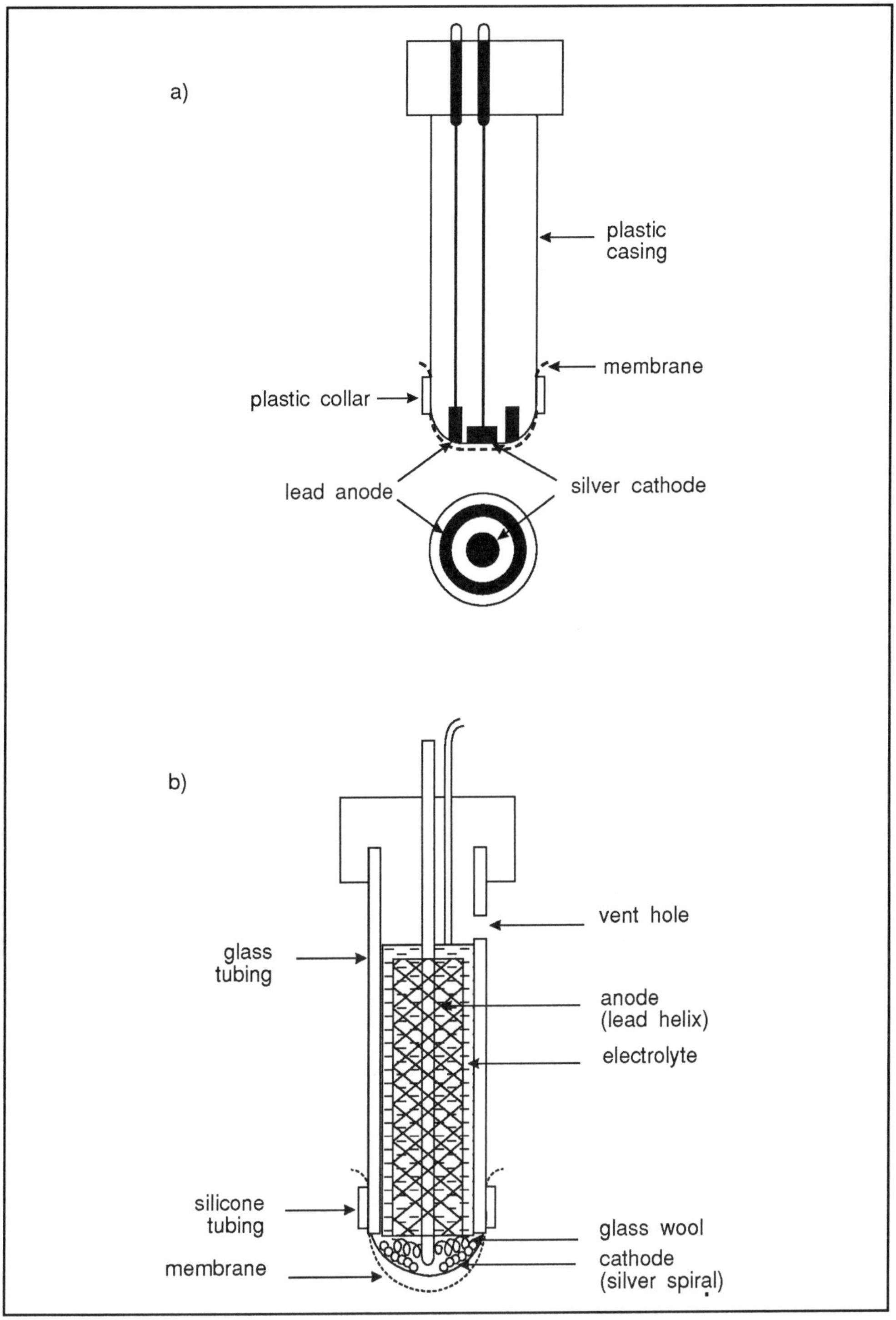

Figure 2.6 Schematic representation of galvanic dissolved oxygen electrodes (see text for details).

2.8 Measuring OTR and k_La

All parameters of Equation 2.11, OTR = k_La (C^* - C_{gL}), have to be known for assessment of the OTR of a system. Methods for measuring C_{gL} and C^* via DOT using oxygen electrodes are easily carried out. Although methods exist for determining k_L and a separately, the results are usually not as easily obtained nor as reliable as that of the determination of the volumetric mass transfer coefficient k_La.

The four most widely used, practical and reliable methods for measurements of OTR and k_La will be treated next.

2.8.1 Oxygen balance method

In this method, the OTR and k_La are determined without the need for disturbing or stopping a fermentation run. It is based on an oxygen mass balance over the bioreactor:

$$\phi_{oxygen,in} = \phi_{oxygen,out} + V_r \, . \, OTR \qquad (E - 2.27)$$

or

oxygen in (inlet gas)	=	oxygen out (outlet gas)	+	oxygen uptake (moles s^{-1}) (reaction, accumulation)

with: ϕ = mass flow of oxygen (mol s^{-1})

V_r = volume of reaction liquid (1 or m^3)

When using Equation 2.27 as such, oxygen accumulation in or removal from gas bubbles in the liquid medium and in the reactor head space should be negligible. If not, corrections for oxygen accumulation or depletion in bubbles acting as oxygen sinks will be necessary. In the case of negligible gas bubble volume and steady state conditions, three oxygen concentrations (in the inlet gas, the outlet gas and the reaction liquid) have to be measured. In addition, the two flow rates of the inlet and the outlet gas should be determined. Knowing V_r and C^*, the OTR can be simply evaluated. Substituting values into Equation 2.11 yields k_La.

precautions in determining OTR and k_La

The oxygen balance method requires precise measurements of gas flow rates and oxygen concentrations in the gas streams. Quite often, the oxygen flow rate and oxygen concentration differences will be small. Therefore, the accuracy of these measurements determines the reliability of the values for OTR and k_La. In addition, the oxygen concentration in the liquid phase should not be place dependent as one assumes a constant DOT all over the reactor. Alternatively, more oxygen electrodes should be inserted in the bioreactor liquid allowing the evaluation of an average DOT value. Such precautions are essential when dealing with bioreactors having dead or badly mixed zones as for example with viscous broths.

2.8.2 Sulphite oxidation method

This method is based on the oxidation of sulphite to sulphate by oxygen in the presence of a catalyst. Usually, divalent cations like Cu^{2+} or Co^{2+} are used as a catalyst. Then,

O_2 gas $\rightarrow$ O_2, dissolved

and

$$2\ SO_3^{2-} + O_2, \text{dissolved} \rightarrow 2\ SO_4^{2-}$$

The sulphite oxidation proceeds much faster than oxygen transfer from gas bubbles into the liquid and continues as long as oxygen is available. Thus, the OTR is the rate controlling step in this reaction. The remaining sulphite in samples withdrawn from the fermenter is reacted with excess iodine followed by back titration with thiosulphate using starch as an indicator. When starting with a known, excessive amount of sulphite, the original amount of dissolved oxygen can be calculated from the stoichiometry of the reaction.

The sulphite oxidation method often yields higher values for OTR and k_La than other methods. It is influenced by pH, oxygen absorbents in the liquid, the type of catalyst, the type of equipment and the purity of the sulphite used.

2.8.3 Static method

In the static method, also called static gasing out, the increase of the oxygen concentration in the medium is followed. As a non-respiring biological system is required for this method, the micro-organisms are killed or cell respiration is blocked by addition of respiratory inhibitors or toxins like azide, malonate or flurocitrate. Fluxing with nitrogen and stirring of this non-respiring system is continued until the oxygen concentration has become zero. Then, aeration is switched on at the required gas flow and stirrer speed. The increase of the oxygen concentration is followed until oxygen saturation ie C* is reached. The change of the dissolved oxygen concentration C in time is described by:

$$\ln(1 - C/C^*) = -k_La \cdot t \qquad \text{(E - 2.28)}$$

Figure 2.7 shows the corresponding graph with the linear relation between ln (1 - C/C*) and t. As an alternative for the oxygen concentration C, the oxygen tension pO_2 (in % saturation) can be inserted in Equation 2.28 and Figure 2.7. Then, C is replaced by pO_2.

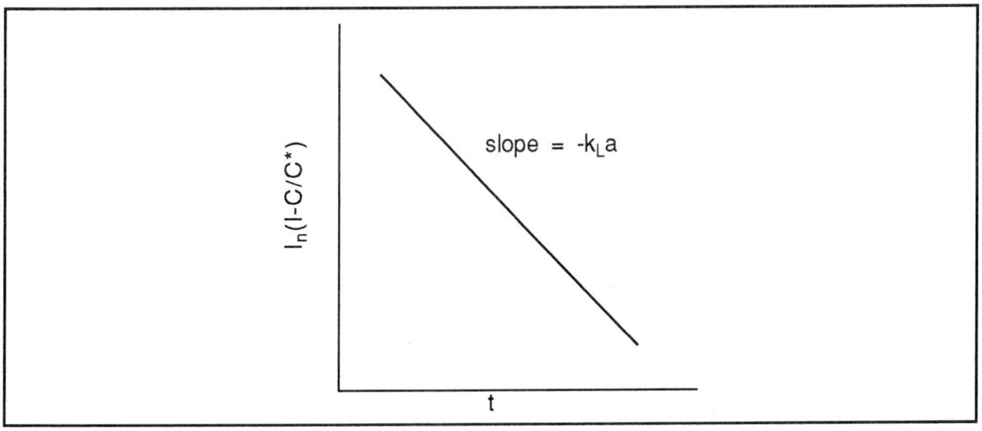

Figure 2.7 Evaluation of k_La in the static method. In this method ln(1 - C/C*) against t is plotted. Note that the concentrations of oxygen (C and C*) can be replaced by oxygen tension (pO_2) in time (t) % saturation.

Π If C is replaced by pO_2 expressed as % saturation in Equation 2.28, what is the value of C*?

You should have realised that since C* is the maximum solubility, then C* must be 100% saturation.

limitations and uses of the static method

When using the static method, one has to take care that no oxygen accumulation or depletion occurs in the gas phase of the fermenter head space or that large amounts of nearly stagnant bubbles develop in viscous broths. This implies that the experimental period should not be too long. Furthermore, one has to bear in mind that the non-respiring biological system may differ from the actual fermentation conditions. Despite these (potential) drawbacks, the static method allows an easy and reasonably accurate determination of the k_La to be expected in the actual process. The OTR during fermentation can be evaluated when C and C* are known from for example oxygen electrode measurements.

2.8.4 Dynamic methods

The main difference between the static method and the dynamic method, also called dynamic gasing out, is that actively respiring cells are present in the dynamic method. This means that the dissolved oxygen concentration is influenced by these cells when aeration is switched on or off.

In the dynamic method, the time course of the dissolved oxygen concentration is followed after the aeration of an actively respiring biological system is stopped and switched on again. The oxygen concentration is allowed to drop until the critical oxygen concentration C_{co} is reached. The critical oxygen concentration is the dissolved oxygen concentration leading to a cell specific growth rate of $\mu = 0.99\ \mu_{max}$. One assumes that the cellular metabolism is not yet influenced at C_{co}. At or above C_{co}, aeration is started again and the dissolved oxygen concentration is monitored until the initial dissolved oxygen concentration C_{io} is re-established. Figure 2.8 shows the time course of the dissolved oxygen concentration in such an experiment.

Throughout the experimental period, the changing dissolved oxygen concentration dc/dt is given by:

$$dC/dt = k_La\,(C^* - C_{io}) - OUR * X \qquad \text{(E - 2.29)}$$

in which:

OUR = oxygen uptake rate (moles g biomass^{-1}. s^{-1})

X = biomass concentration (gl^{-1})

Before stopping aeration at $t = t_1$, the system is in steady state, implying $dC/dt = 0$ and

$$k_La\,(C^* - C_{io}) = OUR * X \qquad \text{(E - 2.30)}$$

or

$$k_La = OUR * X \,/\, (C^* - C_{io}) \qquad \text{(E - 2.31)}$$

The decrease of the oxygen concentration at $t_1 < t < t_2$ after turning off aeration at t_1, is given by:

$$- dC/dt = OUR * X \qquad \text{(E - 2.32)}$$

in which dC/dt is constant when $C > C_{co}$ and X remains the same, indicating that the experimental period should not be too long. This part of the graph allows the calculation of OUR . X as it is the slope of the straight declining line in Figure 2.8a. If the oxygen concentration drops beyond C_{co} the line will become asymptotic and deviate from linearity with time.

Knowing OUR, X and C_{io} the calculation of k_La using Equation 2.30 is possible. However, the accuracy of this calculated value for k_La is rather low and a method involving equations 2.33 and 2.34 described below is usually preferred.

At resumed aeration Equation 2.29 holds again. The increasing dissolved oxygen concentration is described by Equation 2.33 which is a rearrangement of Equation 2.29.

$$C_{io} = C^* - 1/k_La . (dC/dt + OUR * X) \qquad \text{(E - 2.33)}$$

As OUR * X was determined during the period of decreasing oxygen concentration at $t_1 < t < t_2$, a graphical evaluation of k_La is possible when C is plotted against (dC/dt + OUR * X) as is done in Figure 2.8b. Replacement of OUR * X in Equation 2.33 by the term k_La (C^* - C_{io}) from Equation 2.30 and subsequent integration of Equation 2.33 yields:

$$\ln (C_{io} - C / C_{io} - C_2) = -k_La (t - t_2) \qquad \text{(E - 2.34)}$$

at $t > t_2$ (t_2 being the moment of resuming aeration). Then the plot of $\ln (C_{io} - C/C_{io} - C_2)$ versus ($t - t_2$) yields a straight line with slope $-k_La$ (Figure 2.8c). Note that the left term of Equation 2.34 equals zero at $t = t_2$ and that no mathematical solution can be obtained at $t = t_3$ when C has become C_{io} again.

The dynamic method is widely used as it can be applied during a fermentation run under the actual fermentation conditions and requires only a stop watch and an oxygen electrode. The usually small errors in k_La arise from the delay caused by electrode response times and from using an oversimplified mathematical description of the various stages of the dynamic method. Especially electrode characteristics with respect to response times are important and might require correction of the experimental data obtained. Thus, electrodes having response times which are comparable or even slower than the actual changes in dissolved oxygen concentration in the bioreactor liquid might cause serious errors in k_La determinations. One oversimplification of the mathematical representation is neglecting the exchange of oxygen between the gas in the head space of the bioreactor and the liquid medium. Therefore, whenever the experiment lasts a substantial period of time, a complex correction of Equation 2.29 should be made by introducing terms describing the accumulation or depletion of oxygen in this gas phase. To some degree, this so called surface aeration effect can be avoided by filling the fermenter head space with nitrogen gas. Additional accumulation-depletion effects might occur in highly viscous broths when many nearly stagnant gas bubbles are present in the liquid medium.

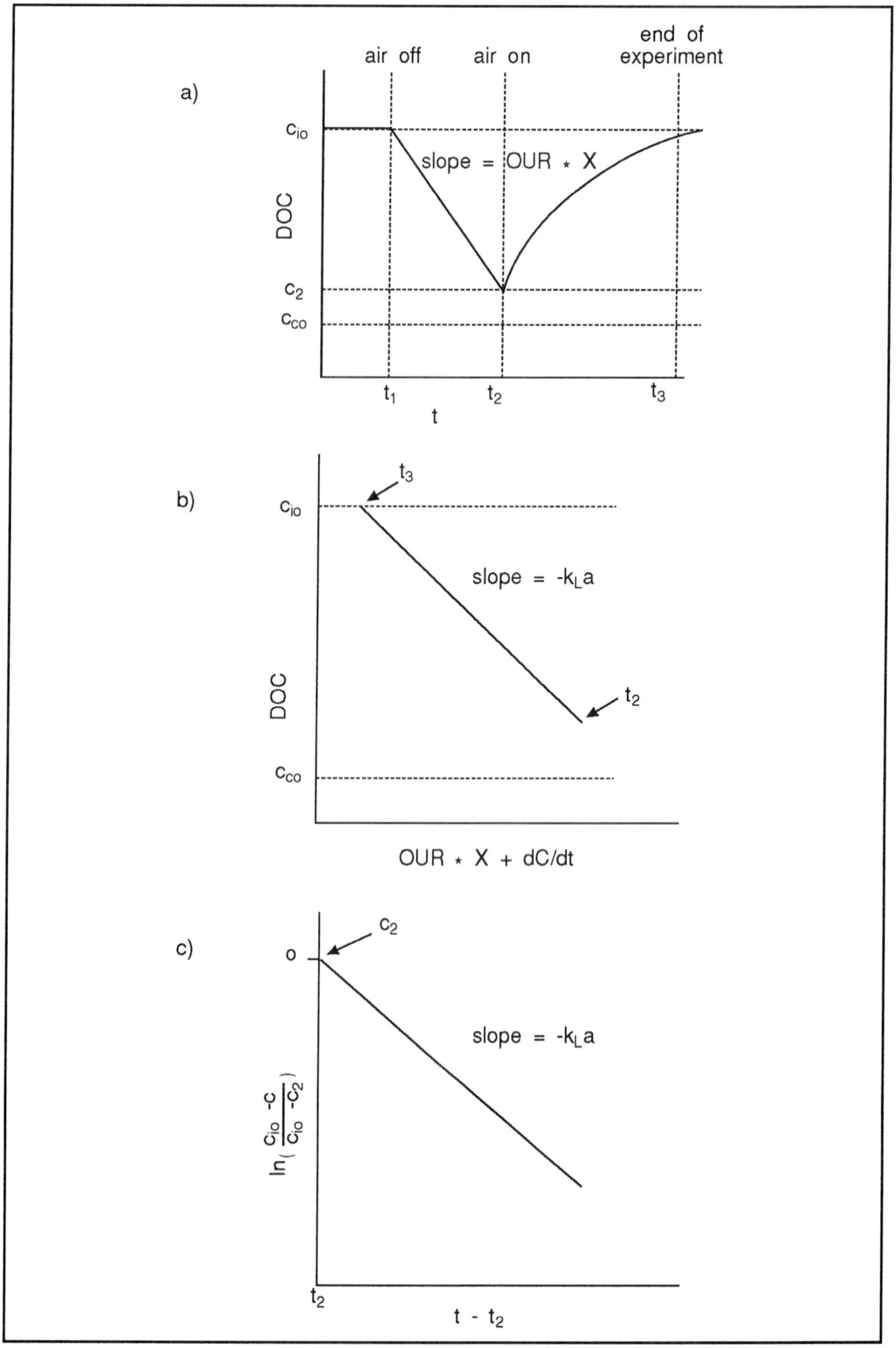

Figure 2.8 a) Time course of dissolved oxygen concentration in the dynamic method, b) the determination of k_La according to Equation 2.33 and c) the determination of k_La according to Equation 2.35. See text for details. DOC = dissolved oxygen concentration, OUR = oxygen uptake rate.

SAQ 2.6

Using the static method, the following time course of pO_2 was measured by a dissolved oxygen electrode.

t	(min)	0	1	2	3	4	5	6	7
pO_2	(%)	0	26	45	59	70	78	83	88

1) Calculate k_La.

2) Would you expect any problems with respect to electrode response times?

SAQ 2.7

Answer true or false to the following statements.

1) k_La values are not greatly influenced by biomass concentration when the biomass is in the form of long filaments.

2) Generally, an increase in impeller speed will increase the k_La value.

3) The Winkler method for determining dissolved oxygen tension enables on-line measurement of oxygen.

4) The tubing method for determining dissolved oxygen concentration requires extensive calibration.

5) The dynamic method for determining k_La is widely used because it can be used while a bioreactor is in use and with little disturbance to its performance.

6) If the experiment shown in Figure 2.8a was extended over a long period (ie $t_1 \rightarrow t_2$ is a long time delay) then the slope of the line between t_1 and t_2 would probably not be linear.

Summary and objectives

In this chapter we have examined gas-liquid mass transfer and have shown that the driving force for diffusion is the concentration difference of the dissolved gas and its maximum solubility. The maximum solubility (C^*) of sparingly soluble gases in liquids is described by Henry's Law. We also explained that the gas transfer rate (GTR) or oxygen transfer rate (OTR) can be adequately described in terms of the volumetric mass transfer coefficient (k_La) and the concentration difference ($C^* - C$). The value of k_La is influenced by a wide variety of parameters and may be evaluated by monitoring the concentration of the dissolved gas after changing the gas flow rates.

Now that you have completed this chapter you should be able to:-

- describe how gas molecules, especially oxygen, are transferred from gas bubbles to cells and to identify the likely rate limiting steps;
- calculate the concentration of dissolved gas using Henry's law and supplied data;
- calculate maximum biomass concentration that can be supported in devices of known k_La values.
- describe and select appropriate methods for measuring O_2, OTR and k_La values.

The conventional stirrer tank reactor

The conventional stirrer tank reactor

3.1 Introduction

One of the oldest and still the most widely applied type of bioreactor in fermentation technology is the Conventional Stirred Tank Reactor (CSTR). The CSTR consists of a cylindrical vessel with liquid height H and diameter T, their ratio depending on the application but usually in the order of 1 to 3 (Figure 3.1). The vessel is normally equipped with four baffles. A rotating stirrer is arranged in the vertical axis of the vessel. In an aerobic process air is introduced through a sparger at the bottom of the vessel and rises freely until it contacts the stirrer.

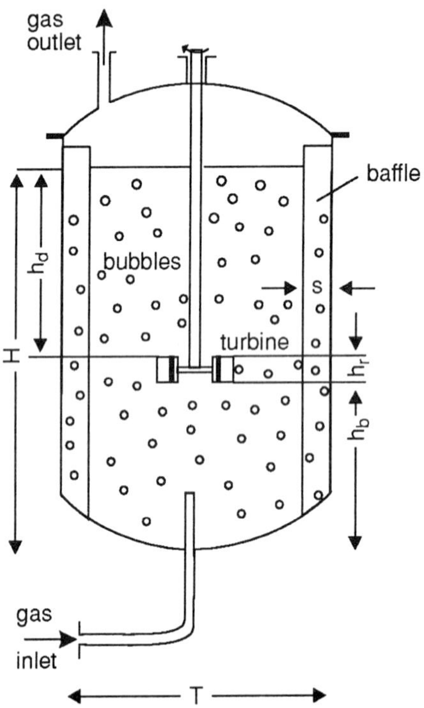

Figure 3.1 General layout of a conventional stirred vessel reactor. (See text for details - redrawn from Brauer, H (1985), Biotechnology Vol 2 Editors H J Rehm and G Reed, VCH, Weinheim, 417-474).

The stirrer plays an important role in the CSTR. It facilitates energy transfer to the fermentation broth in order to achieve mixing and oxygen transfer. This leads to the desired homogeneous conditions in the reactor.

The energy transferred by the stirrer to the fluid is required for a suitable fluid movement in the vessel. Fluid movement is invariably connected with energy dissipation. Mechanical energy is thereby converted into thermal energy. The so-called loss of mechanical energy must be balanced by mechanical energy constantly transferred to the fluid by the stirrer. The function of a stirrer as a energy transfer and mixing element may be described as the ability to achieve and maintain a prescribed fluid movement within a vessel with a minimum of energy input.

The distribution of oxygen consists of two parts. Apart from the distribution of the gas bubbles in the fluid which mainly depends on the fluid movement achieved by the stirrer, the most important function of the stirrer as an element for gas dispersion is the generation of gas bubbles.

types of stirrers

Due to the combination of these functions, the shape, speed and arrangement of the stirrer will be the result of a process of compromise mainly based on practical experience. The stirrers used are in general of a simple construction: disc turbine (Rushton stirrer), vaned disc, open turbine and variable pitch (Figure 3.2).

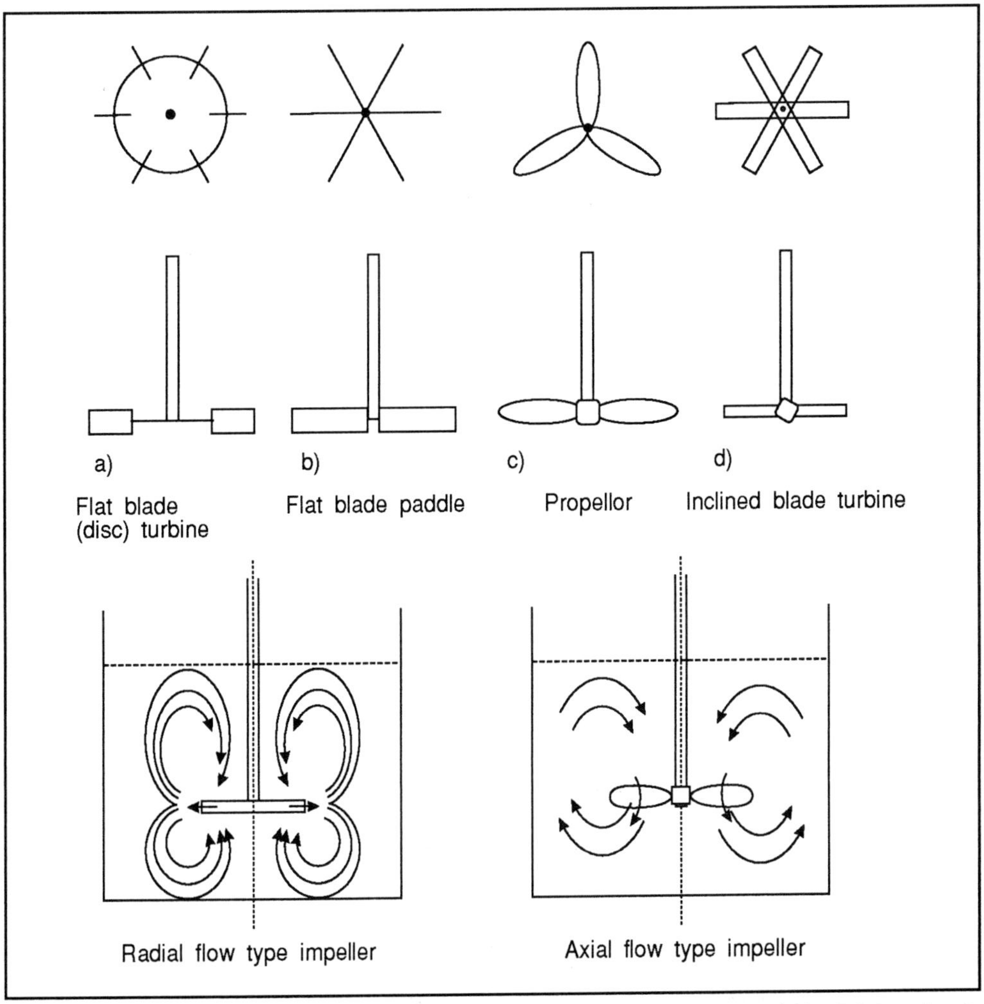

Figure 3.2 Main types of stirrers used in CSTRs. a) disc turbine b) open turbine c) variable pitch impeller d) propeller. Redrawn from Solomons, G L 1965 ' Materials and Methods in Fermentation' Academic Press).

disadvantages of CSTR

The CSTR has some disadvantages over other bioreactors such as the bubble column (Chapter 4): the costs are higher and the shear forces are rather high due to the mechanical stirring. Therefore, the CSTR is less suitable for cells which are shear-sensitive. Although the power input required to reach a certain oxygen transfer rate is for both reactors approximately the same, the power input per volume for a CSTR is often higher. The dissipated power input in stirred bioreactors is of the order of 0.5 - 5 kW m^{-3}. Usually the installed power is usually a factor of 2 higher because of the power loss in motor, gearbox, beltdrive and bearing devices. Since the mechanical energy is converted to thermal energy, heat removal via external cooling loops becomes essential in CSTRs. In some cases a bubble column can not provide the dissipated power needed by the process. In that case a stirred tank reactor has to be used.

Π See if you can list two examples in which a CSTR has to be used because a bubble column may not be suitable.

There are several possibilities but the two most common situations arise when we use viscous broths (filamentous mycelium, viscosity-increasing substrates or products) or when extremely high oxygen transfer rates are required.

In general, the stirred vessel of the conventional design is best suited for (fed) batch operation, and for the fermentative production of relatively inexpensive fine chemicals such as amino acids, vitamins, antibiotics and enzymes.

In the following sections it will be explained that the flow patterns around the stirrer and the liquid properties are very important with respect to liquid mixing, gas dispersion and mass and heat transfer. To give you some idea of the complexity of these processes examine Figure 3.3. Each of the boxes refers to a component involved in the mass transfer processes which occur in bioreactors. Clearly to fully understand the mass processes of bioreactors, or to design and operate bioreactors with predictable performances we need to understand and model each of these components. If you choose any of the parameters described in Figure 3.3 you should be able to appreciate at least in principle if not in detail why each is important. (Why not have a go. We will be examining these in more detail in later sections).

The complexity of these mechanisms makes a quantitative assessment quite difficult. Although the empirical behaviour of model liquids in the CSTR is well established, the complexity of two or more phase fermentation broths makes accurate prediction of industrial systems difficult. When an ample margin is allowed, calculations are sufficiently accurate. However, if narrow margins are required, parameters must be determined experimentally.

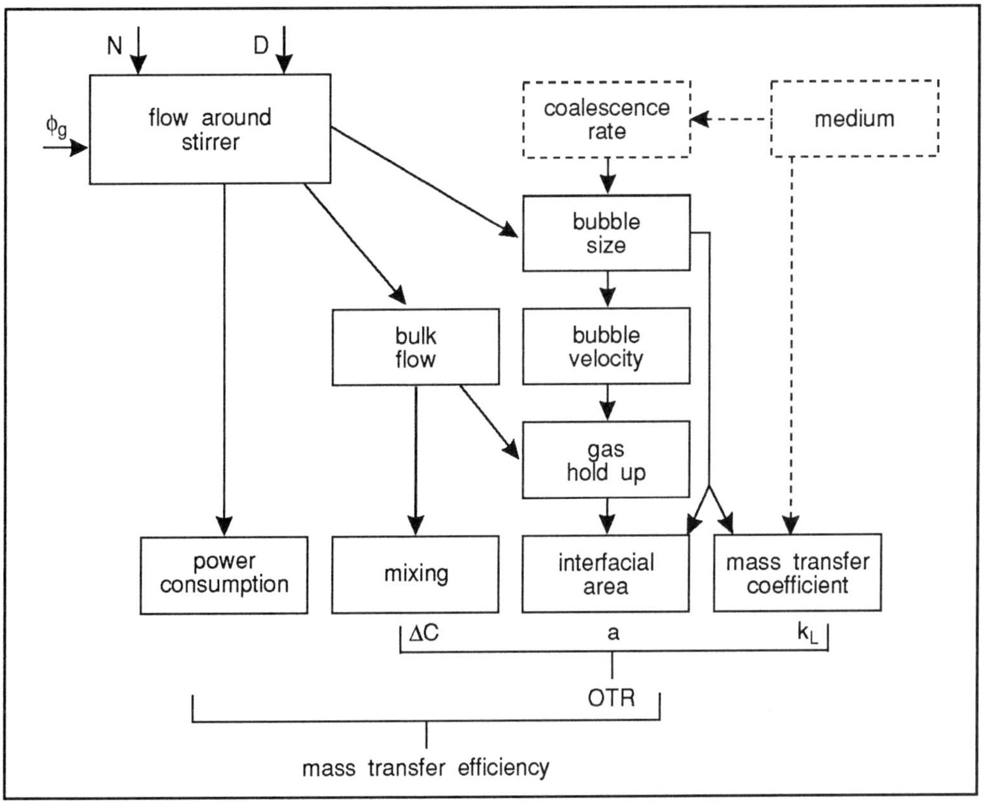

Figure 3.3 Parameters involved in modelling mass transfer in bioreactors. Note N = stirrer speed, D = stirrer diameter ϕ_g = volumetric gas flow rate. Other terms will be described in the text. (Redrawn from T U Delft (1989) 'Bioprocess technologie' Editor; van der Lans RGJM).

3.2 The power consumption

The power dissipation by the stirrer which is essential for mixing and oxygen transfer, is equal to the power consumption. In the turbulent regime (Reynolds number > 10^4) in which power is converted into kinetic energy of the liquid flow, the ungassed power consumption of the radially or axially pumping stirrer P_s is given as:

$$P_s = \phi_p \, . \, 0.5 \, . \, \rho \, . \, \langle v \rangle^2 \qquad \text{(E - 3.1)}$$

in which ϕ_p is the pumping capacity of the stirrer, ρ is the density of the liquid and <v> is the average velocity (we will define this velocity a little later).

significance of Reynolds numbers

If you are unfamiliar with Reynolds numbers we would recommend that you read the BIOTOL text 'Bioprocess Technology: Modelling and Transport Phenomena'. Essentially, if Reynolds numbers are low, then the flow is laminar. High Reynolds number (values >10^4) means that the flow is turbulent and the mixture is well mixed.

For a radially pumping stirrer it is assumed that:

$$\phi_p = \langle v \rangle \, . \, \pi \, . \, D \, . \, h_r \qquad \text{(E - 3.2)}$$

in which D is the stirrer diameter and h_r is the stirrer blade height. When $h_r = \frac{D}{4}$, Equation 3.2 also applies for axially pumping stirrers. In this case:

$$\phi_P = \pi \cdot D^2 \cdot \langle v \rangle / 4 \qquad \text{(E - 3.3)}$$

and assuming that:

$$\langle v \rangle \sim \pi \cdot N \cdot D \text{ and } \langle v \rangle^2 = \langle v^2 \rangle \qquad \text{(E - 3.4)}$$

in which N is the stirrer speed

The power of an axially pumping stirrer in the unaerated case is:

$$P_s = N_P \cdot \rho \cdot N^3 \cdot D^5 \qquad \text{(E - 3.5)}$$

power number

in which N_P is the stirrer power number. The stirrer power number is given by $P/\rho N^3 D^5$.

In the aerated case the gassed power consumption P_g of the stirrer is reduced to about half of the unaerated case, so

$$P_g \approx 0.5 \cdot P_s \qquad \text{(E - 3.6)}$$

This phenomenon is caused by cavity formation behind the stirrer blades as will be discussed in the next section on flow pattern. Many authors have described correlations for the P_g and the P_S ratio. Reuss in 1980 used dimensional analysis to obtain the following correlation:

$$\frac{P_g}{P_s} = 0.0312 \cdot Fr^{-.1.6} \cdot Re^{0.064} \cdot Fl^{-0.38} \cdot \left(\frac{T}{D}\right)^{0.8} \qquad \text{(E - 3.7)}$$

Froude number
Reynolds number
Aeration number
Flow number

in which Fr is the stirrer Froude number (= N^2D/g in which g is the gravitational constant), Re is the stirrer Reynolds number (= ND^2/η in which η is the viscosity) and Fl is the Aeration number or Flow number (= ϕ_g/ND^3 where ϕ_g is the volumetric gas flow rate). T = diameter of the vessel and D = stirrer diameter. (Note the strategy of dimensional analysis is described in the BIOTOL text 'Bioprocess Technology: Modelling and Transport Phenomena').

Although the relationship shown in Equation 3.7 looks complex, it has been shown to fit data from systems with a liquid volume up to $30m^3$ and a T/D ratio of $1.8 < T/D < 3.7$, with confidence limits of 15%.

(If you are interested to follow up how this relationship was established, the key article is Reuss, M, Baypai, RK, Lenz R, Niebelschutз and Papalexiou, A (1980) 6th Internation Fermentation Symposium, London Ontario, Canada paper F-7.2.1).

Hughmark (1980) derived another correlation (Ind Eng Chem Proc Des Dev 19, 638-341). He derived his correlation (Equation 3.8) from 248 sets of literature data.

$$\frac{P_g}{P_S} = 0.1 \cdot \left(\frac{\phi_g}{NV}\right)^{-0.25} \cdot \left(\frac{N^2D}{gh_rV^{0.67}}\right)^{-0.2} \qquad \text{(E - 3.8)}$$

in which V is the volume of the liquid and h_r the stirrer blade width. This relation is claimed to be valid for systems with a liquid volume up to 50 m^3 and a superficial gas velocity up to 0.053 ms^{-1}.

Yet another correlation has been derived by Brown (1981, Instn. Chem. Eng. Symp. Series 64 NI). Brown correlated his experimental data from a production scale reactor with the following empirical interdependency:

$$\frac{P_g}{P_s} = a \cdot \exp(-b * \phi_g) \qquad \text{(E - 3.9)}$$

in which the factors a and b are influenced by the gas flow rate and the stirrer diameter.

Π We do not anticipate that you should remember the specific details of these correlation, but it is worthwhile to know that they exist and to be able to use them as they appear to hold quite well within certain limits. It might be sensible to produce a summary sheet using the following headings.

Author	$\frac{P_g}{P_s}$ ratio =	where
Reuss *et al*	$0.0312\ Fr.^{-1.6}.Re^{0.064}.Fl^{-0.38}\left(\frac{T}{D}\right)^{0.8}$	Fr = Froude number Re = Reynolds number

We would not anticipate that you would need to follow up the original papers in which these relationships were established.

Π We have also used several dimensionless numbers. You may already familiar with many of these. Nonetheless it may be well worthwhile to examine the list of dimensionless numbers provided in the appendix to refresh your memory. If you are unfamiliar with dimensionless numbers we would suggest you read the appropriate section of the BIOTOL text 'Bioprocess Technology: Modelling and Transport Phenomena'. Essentially dimensionless numbers are derived by combining a variety of parameters which influence particular processes. They are essentially ratios between two fundamental features (eg diffusive/convectional transport).

The correlations described above have been used to calculate power consumption at different rates of aeration (ie P_g/P_s) against the flow number (Figure 3.4). The figure shows that the power consumption decreases with increasing aeration. Note the good correlation between Brown's, Hughmark's and Reuss *et al's* derivations.

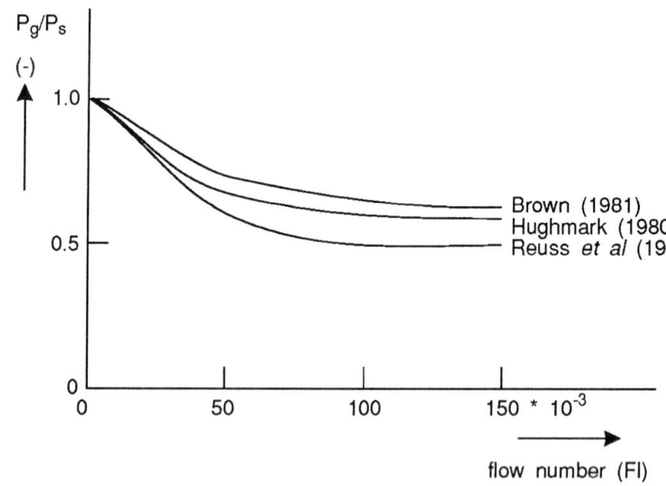

Figure 3.4 Power aeration curve for a vessel with a liquid height (H) to vessel diameter (T) ratio of 0.47 and for stirrer speed (N) = $2.6s^{-1}$ (Adapted from Oosterhuis, NMG 1984, PhD Thesis Th Delft, The Netherlands).

SAQ 3.1

1) A stirrer has a power number (N_p) of value 10. If the stirrer speed (N) is 1 s^{-1}, its diameter D = 1 m and the density of the medium is 1000 kg m^{-3}, what is the ungassed power input (P_s) by the stirrer? (Assume the stirrer is an axially pumping type with a diameter to width ratio of 4:1).

2) What would P_s be if the stirrer speed was increased to $2s^{-1}$?

3) What would be the approximate gassed power consumption (P_g) of the stirrer if the device described in cases 1) and 2) above was aerated?

SAQ 3.2

1) Using Reuss *et al*'s (1980) relationship, what would happen to the P_g/P_s ratio if the diameter of the vessel is doubled?

2) Compare the effects of the volumetric gas flow rate ϕ_g on the $\frac{P_g}{P_S}$ ratio as described by Reuss *et al* (1980) and Hughmark (1980).

3.3 Flow patterns in aerated tank reactors

In an aerated stirred vessel, the main dispersing activity occurs in the immediate vicinity of the stirrer. As a consequence, for a good qualitative understanding of this process, knowledge of the flow behaviour around the stirrer blades is essential. This flow behaviour depends on the stirrer configuration, stirrer speed and gas inflow rate. Here, we will discuss the flow behaviour around a disc turbine stirrer at constant stirrer speed and increasing gas inflow rate.

In an ungassed system the flow behind the blades is dominated by a pair of trailing vortices as shown in Figure 3.5. This results in very intensive shear rates (τ up to 100 N) and distinct under pressure.

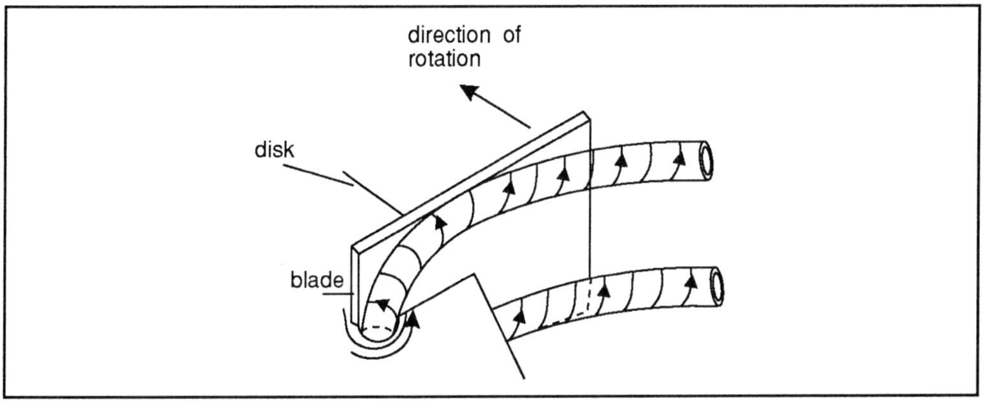

Figure 3.5 Schematic view of the trailing vortex pair behind a turbine impeller blade. Redrawn from Warmoeskerken MMCG (1986) PhD Thesis TH Delft, The Netherlands.

The shape of the cavity behind the stirrer blades changes with the gas inflow rate. With a small gas inflow rate, the gas will be drawn into the centre of the vortex. In this way a so-called ventilated vortex cavity is created (Figure 3.6). Intensive dispersion of the gas occurs at the tip of the cavity where dispersed bubbles enter the fluid bulk.

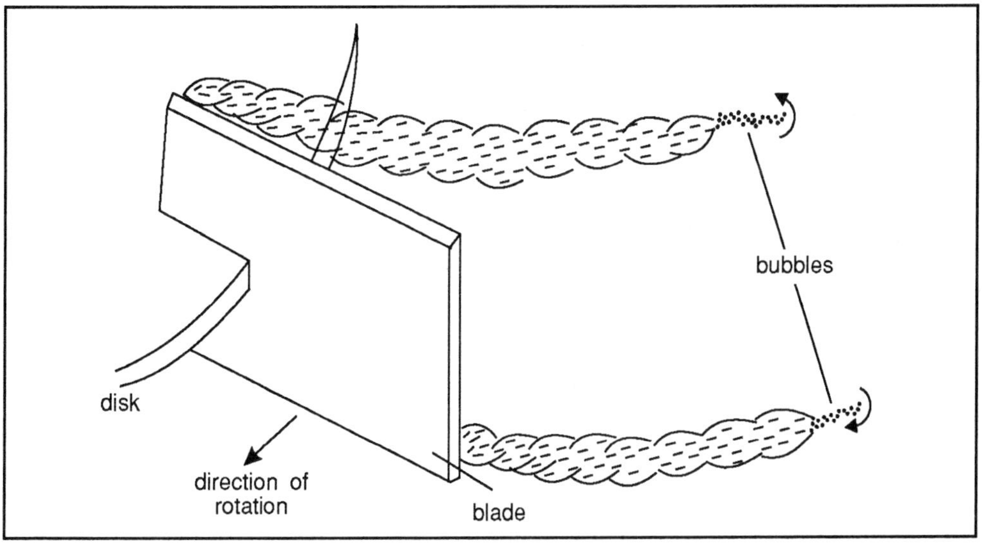

Figure 3.6 Ventilated vortex cavity. Figure shows the cavity shape for small gas feeds. (Redrawn from Warmoeskerken MMCG (1986) PhD Thesis, Delft, The Netherlands). Note that the bubbles are drawn into the centre of the vortex pair behind the blade and the bubbles enter the bulk liquid at the tip of the vortex cavity.

With higher gas inflow rates the vortex completely fills up with gas, the vortex diminishes and the cavity clings to the back of the blades (Figure 3.7). Increasing the gas inflow rate further finally leads to the formation of large cavities (3 large and 3 clinging cavities in case of a 6-bladed turbine, Figure 3.8).

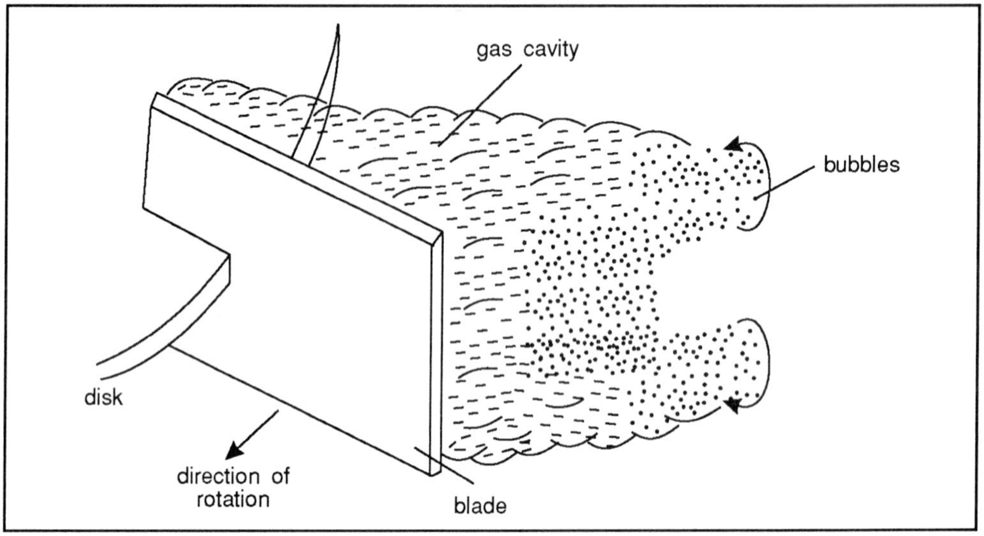

Figure 3.7 Clinging cavity formed behind the stirrer blade when medium gas feed rates are used. (Redrawn from Warmoeskerken, 1986, PhD. Thesis TU Delft, The Netherlands). The cavity in this case is described as a clinging cavity.

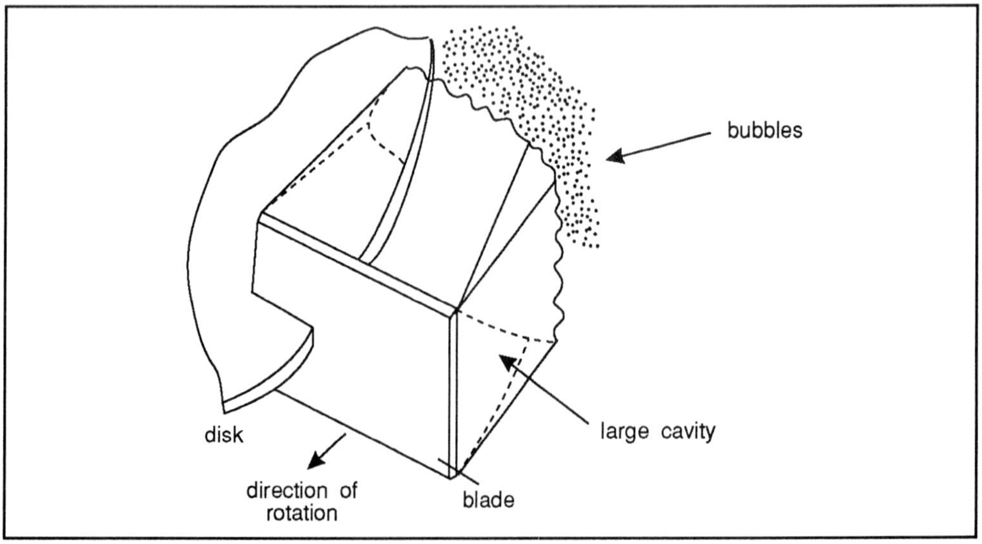

Figure 3.8 Large gas cavity formed behind the blade when large gas feed rates are used. (Redrawn from Warmoesskerken 1986, PhD Thesis, TU Delft, The Netherlands).

flow patterns depend upon stirrer configuration, speed and gas inflow

The overall flow patterns are also dependent on stirrer configuration, stirrer speed and gas inflow rate. Without gas, the liquid flow pattern consists of two circulation loops (see Figure 3.9e). When gas is sparged, different gas circulation patterns can exist as shown in Figure 3.9. At high stirrer speeds and low gas flow rates (Figure 3.9d) the circulation pattern largely resembles that of the non-aerated case (Figure 3.9e). Increasing the gas flow rate or decreasing the stirrer speed changes this pattern gradually until the gas dominates the flow pattern (Figure 3.9b). Now the stirred vessel flow pattern begins to resemble that of a bubble column. At even higher gas throughput or lower stirrer speed the air bubbles will not be distributed at all by the stirrer and the gas escapes in the region of the stirrer axis (Figure 3.9a). Now the liquid circulation direction is exactly opposite to that of the situation depicted in Figure 3.9e.

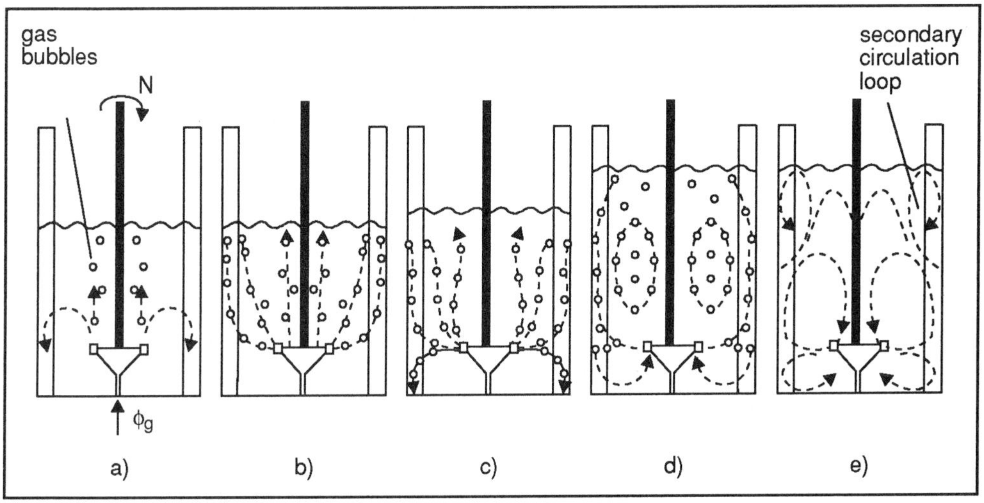

Figure 3.9 Gas circulation patterns in aerated stirred vessels at increasing stirring speed (N) and constant gas feed or constant stirring speed and decreasing gas flow (ϕ). a) represents very high gas flow and/or very low stirrer speed, b) represents high gas flow and/or low stirrer speed, c) represents moderate gas flow and/or moderate stirrer speed, d) represents low gas flow and/or high stirrer speed, e) represents non-aerated state. (Redrawn from Nienow, A.W., Wisdom DJ and Middleton, J.C, 1977, 2nd European. Conference on mixing 30-3 Cambridge, England, Paper FI).

The situation in which gas dominates the flow pattern in Figure 3.9a) has been referred to as flooding.

With reference to Figure 3.9, we can relate the various phenomena to the impeller hydrodynamics. We can consider 3 possibilities:

- impeller flooding as shown in Figure 3.9a (regime I);
- impeller loading without gas recirculation as shown in Figures 3.9b and c. The impeller is loaded with gas and disperses it successfully but the gas is poorly distributed throughout the vessel (regime II);
- impeller loading with gas recirculation as shown in Figure 3.9d (regime III).

flooding depressed by high stirrer speeds

The transition from regime I to regime II is characterised by the release of the trailing gas bubbles from the stirrer blade. With increasing stirrer speed (and constant gas feed) there will be a point when impeller flooding will be depressed. By comparing the discharge velocity of the liquid (expressed by the tip velocity of the blade) with the rise velocity of the bubbles it has been established that to prevent flooding:

$$v_{tip} > 1.5 - 2.5 \text{ m s}^{-1} \quad \text{(E - 3.10)}$$

This condition, however, is insufficient for large gas feeds. A more accurate approach compares the power of the gas P_A (which tends to create a bubble column flow) to the stirrer P_S (which tends to create the circulating flow). To prevent flooding

$$P_A < P_S \quad \text{(E - 3.11)}$$

must be achieved. In the un-aerated case P_S is given in Equation 3.5. Further, it can be established that:

$$P_A \approx \phi_g . \Delta p = \phi_g . \rho . g . H \quad \text{(E - 3.12)}$$

in which Δp is the pressure difference between the bottom and top of the liquid in the vessel. If Equations 3.5 and 3.12 are used together with the definitions of the stirrer Froude number and the Flow number, it results for a standard vessel with turbine stirrer ($T/D = 2.5$, $T/H = 1$, $N_p \approx 6$) that in the un-aerated case

$$Fl < 2.4 \, Fr \quad \text{(E - 3.13)}$$

Remember T = diameter of the vessel, D = diameter of the stirrer, H = liquid height in the vessel, N_p = stirrer power number.

Π Perhaps you would like to derive Equation 3.13 for yourself. Try this and then look at our solution below.

Since $P_A < P_S$ (Equation 3.11) and

$P_A \approx \phi_g . \rho . g . H$ (Equation 3.12) and $P_S = N_p \, \rho \, N^3 \, D^5$ (Equation 3.5)

We can write $\phi_g . \rho . g . H < N_p \, \rho \, N^3 \, D^5$

Thus cancelling p from each side and dividing through by $\frac{N^2 D^3}{g}$

gives $\frac{\phi_g}{N D^3} . H < N_p \frac{N^2 D^2}{g}$

But $\frac{\phi_g}{N D^3}$ = Flow number (Fl) and $\frac{N^2 D}{g}$ = Froude number (Fr) (see text associated with Equation 3.7).

Then $Fl \, H < N_p \, Fr \, D$

or $Fl < Fr \frac{D}{H} N_p$

but $N_p \approx 6$ and since $T/D = 2.5$ and $T/H = 1$ then $H = 2.5 \, D$

Thus $Fl < Fr \frac{6}{2.5}$

or $Fl < 2.4\ Fr$

We have, however, established that in the case of an aerated system that the gassed power consumption P_g of the stirrer is reduced to about half of the un-aerated case (Equation 3.6).

Thus the gassed power consumption of the stirrer is:

$$Fl < 1.2\ Fr \qquad \text{(E - 3.14)}$$

(ie flooding can be prevented providing $Fl < 1.2\ Fr$)

Note that in practice this only applies if the gas flow rate gives a flow number $Fl > 0.05$.

It has also been shown that this result can be extended for non-standard impeller diameters to:

$$Fl < 30\ Fr \left(\frac{T}{D}\right)^{-3.5} \qquad \text{(E - 3.15)}$$

This has been verified experimentally for water systems. Figure 3.10 shows the transition from impeller loading to impeller flooding at different impeller tip velocity and gas flow rates.

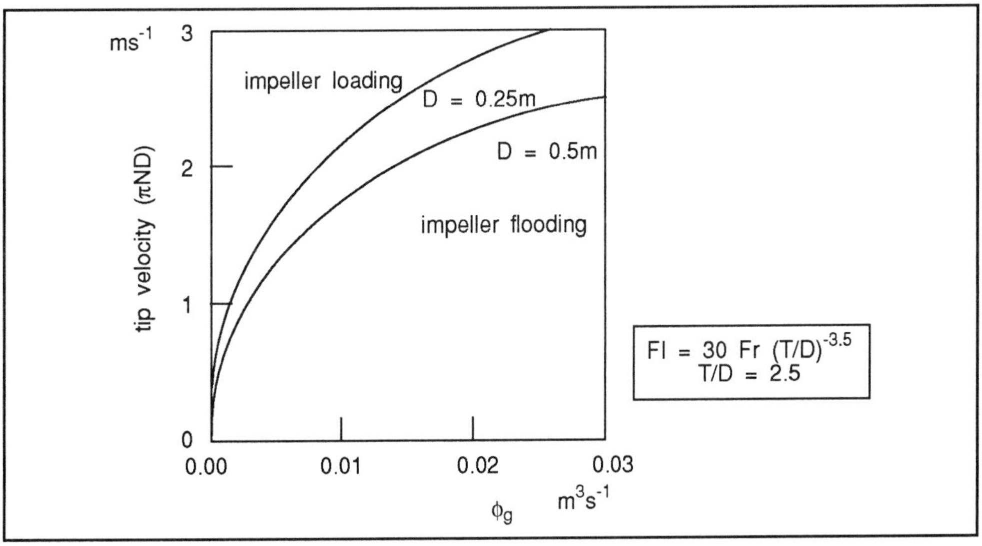

Figure 3.10 Transition from impeller loading to impeller flooding at different impeller tip velocity and gas flow rates (tip velocity = πND, volumetric gas flow rates = ϕ_g). (Redrawn from TU Delft 1989 'Bioprocess technologie' Editor Van der Lans, RGJM. Note that if values of tip velocity and gas flow rates give co-ordinates below the curves, then impeller flooding will occur (see text for details).

Π Using the data presented in Figure 3.10, to determine what impeller tip velocity would be needed to prevent impeller flooding if the gas flow rate was 0.01 $m^3 s^{-1}$ and the diameter of the impeller was 0.5m.

You should have been able to read from the graph that the tip velocity would need to be at least about 1.7 m s^{-1}.

Π Using the same vessel, what would tip velocity need to be if the volumetric gas flow rate was increased to 0.02 $m^3 s^{-1}$?

Again from Figure 3.10, we should be able to determine that the tip velocity should be at least 2.25 m s^{-1}.

flooding reduces mass transfer

Flooding must be prevented in an aerated vessel, since mass transfer will be far less then expected. However, it is only with full gas recirculation regime III (the situation depicted in Figure 3.9d) that mass transfer will be maximal.

The transition from regime II to regime III has been shown to be related to Fl and Fr in the following way:

$$Fl = 13\,Fr^2 \left(\frac{T}{D}\right)^{-5} \qquad \text{(E - 3.16)}$$

Figure 3.11 shows the transition from incomplete to complete gas circulation at different impeller tip velocity and volumetric gas flow rate ϕ_g.

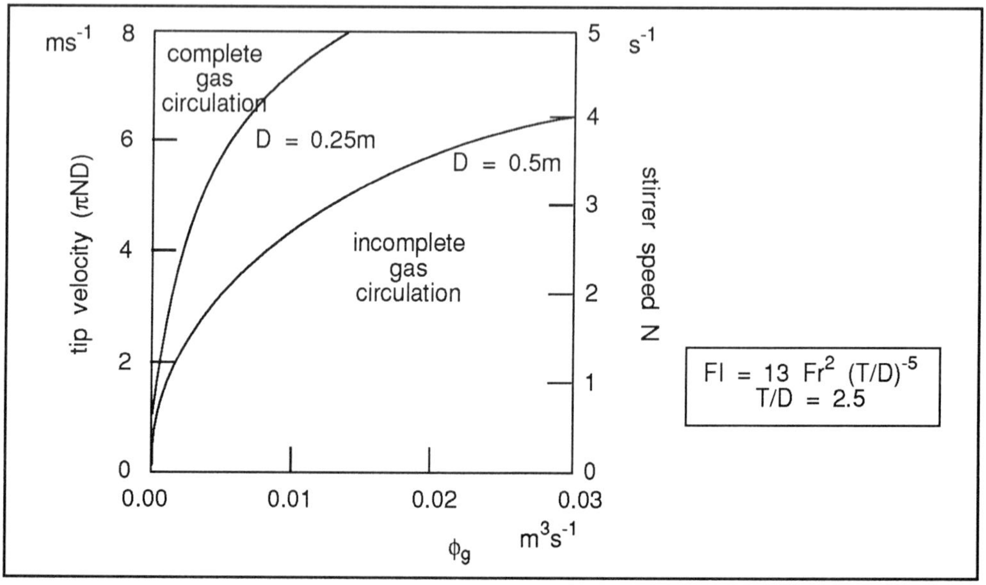

Figure 3.11 Transition from incomplete to complete gas circulation at different impeller tip velocities (πND) and gas flow rates (ϕ_g). (See text for details). (Redrawn from TU Delft, 1989 'Bioprocesstechnologie' Editor; Van der Lans, RGJH).

Thus if we know the values of T and D for a vessel, then in principle we can control the gas flow rate ϕ_g and N to produce maximal mass transfer.

SAQ 3.3

A vessel with a diameter of 1.25 m is fitted with a stirrer of diameter 0.5 m. The vessel is aerated. Which of the following combinations of stirrer speeds and volumetric gas flow rates (ϕ_g) will produce complete gas circulation.

Condition	Impeller speed N (s^{-1})	Volumetric gas flow rate ϕ_g ($m^3 s^{-1}$)
1	6	0.005
2	1	0.02
3	3	0.01
4	4	0.03
5	2	0.015

3.4 Liquid mixing

The aim of the stirring process is to obtain uniform conditions in the working volume of the bioreactor in order to obtain an optimal mass transfer and avoid gradients of any of the components of the medium. In the case of bioreactors we are mainly interested in data for two phase systems: gassed liquids. However, information on such a two phase system is rather scarce. In contrast to this, the single phase system (ungassed liquid) is thoroughly researched. As we can expect single phase (ungassed liquid) and two phase (gassed liquid) behaviour to be related, single phase models are discussed first.

3.4.1 Single phase systems

We can regard the flow through the circulation loops of a CSTR as though it was a plug flow (with dispersion) through a pipe. The circulation time t_c is given by the following relation:

$$t_c = \frac{V}{\phi_c} \qquad \text{(E - 3.17)}$$

in which V is the volume of the circulation loop and ϕ_c is the circulating capacity.

It can be assumed that the circulation capacity is directly related to the pumping capacity of the stirrer ϕ_p:

$$\phi_c \sim \phi_p \qquad \text{(E - 3.18)}$$

Voncken in 1966 established that the dispersion in the loop gives a mixing time t_m as:

$$t_m = 4\ t_c \qquad \text{(E - 3.19)}$$

Equations 3.17 and 3.19 contain empirical parameters of the circulation flow model. We can extend this model to relate mixing time to the operating parameters.

As mentioned earlier the power corresponding with the pumping capacity is (assuming the incoming radial velocity to be zero) is given by:

$P_s = \phi_p \cdot 0.5 \cdot \rho \cdot \langle v \rangle^2$ (see Equation 3.1)

Furthermore, ϕ_p was derived for radially pumping stirrers in Equation 3.2 and Ps for axially pumping stirrers in Equation 3.5. Combining these three equations with Equations 3.17 to 3.19 it follows that:

$$N \cdot t_m \sim 1.2 \cdot \left(\frac{T}{D}\right)^3 \cdot \left(\frac{H}{T}\right) \cdot N_P^{-1/3} \cdot \left(\frac{D}{h_r}\right)^{2/3} \qquad \text{(E - 3.20)}$$

where: T = vessel diameter; D = stirrer diameter; H = liquid height; h_r = stirrer blade height; N_P = stirrer power number; N = stirrer speed.

This equation may look very complex but what it means is quite straight forward. It means that for a Re-number > 5000 (ie turbulent conditions), for which N_p = constant, Nt_m maintains a constant value for a specific stirrer.

Thus for a specific stirrer, doubling N (stirrer speed) halves the value of t_m.

We can, however, simplify Equation 3.20. In many designs D = 4.h_r and for ungassed vessels which are filled with liquid H = 2T. Thus assuming the proportionality to be 1,

$$N \cdot t_m = 3 \cdot \left(\frac{T}{D}\right)^3 \cdot N_P^{-1/3} \qquad \text{(E - 3.21)}$$

It can also be shown that

$$t_m = \left(27 \cdot \frac{\rho T^5}{P_s} \cdot \left(\frac{T}{D}\right)^4\right)^{1/3} \qquad \text{(E - 3.21a)}$$

where P is the power input

These relationships have been checked experimentally. The data are presented in Table 3.1.

Source	Stirrer type	T/D	N_p	Nt_m(exp.)	(calc.)
Voncken (1966)	6 bladed turbine	2.5	6	20	24
Voncken (1966)	6 bladed turbine	3.3	6	35	60
Zlokarnik (1972)	Paddle	2	9.8	10	10
Zlokarnik (1972)	MIG	1.43	0.65	13	10
Zlokarnik (1972)	Propellor	3.33	0.35	150	154
Zlokarnik (1972)	6 bladed turbine	3.33	5	50	64
Hoogendoorn and Hartog (1967)	6 bladed turbine	3	5	32 or 23[1]	16

Table 3.1 Measured and calculated mixing times (see text for discussion). [1] depends on the definition of mixing used.

experimental values agree of calculated values

Although very different stirrer types were used the experimental values agree reasonably with those calculated, particularly when it is realised that mixing time itself is not an objective quantity, but depends on the definition made. Theoretically, good mixing occurs after a infinitely long time. In practice a criterium is used, for example 90% or 95% mixing.

Π Imagine you have a single phase system which under particular conditions gave a mixing time (t_m) of X s. What would the mixing time be if we tripled the stirrer speed (N)? Assume that Re-number > 5000 and that all other conditions remain the same.

The answer is quite straight forward. Under the conditions described, $N.t_m$ remains constant. Thus as we have tripled N, t_m must be one third of its original value. Thus the new mixing time would be $\frac{X}{3}$ s.

Equation 3.21a is very useful in those cases where the power consumption values are known. In this way we can make a very easy first estimate of the mixing times. Mersmann and his colleagues used a similar equation to fit mixing time data and arrived at

$$\frac{t_m^3 P}{\rho \cdot T^5} = 300 \text{ for } 2 < T/D < 3.3 \qquad \text{(E - 3.22)}$$

where P is the power input

larger stirrers reduce mixing times

Finally, the T/D relation in Equation 3.21a shows that for a particular vessel in which a defined amount of energy is consumed, a larger stirrer will lead to shorter mixing times than a smaller one. This, of course, is valid only within limits. All these equations are derived for the turbulent situation (Re > 5000). Large scale industrial plants normally will work under these conditions; small scale bioreactors with a viscous broth definitely will not. This is shown by the data presented in Table 3.2. The data presented in this Table is quite complex so let us work our way through it carefully to make certain we understand what it is telling us.

	N↓	η →	1	100	1000	10000	mPa.s
Volume 0.01 m^3	22.7 (s^{-1})	Re	110000	1100	110	11	
		t_m (s)	12	21	1000	$>10^6$	
T= 0.21 m	12.2 (s^{-1})	Re	59000	590	59	6	
		t_m (s)	20	38	10^4	$>10^6$	
Volume 1 m^3	4.9 (s^{-1})	Re	517000	5170	517	52	
		t_m (s)	48	48	200	10^6	
T= 0.98 m	4.4 (s^{-1})	Re	464000	4640	464	46	
		t_m (s)	56	60	220	10^6	
Volume 100 m^3	1.0 (s^{-1})	Re	$2300 . 10^3$	$23 . 10^3$	$2.3 . 10^3$	$0.23 . 10^3$	
		t_m (s)	240	240	230	2000	
T= 4.57 m	1.6 (s^{-1})	Re	$3700 . 10^3$	$37 . 10^3$	$3.7 . 10^3$	$0.37 . 10^3$	
		t_m (s)	150	150	190	800	

Table 3.2 Mixing times for small and large scale fermenters (see text for details).
Assumptions: H = 2T; 3D = T; 1 stirrer; $N_p = 2.5$. Re > 5000: $t_m = 3\ (T/D)^3\ N^{-1}\ N_p^{-1/3}$

In the legend we have put the ratios of H, D and T which were used to calculate the values reported in the text. Now if you read the left hand column you can see that we have choosen three different vessels of quite different volumes (0.01 m^3; 1 m^3 and 100 m^3). Also reported in this column are the diameters of the vessels. In the next column are the stirrer speeds (N). In the next columns we have reported the Reynolds numbers (Re) and mixing times (t_m) for these vessels if they contain liquid of different dynamic viscosities (η). The dynamic viscosities used are reported at the top of each column. t_m values are reported in seconds.

Let us take the smallest vessel first. Under the conditions described, when the dynamic viscosity (η) = 1 mPas, then the mixing time is 12s. When the viscosity is increased to 100 mPas, then t_m is increased to 21s. When the viscosity is 10000 mPas, then the t_m > 10^6s.

Π What is the t_m value for the 1m^3 vessel when the stirrer speed (N) is 4.4s^{-1} and the viscosity is 1000 mPas?

From the table you should have been able to identify that t_m = 220s.

Note there are a number of other criteria that were used in calculating these values especially at low Re values but we will not deal with them here.

Π Re numbers > 5000 gives turbulent mixing; Re numbers < 5000 gives laminar mixing. Which of these gives the quickest form of mixing? Would you expect the mixing time t_m to increase at low Re values. (Use Table 3.2 to answer this).

turbulent mixing reduces mixing times

You should have expected that high Re numbers (ie turbulent mixing) would give shorter t_m values. The data in Table 3.2 show this very clearly. This has been shown many times experimentally. We have displayed a typical plot of Nt_m against Re number in Figure 3.2. Note that turbulent mixing (Re > 5000), Nt_m is constant for a particular system. At lower Re values, Nt_m rises dramatically.

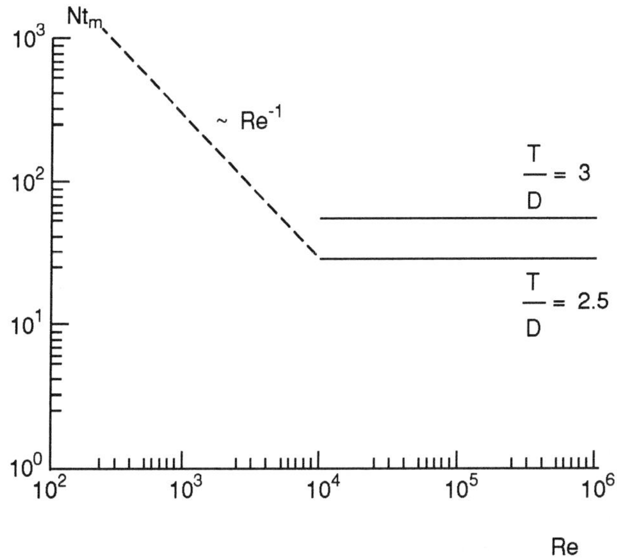

Figure 3.12 Nt_m relation with Re for single phase turbine stirred vessel (see text for details). (Adapted from TU Delft 1989, Bioprocess technologie Editor; Van der Lans, RGJM).

Π Look back to Table 3.2. Can efficient mixing of highly viscous systems be achieved most easily in small laboratory vessels or in large scale reactors?

From the mixing time given in Table 3.2 it is clear that small laboratory scale bioreactors can very easily show an extremely poor mixing behaviour even at rather low viscosities. Therefore, at laboratory scale viscosities > 100 mPas and, on commercial scale, viscosities > 1000 mPas have to be avoided otherwise other reactor types or adapted stirrer systems have to be used.

SAQ 3.4

1) The mixing time (t_m) for a vessel is 20s when the power (P) delivered by a stirrer is 500 W. The T/D ratio for the vessel is 3. What will the mixing time for the vessel be if the power delivered by the stirrer is decreased to 250 W? (Assume that turbulent mixing occurs in both conditions).

2) If the stirrer speed (N) was $20s^{-1}$ when the power (P) delivered to the stirrer was 500 W, what is its speed when the stirrer power is decreased to 250 W.

3.4.2 Two phase systems

The theory presented so far is derived for single phase systems. For aerobic processes, air is an essential component and thus very often there will be a two phase system in the reactor. The reason for discussing single phase systems in such an extensive way is that it provides some insight into the less well studied two phase systems. If we assume that the presence of a second phase (air) does not influence the mixing behaviour, we even can simply apply Equation 3.21a. In an aerated system, however, the power (number) is considerably less than for the single phase case so the mixing times will be somewhat larger.

Experimental data have been produced which show that the increase can be a factor 1.2 to 2.0 at 10% hold-up (see Section 3.5.3) depending on scale of operation and viscosity (we will examine this later).

Some data relating to actual mixing times (t_m) with t_m values calculated from single phase relationships are shown in Table 3.3.

Π Calculate the ratios of t_m (measured) and t_m (calculated) for each of the sets of data. Then come to a conclusion about the ratio of between t_m (measure), t_m (calculated for a single phase).

Author	V(m^3)	P(W)	T(m)	t_m measured	t_m calc single phase[1]
Einsele (1978)	0.14	2100	0.45	15	6
	1	5500	0.86	42	12
	10	25000	1.85	84	26
	20	23000	2.34	51	38
	57	116000	3.31	82	40
Jansen *et al* (1980)	1.9	3800	1.08	29	24
	60	120000	3.45	67	52

Table 3.3 Measured and calculated mixing times in aerated systems (see text for details).
[1]Assumptions: P/V = 2 KW m^{-3} (large scale of Einsele (1978)); H/T = 2.5; Single phase liquid height = 75% of H.

mixing times increased by a factor of about 2 for two phase systems

Although the results do vary, the difference between measured two phase and calculated single phase mixing times is about a factor 2, (ie t_m (measure): t_m (calculated for a single phase is 2:1). This is a useful value to remember.

Therefore, it can be concluded that two phase and single phase mixing times are strongly related and two phase mixing times can be a factor x2 higher than that for single phase systems. For Re < 5000 values as presented in Figure 3.12 Nt_m times a factor 2 might very well be valid.

3.5 Dispersion of gas

bubble distribution and bubble formation

The gas dispersion takes place at the back of the stirrer cavities. All of the gas coming from the sparger, together with a considerable part of the recirculating gas flow, is entrained into these cavities. Equally important to the bubble distribution, which depends on the fluid movement achieved by the stirrer, is the bubble formation. An increased interfacial area leads to an increased mass transfer rate. A large interfacial area requires the generation of small bubbles from a given gas flow rate. Optimum mass transfer conditions may require prescription of bubble diameter. Apart from the stirrer characteristics, the bubble size depends on the liquid properties such as ionic strength, surface tension and viscosity. The liquid phase properties influence the coalescence rate of the bubbles. Tap water and distilled water are regarded as completely coalescing liquids. In other words in those liquids the bubbles will coalesce very quickly. The salt concentration in tap water is approximately 0.02 mol l^{-1}. The addition of higher amounts of salts suppresses coalescence. This has to be taken into account when determining the optimal bubble size.

bubble coalescence

hold up and foaming

The volume fraction of gas in the liquid is called the gas hold-up, which is another characteristic of the gas dispersion. Gas hold-up determines the main surface area and also the effective utilisation of the stirred tank for gas-liquid transfer. Hold-up must, however, be distinguished from foam. Hold-up is characterised by more or less freely moving bubbles present in a continuous liquid phase in quantities < 50% of the total volume. Foam is characterised by more or less stagnant bubbles with very high gas volumes, easily exceeding even 80% of the total volume. The complex character of a fermentation liquid makes prediction of the foaming behaviour rather impossible. Furthermore, hold-up values in stirred vessels are largely influenced by the coalescence characteristics of the liquid. Therefore, the hold-up prediction also becomes difficult. We will, however, discuss hold-up measurements for model systems in later sections. By using the correlations derived from these measurements and by estimating the influence of fluid properties, an indication of the hold-up value to be expected can be given.

3.5.1 Gas mixing

coalescence with the vortex cavity

Although very big bubbles will determine liquid movement in the vessel (flooding), the distribution of small gas bubbles, however, is mainly dependent on the fluid movement achieved by the stirrer. These bubbles will follow the primary and secondary movement of the liquid. At recirculating conditions it has been shown that a large part (10-50%) of the recirculating gas stream coalescences with the cavity behind the stirrer blades, thus mixing intensively with the recently sparged gas. For non-coalescing media this is also the case and even more intensive coalescence with the cavity takes place although the bubbles themselves do not coalesce with each other. This is in agreement with observed results that, for non-coalescing systems providing the stirrer speed is large enough, the gas phase will be well mixed.

3.5.2 Bubble formation

Experimentally determined bubble diameters indicate that very small bubbles are created in a typical stirred reactor. These bubbles have diameters much smaller then 6mm. The reason we have chosen a value of 6mm is that this is the equilibrium bubble size in coalescing media in bubble columns. We will deal with this in the next chapter. Because the bubbles are small we get a dispersion/coalescence behaviour similar to that of a fine bubble sparger in a bubble column. In coalescing liquids the small bubbles will coalesce to finally produce an equilibrium bubble size of about 3-5mm. In non-coalescing liquids the fine bubbles originated at the sparger will remain in the bulk volume. It has also been experimentally confirmed that bubble size in the fluid bulk decreases considerably when salts, known coalescence suppressors, are added. The mean bubble size in coalescing conditions is 2-3mm. At the same stirrer speed under non-coalescing conditions this decreases to about 0.5mm.

SAQ 3.5

1) A foam is a gas:liquid mixture in which [ ] % of the volume of the mixture is a gas.

2) In a coalescing liquid, small bubbles will coalesce to produce an equilibrium bubble size of about [ ] mm.

3) Salts tend to increase/decrease bubble coalescence, thus equilibrium bubble size are increased/decreased.

4) Gas hold up is characterised by freely moving/stagnant bubbles.

3.5.3 Gas hold-up

Gas hold-up, ε, is usually defined as the volume fraction of gas in the liquid:

$$\varepsilon = \frac{V_g}{V_l + V_g} \quad \text{(E - 3.23)}$$

in which V_g = the volume of gas bubbles in the dispersion and V_l = the volume of the liquid.

The hold-up values in water have been measured by many authors who have used various relationships. Some have related hold-up with power per unit value and gas superficial velocity v_s.

Typical of these types of study is the relationship shown in Equation 3.24.

$$\varepsilon = 0.13\left(\frac{P_g}{V}\right)^{1/3} . v_s^{2/3} \quad \text{(E - 3.24)}$$

in which v_s the superficial velocity (= ϕ_g/A in which A is the surface area of the gas bubbles), at 5 and 67 1 scale. All of those who have carried out such studies agree on the general form of the relationship between ε, P_g, V and v_s (that is ε is related to $\left(\frac{P}{V}\right)^{\alpha}$, v_s^{β})

Others have used a relationship between hold-up and the dimensionless Froude, Weber and Flow numbers.

Hughmark has correlated an extensive number of experimental results and found for pure liquids that:

$$\varepsilon = 0.74\left(\frac{\phi_g}{NV}\right)^{1/2} . \left(\frac{N^2 D^4}{g\, h_r V^{2/3}}\right)^{1/2} . \left(\frac{\rho . d_b . N^2 . D^4}{\sigma . V^{2/3}}\right)^{1/4} \quad \text{(E - 3.25)}$$

in which σ is the surface tension in a gas-liquid system, db is the bubble diameter (= 0.0025 m) and a turbine stirrer was used.

These two formulas look very different. However, when Equation 3.25 is applied to the data Equation 3.24 is derived from, it appears that both formulas give nearly the same results. If Equation 3.25 is rewritten for water with the assumption that Np = constant it becomes:

$$\varepsilon = C_1 \cdot \left(\frac{P}{V}\right)^{1/3} \cdot v_s^{1/2} \cdot \left(\frac{D}{T}\right) \cdot D^{-1/6} \qquad \text{(E - 3.26)}$$

where C_1 is a constant

Thus it appears that Equation 3.25 is rather similar to the power per volume relationships described in Equation 3.24. However, when extrapolating the validity of these equations to large vessels (> 10 m^3) Equation 3.25 predicts rather low ε-values, most probably due to the power of -1/6 on D.

In general the ε-values predicted by Equation 3.25 are at about 2 times lower than those predicted by Equation 3.24, a difference that cannot be explained at this moment. It has also been reported that on scales up to 100 m^3 Equation 3.24 is superior to Equation 3.25. It must be noted that for large scales the superficial velocity is not well defined. It may be based on the pressure above the liquid or at the impeller or on a logarithmic mean value. Van der Lans (1985) shows that for a airlift loop reactor with a riser of 0.43 m^3 the logarithmic mean value is best as the scale up parameter for hold-up values.

salts inhibit coalescence of bubbles

The addition of salts above a concentration that corresponds with an ionic strength of about 0.4 mol l^{-1} strongly inhibits the coalescence of bubbles in a dispersion. The resulting smaller bubble size and lower bubble terminal rise velocity (which is not equal to v_s) increases the hold-up considerably. Figure 3.14 is characteristic for this phenomenon. It shows the transition from coalescent to non-coalescent conditions.

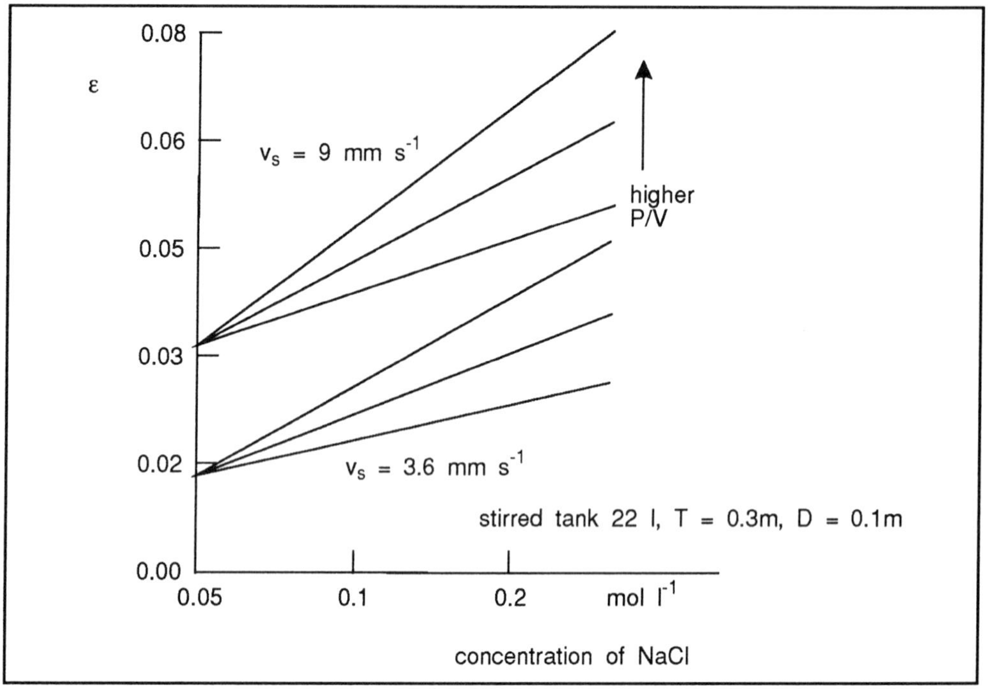

Figure 3.13 Hold up value in relation to salt addition. (Data from Lee and Meyrisk; 1970, Trans, Inst, Chem, Eng 48 T 37).

Π Examine Figure 3.13 carefully. Note that ε is dependent upon the salt concentration. What else influence the ε values reported in this figure.

ε is, of course, influenced by the gas superficial velocity (v_s) and also the power:volume (P/V) ratio. Note that ε can be increased by up to a factor of about 4. No empirical correlation for this increase can be given.

The points to remember are:

- hold up increases with salt concentration;
- it is larger for higher P/V value;
- its relative magnitude is lower for high gas rates.

For fully non-coalescent conditions (when the ionic strength > 0.4 mol^{-1}) the following correlation may be useful for vessels up to 0.7 m^3.

$$\varepsilon = 0.0051 \cdot \left(\frac{P_g + P_A}{V}\right)^{0.57} \cdot v_s 0.24 \qquad \text{(E - 3.27)}$$

in which P_A is the power related to the thermal expansion of the air ($P_A/V = \rho \cdot g \cdot v_s$). Note that the influence of P/V is stronger then v_s. This is in agreement that P/V is representative of bubble size and v_s more closely related to the number of bubbles.

For fermentation liquids it is even more complex because all types of solutes (proteins, salts, antifoaming agents, alcohols) and culture conditions (pH, temperature) influences coalescence behaviour and thereby the hold-up value. Figure 3.14 gives an impression of the relationship between ε and stirrer speed with different additives. Note that all the relationships shown in Figure 3.14 are at the same v_s values.

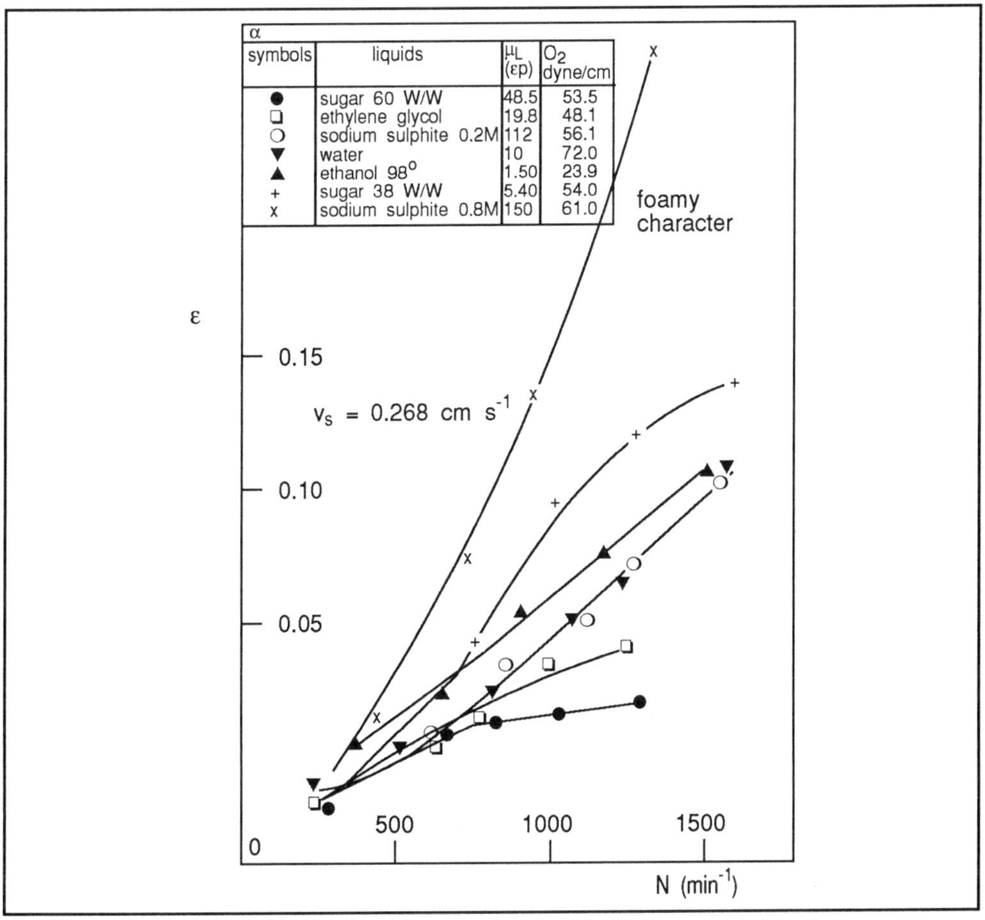

Figure 3.14 Some influences of additives on gas hold-up.

hold up in fermentation broths is difficult to predict

From this description, it will be understood that for complex liquids no relationship can be found. Nevertheless some prediction can be made. Pure water is a coalescing liquid. This means that normally the fermentation liquid will be less coalescing then water. Hold-up values will be higher than those predicted by Equation 3.27. The values predicted by this equation can be regarded as minimum values. Only for extremely coalescing liquids can the hold-up become somewhat less. Data in the literature indicates that above a viscosity of 20 mPas coalescence is strongly promoted. As a result of the flow around a large bubble, smaller bubbles are encaptured in its wake and will then coalescence. The presence of a large number of small bubbles has a relatively small counteracting influence.

SAQ 3.6

1) If the liquid volume in a reactor is 10 m^3 and the gas volume is 2 m^3, what is ε?

2) Will the P/V ratio have a greater or lesser effect on ε than v_s when a) the ionic strength of the medium is greater than 0.4 mol l^{-1}. b) the medium is virtually pure water?

3.6 Mass transfer

At first sight there is not much consistency in the literature on mass transfer in stirred vessels. One can assign various reasons for this. Firstly many different methods of measuring oxygen and k_La-values have been used which in some instances have given erroneous values (see Chapter 2).

major factors which influence k_La

In addition k_L and a can be affected in several ways. The most important variables are power consumption, gas superficial velocity and liquid phase properties which influence the coalescence behaviour (Figure 3.3).

Different values can be found for k_La under seemingly identical conditions, because the various variables are not always selected in a consistent manner or may be latently present. To get an insight into the k_La-value for fermentation broths, we will follow the same path as for the hold-up data:relationships for the coalescing model fluid (tap or distilled water) and those for ionic solutions are derived first (Section 3.6.1) and then fermentation liquids will be discussed (Section 3.6.2). Last but not least, some attention will be given to the influence of vessel geometry parameters (Section 3.6.3).

3.6.1 k_La relations in low viscous media

Power per volume and gas superficial velocity are major correlation coefficients for k_La. Therefore, equations of the following type are frequently found in the literature:

$$k_L.a = K . \left(\frac{P_g}{V}\right)^{\alpha} . (v_s)^{\beta} \qquad \text{(E - 3.28)}$$

in which K, α and β are constant. So when P is omitted, an equation like that of 3.29 can usually be adopted to the reactor parameters N and D:

$$k_L.a = K . N^{\alpha 1} . D^{\alpha 2} . (v_s)^{\beta} \qquad \text{(E - 3.29)}$$

k_La related to stirrer speed, stirrer diameter and superficial gas velocity

In other words we can usually relate k_La to the stirrer speed, the diameter of the stirrer and the superficial gas velocity. It is also usually possible to convert Equation 3.28 to the form shown in Equation 3.27. Thus k_La may be related to gas hold-up.

Literature values found for α and β show a fairly great variation: $0.4 < \alpha < 1$ and $0 < \beta < 0.7$. It is not unusual for the K value to remain unmentioned.

In the German literature, however, it is practice to use an equation in a dimensionless form:

$$k_L \cdot a \left(\frac{\nu}{g^2}\right)^{1/3} = f\left(\frac{P_g}{V \cdot \rho \cdot (g^4\nu)^{1/3}}, \frac{v_s}{(g\nu)^{1/3}}, \frac{\nu}{D}, \frac{\sigma}{\rho(g\nu)^{1/3}}\right) \qquad \text{(E - 3.30)}$$

in which ν is the kinematic viscosity and σ is a material dimensionless number, unknown but introducing coalescence behaviour. For water systems (the salt concentration in tap water is approximately 0.02 mol l^{-1}, which causes only an increase by 5-10% in k_La over distilled water) this can be simplified to:

$$k_L \cdot a \left(\frac{\nu}{g^2}\right)^{1/3} = k \cdot \left(\frac{P_g}{\nu \cdot \rho(g^4\nu)^{1/3}}\right)^{\alpha} \cdot \left(\frac{v_s}{(g\nu)^{1/3}}\right)^{\beta} \qquad \text{(E - 3.31)}$$

Since ρ, g and ν will hardly change this expression is similar to Equation 3.27 and 3.28. It is, however, less applicable.

influence of temperature on k_La

Interpretation of the literature is, furthermore, hampered by two factors. Firstly, k_La-values determined by the method sulphite are often not reproducible. Secondly, the k_La-values are temperature-dependent. It appears that k_La increases according to the relationship:

$$k_La = k_L20a \cdot \theta^{(T-20)} \text{ with } \theta = 1.020\text{-}1.024 \qquad \text{(E - 3.32)}$$

with T in °C. Thus k_La increases by about 2.5% /°C over the temperature ranges normally used. Therefore, the k_La-values given below have been recalculated for a temperature of 20°C. Most data with the diameter of the vessel << 0.1 m have to be omitted, as well as data from measurements done by the sulphite method is unreliable without extensive further data on reaction kinetics.

Let us first consider the case of low viscosity media. The various authors providing k_La values have used rather different values of K, α and β. Nevertheless, a general relationship can be produced from the published data:

$$k_La^{20} = 0.0026 \cdot \left(\frac{P_g}{V}\right)^{0.4} \cdot v_s^{0.5} \qquad \text{(E - 3.33)}$$

This holds for pure water and for volume up to 2.6 m^3 and $500 < P_g/V < 10000$ and v_s up to 4 cm s^{-1}. The deviation of the measured values is within 20-40%. Note that in Equation 3.33, and also in the following equations, the dissipated gassed power P_g has to be used.

An alternative correlation has been described by Henzler. He used Equation 3.30 and found that:

$$\frac{k_La}{v_s}\left(\frac{\nu^2}{g}\right)^{1/3} = 7.5 \times 10^{-5}\left(\frac{P_g}{V} \cdot \frac{1}{v_s \rho g}\right)^{0.43} \qquad \text{(E - 3.34)}$$

He included volumes up to 900 m^3.

Equation 3.33 can be modified to:

$$k_La = 0.027 \left(\frac{P_g}{V}\right)^{0.43} . v_s^{0.57} \qquad \text{(E - 3.34a)}$$

for $g = 9.8\ m\ s^{-2}$, $\nu = 10^{-6}\ m^2\ s^{-1}$, $\rho = 10^3\ kg\ m^{-3}$

The similarity between Equations 3.33 and 3.34a is clear and even for extreme values of P/V and v_s, the predicted k_La-values vary within 20% of the actual measured values. It can thus be concluded that for Equations 3.33 and 3.34a predict and measured values agree very well. The dimensionless Equation 3.27 can be used but usually for simplicity Equation 3.33 or 3.34a are preferable.

k_La data has also been generated for strong ionic (non-coalescing) solutions in water. In this case:

$$k_La^{20} = 0.0016 . \left(\frac{P_g}{V}\right)^{0.7} . v_s^{0.2} \qquad \text{(E - 3.35)}$$

and

$$\frac{k_La}{v_s}\left(\frac{\nu^2}{g}\right)^{1/3} = 8.3 \times 10^{-5}\left(\frac{P_g}{V} . \frac{1}{v_s . \rho . g}\right)^{0.71} \qquad \text{(E - 3.36)}$$

to be transformed to:

$$k_La = 0.0023 \left(\frac{P_g}{V}\right)^{0.71} . v_s^{0.29} \qquad \text{(E - 3.36a)}$$

(Conditions were $500 < P_g/V < 10000\ W\ m^{-3}$, v_s up to 4.7 cm s^{-1}, volume up to 4.4 m^3).

Again both these relationships (3.35 and 3.36a) are very similar and predict the same k_La-values within 20%. On comparison with Equation 3.33 and 3.34 for pure water, it will be found that the α and β (Equation 3.28) differ markedly, as can also been seen in Figure 3.15 where ionic solution and pure water results are given. At higher P/V values, k_La-value can be greatly increased.

The graph shows the relationships between k_La and P_g/V at different superficial gas velocities (v_s).

So it can be concluded that the results presented in the literature show much agreement with theoretical calculation and the observed accuracy of, say 20-40% is regarded as acceptable. For a given determination to be more accurate, the k_La in the vessel and fluid concerned should be measured as exactly as possible. The question is whether this can be really be achieved bearing in mind the limitations of the experimental methods. It might be better to be satisfied with this margin of accuracy.

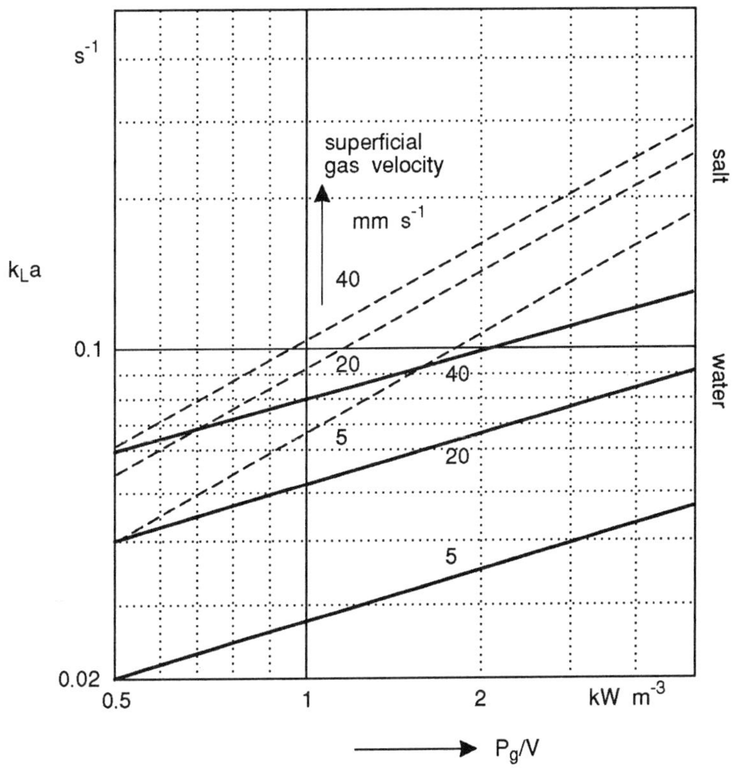

Figure 3.15 Correlation equations at several superficial gas velocities for pure water and ionic solutions (TU Delft 1989 Bioprocesstechnologie, Ed Van de Lans, RGJM).

Π In this section we have given several relationships between k_La and P/V and v_s. It would be a useful form of revision to produce yourself a summary chart of the following type. We have started it for you.

Pure water

Relationship	Conditions/Comments
$k_La = K\left(\frac{P_g}{V}\right)^{\alpha}(v_s)^{\beta}$	General relationship
$k_La = K\,N^{\alpha 1}\,.\,D^{\alpha 2}\,(v_s)^{\beta}$	General relationship
$k_La^{20} = 0.0026\left(\frac{P_g}{V}\right)^{0.4}.\,v_s^{0.5}$	V up to 2.6m^3, $500 < P_g/V < 1000$ v_s upto 4 cm s^{-1}
$k_La = 0.027\left(\frac{P_g}{V}\right)^{0.43} v_s^{0.57}$	V up to 900m^3 $\rho = 10^3$ kg m^3

So far we have dealt with water and ionic solutions. A distinction between water with and without dissolved salts is, however, essential. By raising the ion concentration in solution,k_La is increased considerably. Once a limit of about a concentration of 0.4 mol l^{-1} has been reached, the increase is much smaller. The k_La's for ionic solutions are more dependent on P/V than those for pure water. This was also seen to be true for gas hold-up (see Figure 3.13).

affects of higher alcohols, silicone oil on k_La values

For low viscous fermentation liquids the situation becomes rather complex. Not only is the influence on bubble size (and thus interfacial area), in a comparable way to that described for gas hold-up, important but also the influence of additives on the k_L-value. For instance, higher alcohols are known to be strong coalescence-suppressors but at the same time lower the k_L-value in such way that the mass transfer is relatively unaffected. Experimentally it has been demonstrated that the presence of only a small amount (say 2 ppm) of silicone oil (often used as a anti-foam agent) may be disastrous, since the k_La-value may drop by as much as a factor 5 or more. The effect is larger for smaller P/V-values. This effect results from the extra hindrance that may be exerted on mass transfer by a mono-layer of oil or alcohol on the bubble surface. With a controlled addition of an anti-foam agent, one may expect that k_La will be in between the extremes given in Figure 3.15.

A strategy can be developed to predict k_La on plant scale accurately for a specific broth from pilot scale measurements. Firstly, biological parameters like the cell density are determined. When the cell density is known, the oxygen uptake rate can be estimated and used in the OUR equation (OUR = k_La C^*-C_l) where C^* is the oxygen saturation of the liquid and C_l the actual oxygen concentration of the liquid). Remember that OUR = oxygen uptake rate (see Chapter 2). If C_l is known ($C_l > X\%$; the usual range being $20 < X < 80\%$ of air saturation) k_La can be determined. It is then possible to determine P_g/V at a specific v_s from Figure 3.13. With the help of Equation 3.5, N_p and eventually D can be determined. The pilot plant can be used to establish the OUR equation (in fact the medium dependency) or Equation 3.28. A disadvantage of this procedure is that there can be scale-up effects.

The measured line will almost certainly lie in between the water and water with salts in solution line (Figure 3.15). The slope will be in between that of these two curves. Each broth and even a broth at various stages of cell growth, will need its own k_La correlation, to be established in this way.

SAQ 3.7

1) k_La for a system has been measured at being 10 s^{-1} at 20°C. What will the k_La value likely to be at 30°C?

2) Write down the general form of relationship between k_La, $\frac{P_g}{V}$ and v_s.

3) If the oxygen uptake rate = 1 g l^{-1} s^{-1} when the concentration of oxygen in solution is 90% saturation, what is the k_La value for the system? What further piece of information do you need to be able to calculate k_La in a meaningful way.

3.6.2 The influence of viscosity

There is no general agreement about the influence of viscosity in the range of 1-10 mPas. Much more agreement exists on the influence of viscosity above 50 mPas. Various correlations between k_La and η have been reported in the literature. Thus:

many correlations between k_La and η have been established

Deindorfer and Gaden (1955) and Deindorfer and West (1960) report $k_La \sim \eta^{-0.9}$, Yagi and Yoshida (1975): $k_La \sim \eta^{-0.4}$ and Hocher et al (1980): $k_La \sim \eta^{-0.69}$, all indicating a distinct negative influence of viscosity on k_La.

These experimental results have been correlated for two types of viscous solutions. One containing carboxymethyl cellulose (CMC), the other glucose. These correlations are:

$$\text{CMC solution: } \frac{k_La}{v_s}\left(\frac{v^2}{g}\right)^{1/3} = 3.6 \times 10^{-4} \cdot \left(\frac{P_g}{V} \cdot \frac{1}{v_s \cdot \rho \cdot g}\right)^{0.55} \quad \text{(E - 3.37)}$$

$$\text{Glucose solution: } \frac{k_La}{v_s}\left(\frac{v^2}{g}\right)^{1/3} = 9 \times 10^{-5} \cdot \left(\frac{P_g}{V} \cdot \frac{1}{v_s \cdot \rho \cdot g}\right)^{0.7} \quad \text{(E - 3.38)}$$

inadequacy of correlations between n and k_La for viscous solutions

Other liquids delivered values in between these two sets. The difference in k_La under the same conditions predicted by these two formulas can be by as much as a factor 10. This indicates that further research is needed to gain insight into the important phenomena involved in viscous gas-liquid mass transfer.

A problem is the inadequacy of some of the measuring methods known. The sulphite method (Chapter 2) should not be used and the popular dynamic method cannot be applied at high viscosities due to the presence of very small bubbles. The CMC results cited above might also have been influenced by this phenomenon. Apart from this, it is generally accepted that in highly viscous non-Newtonian fermentation broths, mass transfer is almost completely limited to the stirrer region. In what ways this influences mass transfer formulas such as Equations 3.37 and 3.38 is not known. The equations overall P/V in these formulas will certainly need correction depending on the stirrer volume. How this has to be done is not yet known.

3.6.3 The influence of vessel geometry parameters

k_La often related to N and D

A large variety of stirrers such as turbines, paddles, propellers, rods and self-sucking agitators has been used. Most authors do not correlate k_La with P/V but with various N and D combinations. On evaluation of their results, it becomes clear that P/V correlations are not sensitive to the choice of a disperging stirrer. When it is unimportant which type of stirrer is used and only the power dissipated in the fluid is essential, it may be expected that the number of stirrers is also immaterial. This has been confirmed experimentally. It must be added here that on some occasions a fluctuating difference (up to 40%) in the amount of mass transfer occurs but this may be related to experimental accuracy.

For H > > T and for many stirrers, gas phase modelling becomes important and has to be included in the calculation of the final oxygen transfer rate.

Frequently, a sparger is used under the lower stirrer. Examples are ring spargers and open pipes. It has been demonstrated that bubbles from the sparger collect under the stirrer and are invariably dispersed via the cavities. The choice of sparger has little effect on the mass transfer.

As the diameter of the sparger relative to that of the stirrer becomes greater than the stirrer diameter, the mechanism of dispersion undergoes a change. Thus, in the case of ring spargers having a diameter larger than that of the stirrer, fewer bubbles enter the vessel via the stirrer. This diminishes dispersion and leads to a relatively higher value of power consumption. It may be expected that, as a consequence, mass transfer conditions become less favourable. Thus we can conclude that spargers larger in diameter than the stirrer are less suitable.

No direct correlations with P/V have so far been reported for the influence of fluid height. But, at a given stirrer speed (and, thus, a practically constant P), k_La ~ T/H (ie ~ 1/V).

very tall vessels may be split into two zones

The influence of the fluid height can be described in terms of P/V, yet only within a defined region, as it assumes a well mixed gas phase and invariable k_La over the contents of the vessel. At extremely high H/T ratios and one stirrer near the bottom, the contents of a vessel can be more properly described as a well-stirred area in the lower part along with a bubble column (see Chapter 4) in the upper part. Eventually, the mass transfer in the bubble column will become independent on P/V dissipated in the well-stirred area. For one stirrer in very viscous systems with mass transfer only occurs in the stirrer region, no transfer at all will take place in the upper part of the vessel, whatever height it may have. Results even indicates similar behaviour for low viscous systems in larger reactors ($\approx 20\ m^3$). Therefore, one has to be careful with the prediction of k_La for fluid heights > T, particularly when only one stirrer is employed.

As stated above neither the type nor the number of stirrers have any influence on k_La versus P/V. It may thus be said that the stirrer position is also of no importance. However, complications may occur with regard to the flow phenomena as shown in Figure 3.9. Even before flooding occurs there can be a lack of gas recirculation in the lower compartment. However, this is only a theoretical problem since in most industrial applications the stirrer is positioned at about a stirrer diameter above the bottom.

If the distance between stirrer and bottom is too small, ie smaller than the stirrer diameter, dissipated power decreases. This diminishes mass transfer. Unfortunately, experimental data or any correlation with P/V are lacking.

Lastly, if the stirrer is positioned close to the surface, air can be entrained. At the same time power consumption decreases. This causes a decrease in k_La, but the relation k_La -> P/V still stands.

Π List the important conclusions that can be made about mass transfer in non-viscous systems based on the above discussion and compare it with our conclusions described below.

It can be concluded from the correlation of k_La with P/V, that there is no influence of stirrer geometry or number of stirrers on mass transfer in non-viscous systems. In the cases of full recirculation, the stirrer position is rather immaterial. By placing the stirrer one stirrer diameter away from the bottom, an optimal power consumption value and mass transfer are ensured.

At a fluid height in the range of $0.5 < T/H < 1.5$, P/V appears to be a variable with which k_La can be correlated. The sparger should not be fitted at a distance from the centre larger than the stirrer radius.

In general, it may be concluded that, provided that the vessel is properly stirred (which means absence of flooding or surface air entrainment and the presence of a suitable gas sparger) and allowance is made for the ions in the solution, P/V and v_s are sufficient to described k_La for non-viscous fermentation broths.

3.7 Heat transfer

heat produced close to the stirrer

All the mechanical energy of the stirrer transferred to the fluid in order to maintain fluid movement is dissipated into heat. The consequence is that heat production takes place in a space close to the surface of the stirrer. Whatever the initial conditions of heat transfer are, after a certain time, the dissipated mechanical energy must be transferred from the fluid through the walls of the vessel to an outside medium.

(Note that heat transfer processes are described in the BIOTOL text 'Bioprocess Technology: Modelling and Transport Phenomena' - if you require more background information about heat transfer processes).

Heat transfer data are known mainly for unaerated systems. The question is to what extent aeration influences heat transfer. Depending on circumstances aeration can improve as well as deteriorate the heat transfer coefficient α. This depends on the stirrer speed and might be related to flooding. Generally the influence of gassing rate is only minor in relation to the other variables, thus the data for non-aerated systems can be used. Relations can be found in standard literature. For standard vessels, the average heat transfer coefficient α to the wall follows from:

$$\frac{\alpha \cdot T}{\lambda} = 0.75 \cdot \left(\frac{N \cdot D^2}{\nu}\right)^{2/3} \cdot Pr^{1/3} \cdot \left(\frac{\eta_w}{\eta}\right)^{0.14} \quad \text{(E - 3.39)}$$

in which λ is the heat conductivity, Pr the Prandl number and η_w the viscosity of water. Other types of stirrers and other cooling devices (pipes and, in large reactors, coils because the cooling surface of pipes is not large enough) will differ somewhat in heat transfer. The correction term given for the wall viscosity is doubtful since the presence of air bubbles will interfere with the wall layer.

3.8 Modified stirrer tank reactors

As described above in the section on mass transfer in stirred tank reactors some problems can arise when only one stirrer is used. In a viscous, non-newtonian system mass transfer only takes place in the stirrer region and not in the upper part of the vessel. This also holds for low-viscous fluids when the liquid height H is larger than the vessel diameter T. Another problem of the stirred tank reactor is that flow patterns are not well defined. It is for these reasons that, in some cases, the stirrer tank reactor has been modified. These are discussed in the following sections.

3.8.1 The multi-stage CSTR

Apart from adapting a stirrer to achieve a better mixing, the use of multiple stirrers is frequent when H/T ratios > 1 (Figure 3.16). Each stirrer can be considered to be in one mixing cell. Each mixing cell can be analysed as a separate stage in the reactor which is connected to other stages by the gas and liquid flows.

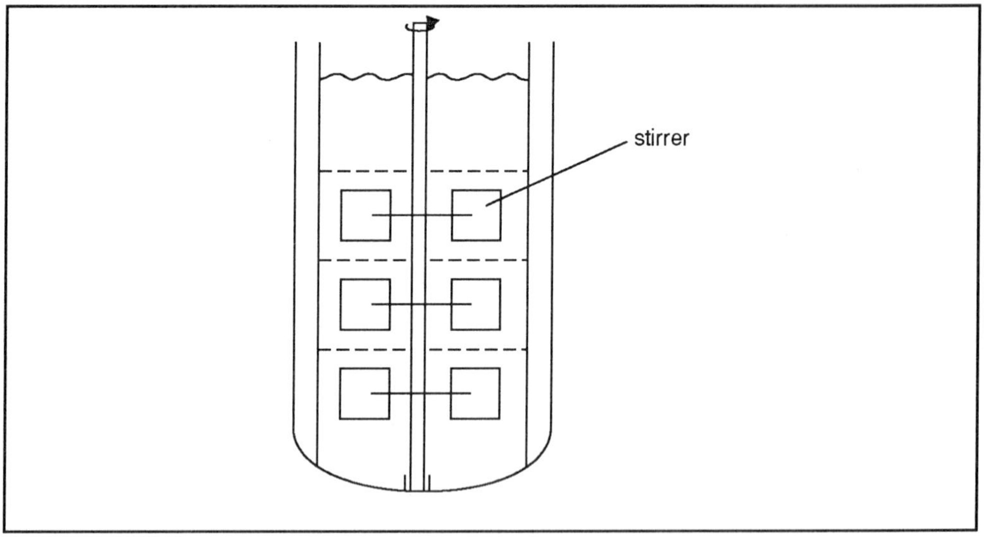

Figure 3.16 General layout of a slender fermenter. This one has a three stirrer arrangement (see text).

The influence of the number of stirrers on the mixing time is rather unknown. At Re < 5000 a clear compartimentation begins to form. This leads to an increase in mixing time at decreasing Re-numbers. At Re < 500 the mixing time is nearly infinitely long. Liquid is transported by bubbles moving through the interface of each stage, thus increasing mixing. No data on large scales are known. For Re > 5000 it seems logical to apply Equation 3.20 for each compartment. For a limited number of compartments, the total mixing time will be in between the mixing time per compartment and the product of the number of compartments and this mixing time. It must be emphasised that it is not the total mixing time that is the characteristic parameter, but rather the time needed to mix each of the compartments.

3.8.2 CSTR with internal draft tube

It is also possible to insert a draft tube in the CSTR. In this case the stirrer, usually of the propeller type, can be placed at the bottom or at the top of the draft tube (Figure 3.17). By the action of the propeller a better defined flow pattern exists than in a CSTR without draft tube.

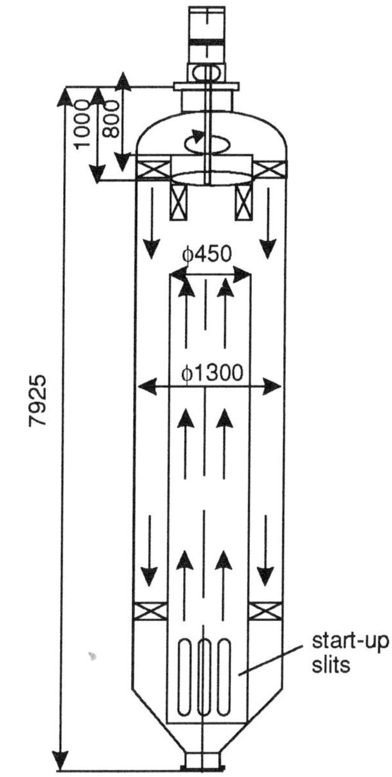

Figure 3.17 Continuous stirred tank reactor with internal draft tube (see text). (After Blencke H, 1985 Biotechnology, Vol 2, Eds H J Rehm and G Reed, VCH, Weinheim, 471-474).

SAQ 3.8

Answer true or untrue to the following:

1) For a CSTR with a non-viscous liquid, the number of stirrers has little influence on the mass transfer.

2) The lower the stirrer is placed in the vessel, the greater the power dissipated.

3) k_La is directly related to the height/diameter ratio of a CSTR.

4) To be effective, a ring sparger should have a smaller diameter than the stirrer.

Summary and objectives

In this chapter, we have focused our attention on the most widely used type of bioreactor, the Conventional Stirred Tank Reactor. Particular attention was paid to consideration of the major parameters involved in mass transfer within such bioreactors. This power consumption, flow patterns, liquid mixing and the dispersion of gases were examined and relationships between these factors and transfer processes were established. In the final part of the chapter, some modifications of stirred tank reactors were considered briefly.

Now that you have completed this chapter you should be able to:

- list the parameters that need to be considered in modelling mass transfer processes in bioreactors;
- be able to calculate the power input of gassed and ungassed systems from data relating to stirrer power number, stirrer speed and stirrer dimensions;
- use relationships between $\frac{P_g}{P_s}$ ratios and gas flow rate ϕ_g;
- use data to calculate impeller tip velocities required to prevent flooding;
- calculate mixing times from supplied data relating to reactor dimensions and stirrer power number;
- describe the effect of ionic strength on gas hold-up;
- use supplied data relating to P_g/V ratio and superficial gas velocity (v_s) to calculate k_La;
- calculate k_La at a variety of temperatures from data relating to k_La at 20°C;
- demonstrate knowledge of the effects of the number of stirrers, stirrer position, sparger dimensions and viscosity on k_La values.

Bubble column, airlift reactors and other reactor designs

Bubble column, airlift reactors and other reactor designs

4.1 Introduction

emphasis on bubble size and gas hold-up

This chapter has a similar structure to the previous chapter except we will be dealing with different types of reactors. We will first give a general description of bubble column and airlift reactors and then go on to discuss flow patterns in these reactors. Important in this discussion will be liquid and gas mixing processes with particular emphasis on bubble size and gas hold-up. In the final part of the chapter, we will give a brief introduction to new developments in reactor design.

4.1.1 General features of bubble column and airlift reactors

Although the overall energy input of bubble columns and airlift reactors is approximately similar as that for the CSTR, the former are cheaper and easier to scale-up. Furthermore, the design avoids heat generation by mechanical agitation. The basic construction of a bubble column in its simplest form is schematically shown in Figure 4.1. It is a cylindrical vessel with liquid height H and diameter T. The diameter of the cylindrical reactor is usually enlarged at the top to facilitate bubble disengagement and foam destruction. The air is distributed at the reactor bottom by means of various kinds of spargers, and rises upwards in the form of bubbles concurrently or counter currently to the liquid. As the bubbles, rising faster than the surrounding liquid, carry liquid upwards in their wake, a liquid circulation pattern is established which causes good mixing intensity in the liquid phase.

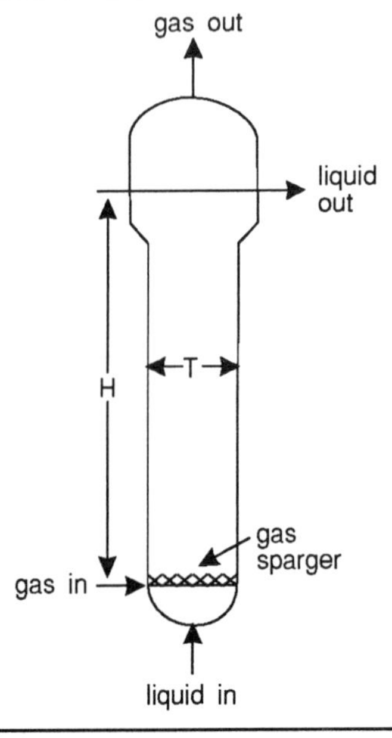

Figure 4.1 Conventional bubble column. T = diameter, H = liquid height.

The most important process in a bubble column is the formation of gas bubbles at the gas sparger. As mentioned earlier, smaller bubbles give a larger area for mass transfer between the gas and liquid phases. To achieve this, a large variety of sparger designs have been developed. A few are shown in Figure 4.2. However, the diameter of the gas bubble at the sparger is not necessarily the same as the bubble diameter in the bulk of the column, because the bubbles from the sparger undergo a coalescence and/or a (re)dispersion process. We will not discuss the merits or otherwise of the various types of spargers at this stage.

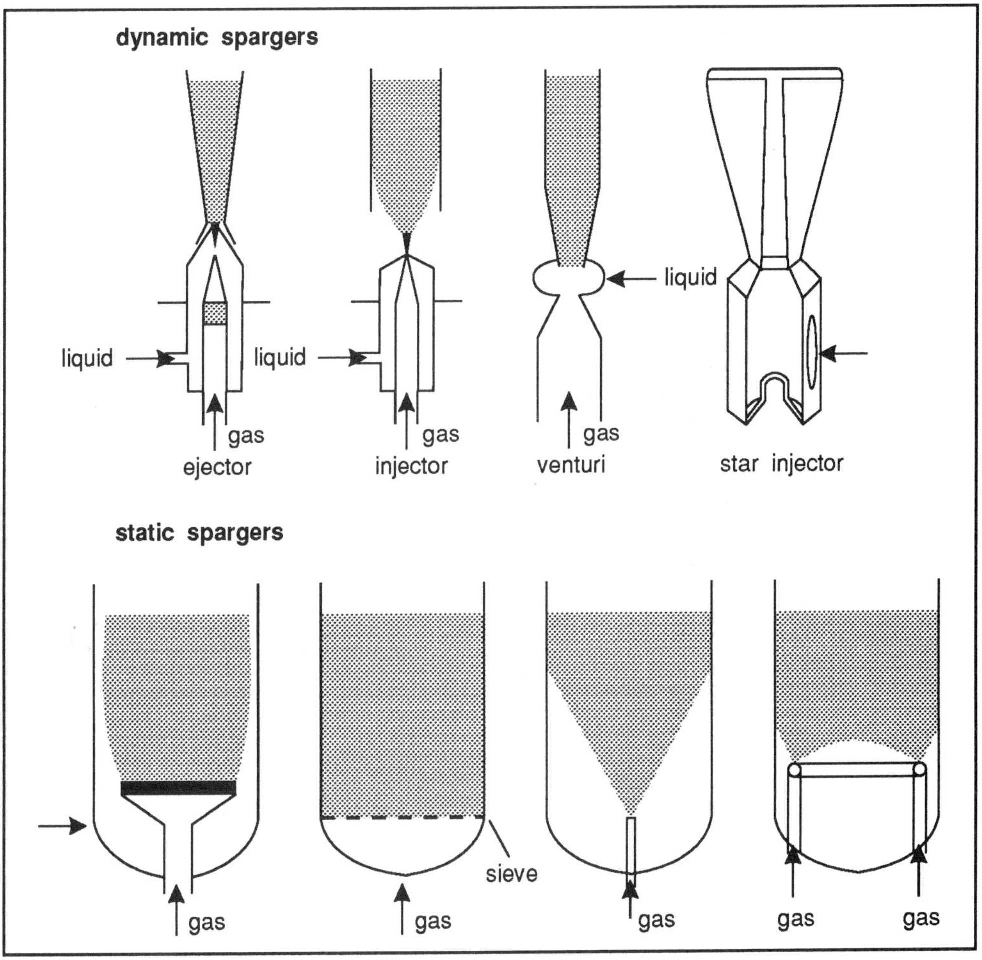

Figure 4.2 Static and dynamic gas spargers. (Redrawn from Deckwer, W D, 1985, 'Biotechnology' Ed H J Rehm and G Reed, VCH, Weinheim, 2, 445-464.

Contrary to the stirred tank reactor, bubble columns are usually used for low viscous broths and to process large volumes. They are also suitable for shear sensitive organisms due to the lack of mechanical stirring.

airlift loop reactors

In order to meet other specific process requirements there has been a tendency to modify the bubble column. This has led to a variety of bubble-column-like reactors. For example the need for better liquid circulation has stimulated the development of the airlift reactors with an internal or external liquid circulation loop (Figure 4.3). These utilise the specific gravity difference between the air-rich part of the liquid in the riser and the denser air-depleted part of the liquid in the downcomer to obtain continuous liquid movement. Due to the high circulation-flow rate, efficient mixing and oxygen transfer is combined with a controlled liquid flow in the absence of mechanical agitators.

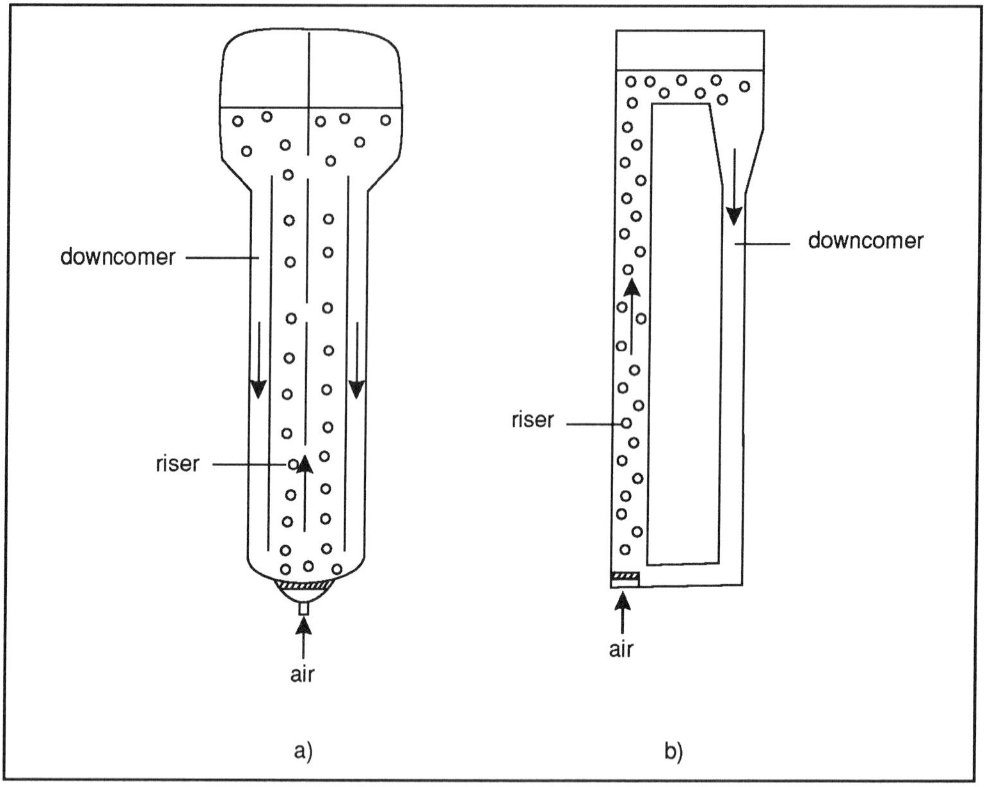

Figure 4.3 Airlift reactors a) airlift with internal loop, b) airlift with external loop.

airlift with sieve plates

The fact that better mass transfer was needed has also resulted in production of bubble columns which are comparable with the multi-stage CSTR: Instead of the multiple stirrers, gas is redispersed at various levels through sieve plates.

Both bubble columns and airlift reaction have found widespread application in large scale biotechnological processes such as waste water treatment, production of bakers' yeast, single cell protein (SCP) production, citric acid fermentation, etc. However, the popularity of the bubble column (or variants) is also increasing for production of high cost products.

4.1.2 The important factors which influence mass transfer in a bubble column

importance of bubble formation, coalescence and redispersion

This chapter will be mainly based on the description of the simple bubble column with a low viscous liquid, emphasis being laid on the mechanisms of the gas dispersion. These mechanisms include bubble formation at the sparging device, coalescence and redipersion and the resulting flow characteristics. A better understanding of such mechanisms leads to a better insight into the resulting liquid mixing, and heat and mass transfer. The phenomena which determine the mass transfer efficiency are given in Figure 4.4 as schematic guidelines.

Π Examine Figure 4.4 carefully. You will see that many of the factors we discussed in the previous chapters relating to mass transfer are also important in bubble column devices. You will see that at the bottom we have listed the key parameters k_La C^* and C_l. Can you write down how these are related to the rate of mass transfer of oxygen?

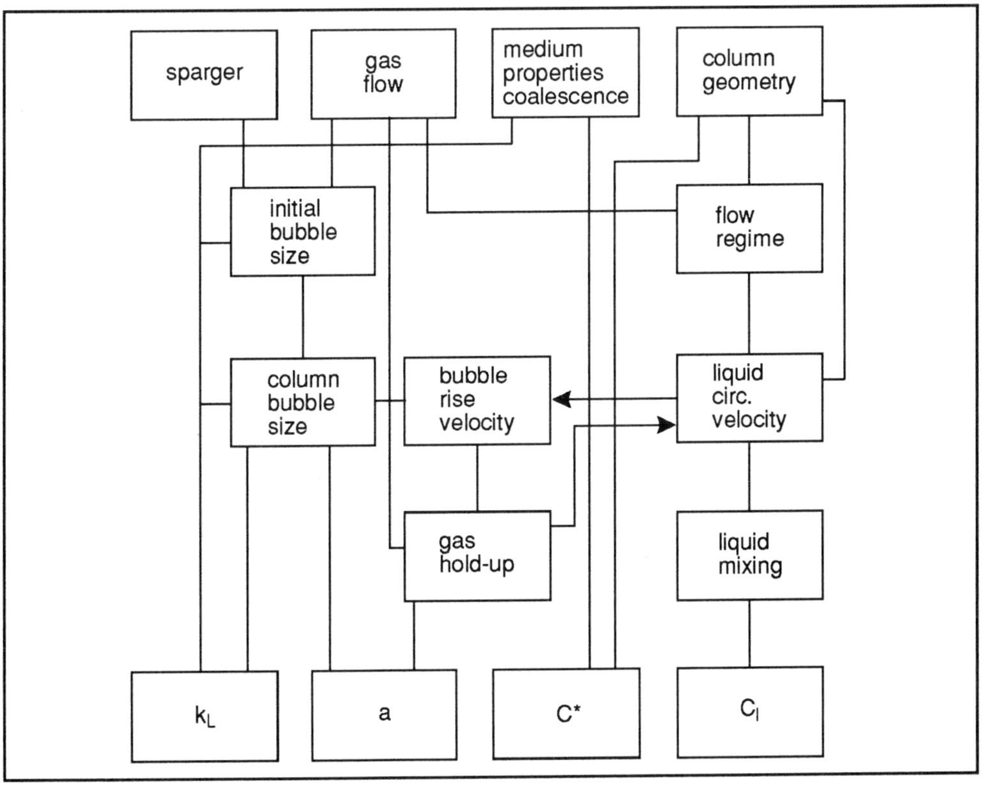

Figure 4.4 Interrelated processes in a bubble column.

We hope that you will recall from the two previous chapters that the rate of mass transfer of oxygen = k_La (C^*-C_l) where k_L = liquid mass transfer coefficient, a = specific surface area per unit volume of liquid, C^* is the saturation concentration of oxygen in the liquid, C_l = actual concentration in the liquid.

It should become clear from the discussion given below that liquid mixing, mass and heat transfer can be described by a series of simple correlations providing the mechanisms underpinning these processes are understood.

4.2 Flow patterns in aerated bubble columns and airlifts

homogenous flow regime

heterogenous flow regime

The bubbles generated at the sparger rise in a so-called bubble swarm. Bubble swarm behaviour in a bubble column is mainly determined by the gas superficial velocity. At low v_s values there is a fairly homogeneous distribution of rising bubbles through the bubble column area. This is called the homogeneous flow regime. At high gas superficial velocities (in a pure liquid) a number of very large bubbles exist due to coalescence. This results in an increase in bubble rise velocity and has a self-magnifying effect. As the large bubbles occur mainly in the centre region, this leads to large scale circulation patterns of liquid as well as gas. This is called the heterogeneous flow regime. Transition from homogeneous to heterogeneous flow occurs at about $v_s = 0.03$ m s^{-1}. The exact value is very dependent on liquid properties and column and sparger configurations.

Under heterogenous flow regime, circulations arise from higher (time-averaged) hold-up values in the centre of the column leading to an (time-averaged) upward velocity at the centre of the column. Although there is an upward velocity along the whole axis, circulations do not cover the column length. The momentum transport between upward and downward liquid currents leads to a break-up of the large circulation into smaller ones (Figure 4.5). These appear to be of the order of the column diameter.

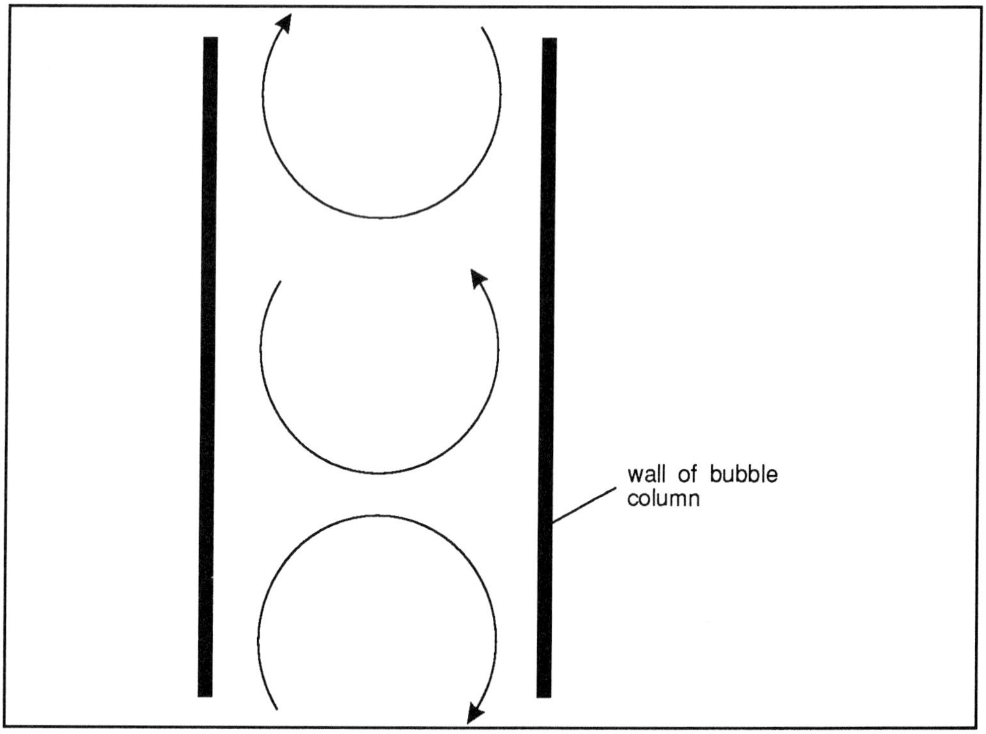

Figure 4.5 Generalised flow patterns in a bubble column without a draught tube. The circulations occur with gas superficial velocities above about 0.03 ms^{-1} (see text).

A simple way to stabilise circulation flow over the column length is the introduction of a draft tube (Figure 4.6), which in effect is the principle of an airlift reactor. However, we should remember unhampered circulation can be expected to improve mixing characteristics.

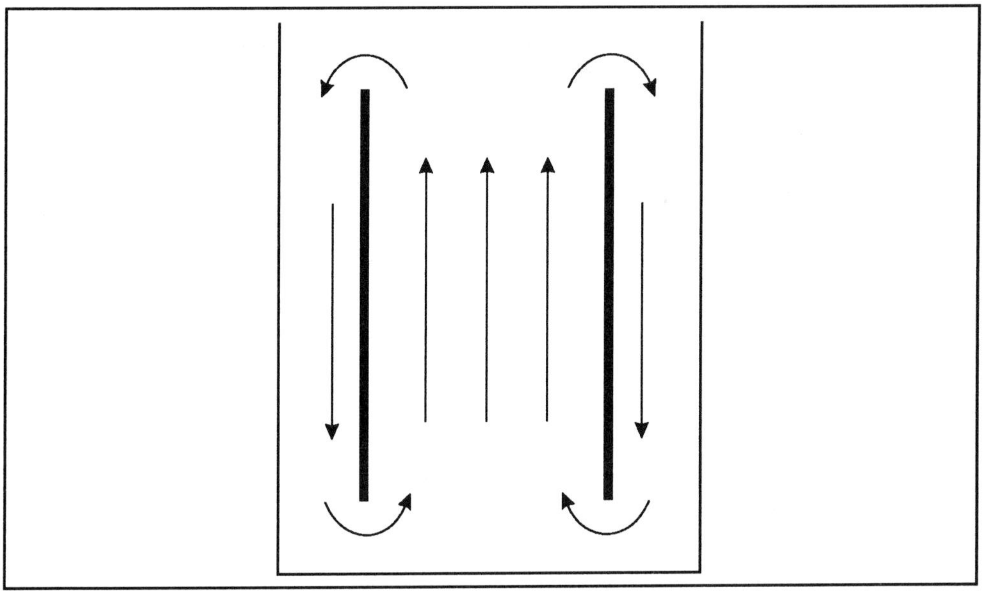

Figure 4.6 Generalised flow patterns in a bubble column with a draught tube (see text).

4.3 Liquid mixing

4.3.1 Liquid mixing based in fluid velocities

liquid dispersion coefficient

Liquid mixing in a bubble column is due to several processes. Under the homogeneous flow regime the bubbles themselves contribute to mixing because of liquid transport in the bubble wake as well as transport due to velocity differences around the bubble. This mixing process can be regarded as a kind of diffusive mixing to be characterised by an effective liquid dispersion coefficient D_l. We will deal with this approach later.

liquid circulation velocity v_{lc}

Under heterogeneous flow conditions, a very different mixing process is accomplished through the action of large circulating flows. From a mathematical approach based on a flow model with circulatory flow of vessel diameter T, a liquid circulation velocity v_{lc} can be calculated.

We will not go into the mathematical derivation here, but the relation described in Equation 4.1 has been shown to be consistent with experimental data.

$$v_{lc} = 1.18\,(g \,.\, T \,.\, (v_s - \varepsilon \,.\, v_{bs}))^{1/3} \qquad \text{(E - 4.1)}$$

in which: v_{bs} is the rise velocity of a single bubble relative to the liquid; v_s = gas superficial velocity; ε = gas hold-up; g = acceleration due to gravity; T = vessel diameter.

A major disadvantage of this equation in engineering calculations is the inclusion of the dependent variables ε and v_{bs}. Complicated methods of obtaining the values of ε and v_{bs}

have been described. However, from actual ε and v_{bs} data it can be argued that under heterogeneous flow conditions $v_s > \varepsilon v_{bs}$, and Equation 4.1 reduces to:

$$v_{lc} = 0.9\,(g\,.\,T\,.\,v_s)^{1/3} \qquad \text{(E - 4.2)}$$

in which only the independent variables T and v_s are present.

v_{lc} has been measured and compared with v_{lc} values predicted by Equations 4.1 and 4.2. The results are shown in Figure 4.7.

Π Use Figure 4.7 to determine the limits of v_{lc} in which the experimental and calculated value are consistent with each other.

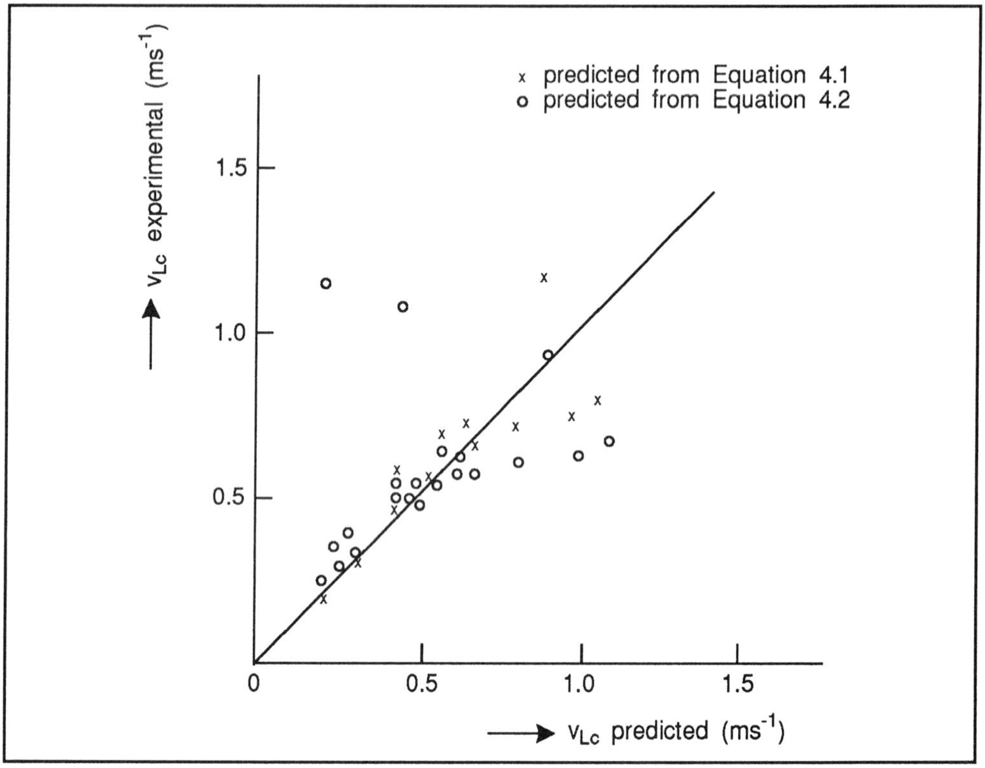

Figure 4.7 Comparison between liquid velocities calculated from Equation 4.1 and Equation 4.2 and measured velocities (data from Joshi (1980) Trans Instn Chem Eng 58 155-165 TU Delft 1989, 'Bioprocesstechnologie' Ed Van der Lans RGJM.

From the data in Figure 4.7, you should have realised that up to $v_{lc} = 0.7$ m s^{-1}, the agreement of both equations with experimental data is largely satisfactory. Experimentally it has been shown that Equation 4.1 is adequate in the range of v_s = 0-0.4 m s^{-1} and T = 0.1-7.5m.

In the above discussion, v_{lc} was assumed to be dependent only on v_s and T. However, it has been shown v_{lc} increases as the distance from the sparger becomes larger. These increases are of the order of 10-50% and it is not clear whether they are due to bubble expansion and/or entrance effects. In any case, it indicates that velocity prediction with inaccuracies less than 10% is unnecessary because of a 'natural' variation of bubble size. Liquid superficial velocities (v_l) only influence v_{lc} for $v_l >> v_s$. Thus providing $v_l >> v_s$, then Equation 4.1 and 4.2 can be used.

parabolic upward liquid velocity

It has also been shown that the (time averaged) upward liquid velocity profile is more or less parabolic and that the mean upward velocity is between 0.6 and 0.9 times the centre velocity, 0.8 times being an adequate estimate.

An important relationship in bubble columns is that between the liquid mixing flow rate (ϕ_l) and the gas flow rate (ϕ_g). It has been established that the following relationship holds:

$$\phi_l = 0.30 \,.\, T^{5/3} \,.\, \phi_g^{1/3} \,.\, g^{1/3} \quad \text{(E - 4.3)}$$

liquid mixing determined by smallest linear measurement

We have, however, to be a little cautious of this relationship. It only applies if the vessel diameter (T) is smaller than the liquid height (H). If $H/T << 1$ then T in Equation 4.3 has to be replaced by H + 0.07 (shown experimentally). Thus it appears that liquid mixing in bubble columns is determined by the smallest linear dimension of the column. Thus if $H/T > 1$ then we use the term T, and if $H/T < 1$ then we use H.

4.3.2 Liquid mixing based on effective dispersion coefficient

effective dispersion coefficient

In much of the literature on liquid mixing in bubble columns mixing data are presented in terms of an effective dispersion coefficient. By **effective** dispersion coefficient, we mean dispersion in a column with bubbles as opposed to dispersion in a column. Clearly the presence of the bubbles influences dispersion. We will meet an analogous situation when we discuss immobilised systems in Chapter 6. In this latter case, dispersion is influenced by the matrix material and will not be the same as in free solution. Experimental data has let to the following correlation being established:

$$D_l = 0.37 \,.\, T \,.\, v_{lc} \quad \text{(E - 4.4)}$$

Combined with Equation 4.2 this results in:

$$D_l = 0.33\,(g \,.\, T^4 \,.\, v_s)^{1/3} \quad \text{(E - 4.5)}$$

Note, similar equations have been derived from turbulence theory. The only difference is that the constant factor is 0.35 instead of 0.33. Considering the scatter of experimental data, this is quite an excellent correlation. We would prefer you to use the correlation based on the experimental data.

Another means of describing the mixing process in provided by the Peclet number Pe. For simple tube flow the relation between the Peclet number and the number n of theoretical mixers is:

$$0.5\,Pe = \frac{v_{lc} \,.\, H}{2 \,.\, D_l} = n - 0.5 \quad \text{(E - 4.6)}$$

Note that a 'theoretical mixer' is an ideally mixed compartment.

It has been calculated that:

$$n = \frac{H}{0.8 \,.\, T} \quad \text{(E - 4.7)}$$

Assuming that n >> 1, Equations 4.4, 4.6 and 4.7 lead to:

$$D_l = 0.36\,(g\,.\,T^4\,.\,v_s)^{1/3} \qquad \text{(E - 4.8)}$$

This is in surprisingly good agreement with the other correlations. It also shows the simple background of the dispersion data, ie a column consists of n ideally mixed compartments, with n given by Equation 4.7, coupled with an exchange flow between the compartments. These relationships are sufficient for engineering purposes.

It is unusual to express the mixing in columns by means of a mixing time. Yet we want to know at least an order of magnitude, to compare the characteristic times of the process. It is possible to use the liquid circulation velocities to derive these equations. In this way, assuming that the interaction between the compartments is ideal, Equation 4.2 leads to:

$$t_c = \frac{H}{0.9\,.\,(g\,.\,T\,.\,v_s)^{1/3}} \qquad \text{(E - 4.9)}$$

where t_c = circulation time.

Unfortunately there is no general agreement on what exactly is defined as mixing time (eg 90% mixed, 95% mixed etc). Depending on the definition used for mixing time, t_m equals 2-5 times the circulation time.

4.3.3 Influence of liquid viscosity and surface tension

The influence of liquid viscosity and surface tension on liquid mixing has not been investigated extensively. In very turbulent systems, liquid viscosity and surface tension do not have much effect on liquid mixing. At very large viscosities an influence can be expected, although bubble columns are not normally used under these circumstances because mass transfer rates are low. We will discuss this in Section 4.5.

In the above consideration the dispersion coefficient has been regarded as gradually increasing with gas velocity. This is true in so far as the flow is heterogeneous in the column. At low velocities the flow is homogeneous and it appears that precisely when a transition to heterogeneous flow occurs, the dispersion coefficient, although small, reaches a maximum. In this gas flow region (v_s < 0.03 m s^{-1}) the dispersion coefficient can be estimated from Equation 4.8. Channelling can also occur, particularly at very low air velocities with one sparger. It may occur at all air velocities, if the column is slightly tilted.

A word of caution: a review of flow velocities and dispersion data from real process leads to rather consistent relations. However, scatter of data is large (10-50%) and deviations can occur due to particular geometry and flow phenomena. Therefore, whenever specific geometric data, fluid behaviour and gas velocities are used, a critical evaluation is needed to rule out the possibility of deviations from the above relations.

SAQ 4.1

1) What is the liquid circulation velocity in a bubble column, if the gas superficial velocity is 0.1 m s^{-1} and the diameter is 1m? Report your answer in terms of g.

2) If the gas flow rate into a bubble column is 1 m^3 s^{-1} and the diameter of the vessel is 1m, what is the liquid mixing flow rate? Again report your answer in terms of g.

3) A bubble column has a liquid height of 10m and a diameter of 1m. How many ideally mixed compartments are there in the vessel?

4) What is the liquid circulation time if the column diameter is 1m and liquid height 10m and the gas superficial velocity is 0.1 m s^{-1}? Report your answer in terms of g.

4.4 Gas dispersion

importance of bubble size

Equally important to gas mixing, which is mainly dependent of the flow regime, is the bubble size. Bubble size is a key parameter in bubble column reactor design. As we have already learnt an increased interfacial area (a) leads to an increased mass transfer rate. Bubble size is determined by the initial size created at the sparger and the coalescence-redispersion of the bubbles in the bulk. Depending on those coalescing/non-coalescing properties of the liquid the bubble diameter at the sparger can differ significantly from the ultimate bubble diameter in the column. Therefore, in this section both, bubble formation at the sparger and bubble formation at a distance, will be discussed. The bubble formation at the sparger can be realised by means of a variety of spargers. Here we will make a division into three groups:

- single orifices and perforated discs;
- porous discs;
- two phases injectors.

Another characteristic of gas dispersion is the gas hold-up, which determines the effective utilisation of the bubble column volume in respect of oxygen mass transfer. The gas hold-up is mainly dependent on the liquid properties and the bubble velocity. The mechanisms and formulae given in this section afford insight into the estimation of the gas mixing, bubble size and gas hold-up.

4.4.1 Gas mixing

contrast between homogenous and heterogenous flow regimes

Many observations show that in the homogeneous flow regime, where $v_s <$ 0.02-0.04 m s^{-1}, for water, the gas is in plug flow with a very small and constant effective gas dispersion coefficient D_g of about 10^{-2} m^2 s^{-1}. If the heterogeneous flow regime is reached the gas mixing increases very rapidly as a result of large liquid circulatory flows which drag down the rising bubbles. In this regime the effective gas dispersion coefficient for air-water is correlated by:

$$D_g = 78\,(v_s * T)^{1.5} \quad \text{(E - 4.10)}$$

From Equation 4.10 it appears that at large bubble column diameters D_g increases sharply. If Equation 4.10 is used to calculate the Pe-number it can be shown that in small diameter columns with high H/T ration (> 10) the gas mixing is of the plug flow type contrary to large diameter columns with low H/T ratio (< 3) and large superficial gas velocities, in which the gas phase is ideally mixed. It is evident that this phenomenon is especially important in scale-up procedures for bubble column reactors. It may be noted here again that the transition from the homogeneous to the heterogeneous regime depends on geometric parameters (sparger type and location) and viscosity of the medium. No design rules based on the transition value of v_s exist.

4.4.2 Bubble size

Sauter mean diameter d_{vs}

To characterise bubble size distribution we will use the Sauter mean diameter or volume-surface mean diameter d_{vs}. This is given by:

$$d_{vs} = \frac{\Sigma d^3}{\Sigma d^2} \qquad \text{(E - 4.11)}$$

This is widely used as the most representative average diameter for mass transfer calculations. In the literature, you may find many different definitions of mean bubble diameters. These may lead to quite different values in subsequent correlations.

Because of the wide spread use of d_{vs}, we will focus on this term for subsequent correlations.

Π Before embarking on a discussion of these correlations, let us calculate d_{vs} from some data. Let us assume that we have measured some bubbles and they have diameters described in the table below. Use this data to calculate d_{vs}. Note that we have used only a few bubbles. Obviously the more bubbles that are measured, the more accurately the calculated d_{vs} value will represent the population. In this case, however, we did not want you to do a long and tedious calculation. Use the table to complete your calculation.

Bubble number	Bubble diameter (d) (mm)	d^2	d^3
1	5		
2	4		
3	5		
4	6		
5	4		
6	3		
7	4		
8	6		
9	8		
10	3		
		Σd^2	Σd^3

Your answer should come to 5.714 mm. (First you need to calculate all the d^2 values and add them together to give Σd^2. This should yield 252. Then calculate all the d^3 and add them together to give Σd^3. This should yield 1440. Thus $d_{vs} = \frac{\Sigma d^3}{\Sigma d^2} = 5.714$ mm).

Bubble size at the sparger

With an orifice or perforated disc gas is introduced into the column through one or more holes with a diameter larger than about 1 mm. For bubble formation at increasing gas flow rates four regimes need to be distinguished:

- separate bubble formation;
- chain bubbling;
- jet regime;
- turbulent regime.

separate bubble formation

For the separate bubble formation regime, d_{vs} is dependent on the equilibrium between surface tension and buoyancy forces. d_{vs} can be calculated from:

$$d_{vs} = 1.7\left(\frac{\sigma \, . \, d_o}{\Delta\rho \, . \, g}\right)^{1/3} \qquad \text{(E - 4.12)}$$

σ = surface tension of gas liquid interface; $\Delta\rho$ = difference in the densities of gas and liquid.

This equation is valid for $d_{vs} > d_o$ where d_o is the diameter of the sparger orifice. In air-water systems this means that $d_o < 6$ mm.

chain bubbling

At increasing gas flow rates, bubble separating frequency increases so that bubbles leave the sparger as a continuous chain. Here the diameter is frequency determined. From a large number of data it can be shown that the following simple relation is appropriate:

$$d_{vs} = 1.17 \, v_o^{0.4} \, . \, d_o^{0.8} \, . \, g^{-0.2} \qquad \text{(E - 4.13)}$$

where v_o = gas velocity at the sparger orifice.

More sophisticated relations have been reported. Equation 4.13, although simple, is adequate for our purposes.

jet regime

When gas flow rates are further increased, bubble formation becomes unstable. A transition is found to occur for low viscosity systems at $2000 < Re_o < 10000$, in which Re_o is the Re-number for flow through the orifice. In such a case the dependence on gas flow rate is only marginal and it can be stated that for the jet regime:

$$4 \text{ mm} < d_{vs} < 6 \text{ mm} \qquad \text{(E - 4.14)}$$

In Figure 4.8 d_{vs} values are given for a number of sparger diameters.

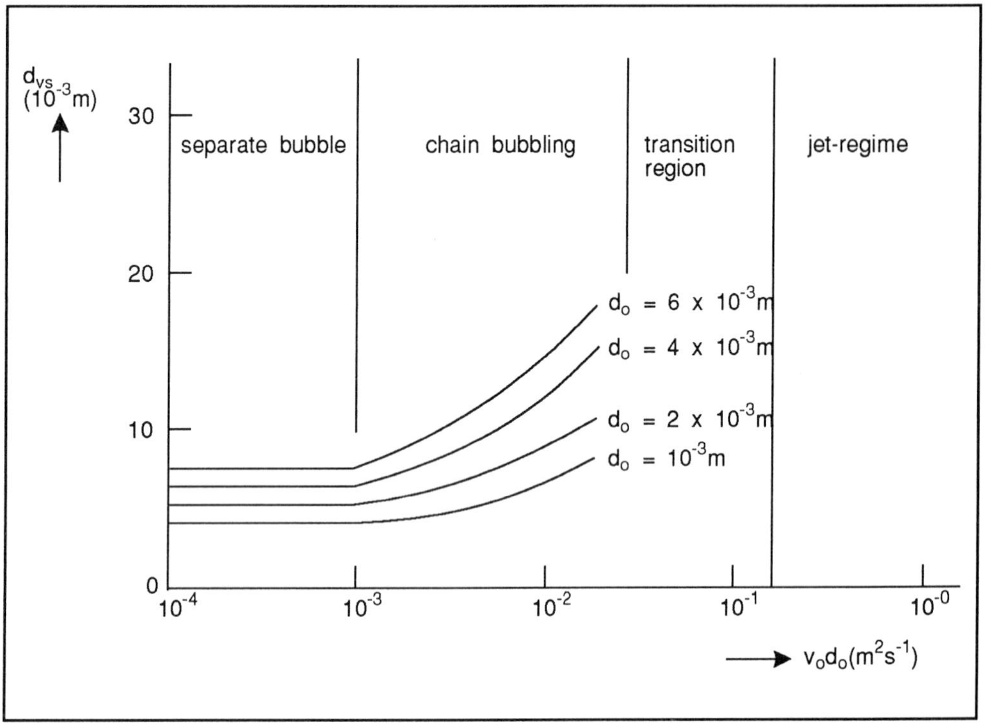

Figure 4.8 Bubble formation from an orifice in the air-water system. (TU Delft, 1985, 'Bioprocesstechnologie' Editor; Van der Lans, RGJM).

Π Examine the data presented in Figure 4.8 carefully and decide how to obtain small bubbles.

Obviously, the smaller the diameter of the sparger orifice, the smaller the bubble.

To obtain really small bubbles it is desirable to operate in the separate bubble regime. However, it can easily be established that only low superficial gas velocities can be used to achieve this. Therefore, orifices will nearly always need to be operated in the jet regime. If bubble diameters < 4 mm are desired, a large number of orifices are required and this is usually achieved by using porous discs. These are discussed in the next sections. It must be stressed that d_{vs} of bubbles from this type of spargers is not influenced by the coalescing/non-coalescing properties of the medium. This is in contrast to the other type of sparger (eg ejector) as we will show in a later section.

porous discs produce small bubbles

As described above, an orifice sparger leads, in practice, to bubbles of 4-6 mm diameter. For mass transfer, smaller bubbles are far more effective. This has led to the development of spargers which can generate extremely small bubbles. One of these is the porous disc. Just above the porous disc the multitude of very small bubbles collide with each other and coalesce to a size dependent on the coalescence hindering properties of the medium. The d_{vs} in non-coalescing (which means strong coalescence hindering, resulting in non-coalescence in a free turbulent flow) media can be calculated from:

$$d_{vs} = 0.0061\,(v_o\,.\,d_o)^{0.1}\,.\,d_o^{0.08}\,.\left(\frac{\sigma}{\sigma_w}\right)^{0.38} \qquad \text{(E - 4.15)}$$

where σ_w is the surface tension of water.

In coalescing media d_{vs}, measured at a short distance above the sparger the relationship is:

$$d_{vs} = 0.0104\,(v_o\,.\,d_o)^{0.16}\,.\,d_o^{-0.07}\,.\left(\frac{\sigma}{\sigma_w}\right)^{0.41} \qquad \text{(E - 4.16)}$$

Figure 4.9 shows these relations.

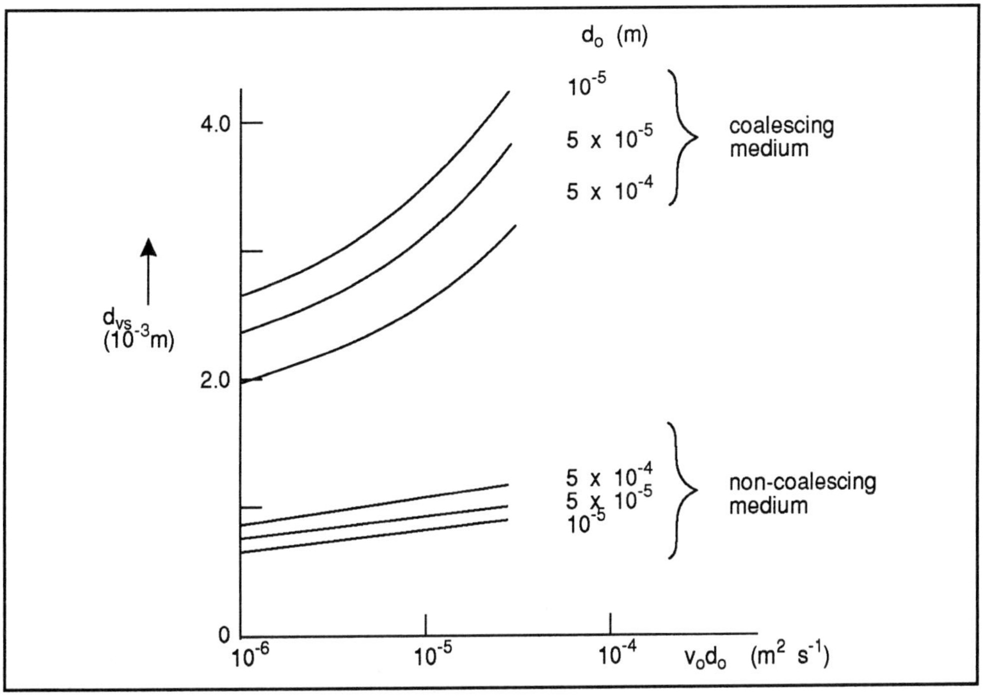

Figure 4.9 Bubble formation from a porous plate in the air-water system. (TU Delft, 1985, 'Bioprocesstechnologie' Editor; Van der Lans, RGJM).

Π What are the range of bubble sizes produced from porous discs in non-coalescing and in coalescing media.

As can be seen from Figure 4.9 in non-coalescing media $d_{vs} < 1$ mm. For coalescing media the bubbles are larger, ranging from 2-4 mm. It is also apparent that the bubbles rapidly become larger at greater air flow rates.

alternative to porous discs

Because porous discs are frequently subject to clogging, alternative small bubble spargers have been developed. This has resulted in two phase ejectors/injectors being developed. These are illustrated in Figure 4.2. In this type of sparger, gas and liquid are concurrently pumped through a small nozzle. Due to the extreme energy dissipation in this nozzle, very small bubbles are generated.

In non-coalescing media bubbles with d_{vs} = 0.5-1 mm are produced. These results closely resemble the bubble formation at the porous disc sparger. Therefore we may conclude that ejectors/injectors generate bubbles of d_{vs} = 4-6 mm in coalescing media.

In summary, the following conclusions about the initial bubble size at the sparger can be drawn. From Figure 4.8 it is evident that to generate bubbles with $d_{vs} << 4 - 6$ mm very small orifice diameters and low orifice gas velocities are required. In practice, however, gas velocities are large in the orifice and orifice diameters are mostly much larger than 1 mm in order to prevent clogging. This means that an orifice sparger usually leads to a bubble diameter of d_{vs} = 4-6 mm irrespective of medium properties (coalescing/non-coalescing) of the broth. In bioreactors, this is especially important because coalescing or non-coalescing properties of the broth can change dramatically due to microbial activity.

The lower d_{vs} values (less than 1 mm) can only be reached using porous disc type or ejectors/injectors and exclusively for a non-coalescing fluid or extremely low superficial gas velocities in a coalescing fluid.

energy costs of sparger types

Finally it should be noted that the energy needed to operate ejectors can be considerable. For example, the liquid pumping through ejectors can consume as much as 50% of the gas compression energy. Gas dispersion through an orifice requires not much more energy than that due to the hydrostatic pressure.

SAQ 4.2

Answer true or false to the following.

1) In low viscosity system, if the Re-number for flow through a sparger orifice is in the range of 2000-10000, then the bubbles produced will have d_{vs} values in the range 4-6 mm.

2) Separate bubbles will be produced at a sparger providing the product of the gas velocity at the sparger and the sparger orifice diameter does not exceed about 10^{-3} m^2 s^{-1}.

3) Typical bubble sizes produced by a porous disc in non-coalescing medium are of the order of 0.5-1 mm.

4) Two phase ejectors and injector type spargers produce bubbles of similar dimensions as porous discs.

Bubble size at a distance from the sparger

The bubble generated at the sparger can either coalesce to larger bubbles or be dispersed to a smaller diameter bubble. In coalescing media d_{vs} can be calculated from:

$$d_{vs} = 4.15 \cdot \frac{\sigma^{0.6}}{(P/V)^{0.4} \cdot \rho^{0.2}} \cdot \varepsilon^{0.5} + 9 \times 10^{-4} \qquad \text{(E - 4.17)}$$

On the assumption that $P/V = \rho.g.v_s$ and $\varepsilon = v_s/v_{bs}$ with $v_{bs} = 0.25$ m s^{-1}, this leads for an air-water system to a d_{vs} value of about 6 mm. This is in accordance with the d_{vs} values for the jet regime of Figure 4.8. The d_{vs} value of about 6 mm is rather insensitive to v_s variations as the v_s influences in ε and P/V largely cancel out. Experimental confirmation of this for bubble columns has been shown in a large number of published works. It should be noted that Equation 4.12 is derived for stirred vessels in which circulation is very intensive and the chance of collisions between bubbles is high. In bubble columns at very low superficial gas velocities chances of collision may be low and this results, for porous discs, in bubbles smaller than those calculated by Equation 4.12. With a porous disc in air-water d_{vs} increases sharply with gas superficial velocities. Measurements indicate that Equation 4.17 is applicable for $v_s > 0.01\text{-}0.02\ m\ s^{-1}$.

In non-coalescing media for cases where the initial bubble diameter is larger than the bulk bubble diameter the ultimate bubble diameter d_{vs} can be calculated according to the relationship.

$$d_{vs} = 1.93 \cdot \frac{\sigma^{0.6}}{(P/V)^{0.4} \cdot \rho^{0.2}} \qquad \text{(E - 4.18)}$$

Figure 4.10, d_{vs} is calculated for air-water systems with coalescing or non-coalescing properties for different P/V values. It is assumed that $P/V = \rho.g.v_s$, which is a satisfactory estimate for normal height values. Thus we are able to plot v_s to represent P/V.

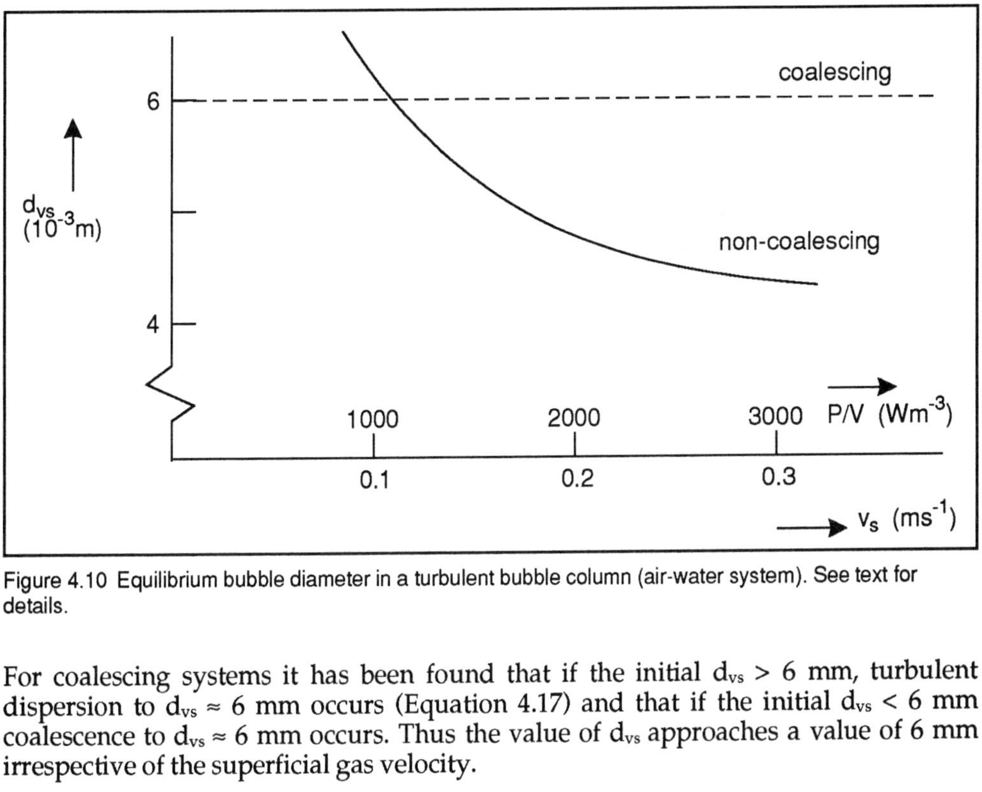

Figure 4.10 Equilibrium bubble diameter in a turbulent bubble column (air-water system). See text for details.

For coalescing systems it has been found that if the initial $d_{vs} > 6$ mm, turbulent dispersion to $d_{vs} \approx 6$ mm occurs (Equation 4.17) and that if the initial $d_{vs} < 6$ mm coalescence to $d_{vs} \approx 6$ mm occurs. Thus the value of d_{vs} approaches a value of 6 mm irrespective of the superficial gas velocity.

If non-coalescing media are used it can be concluded that if the initial d_{vs} > bulk d_{vs} turbulent dispersion occurs until the bubble from the sparger has reached the bulk value (Equation 4.18) and if the initial d_{vs} < bulk d_{vs} the bubble diameter from the sparger will not change.

It may be noted here that practical fermentation fluids will generally not be completely non-coalescing. In such circumstances we might consider we are working with a coalescence suppressing medium which still gives a bubble diameter between the sparger value and 6 mm. In practice this might be the case for many fluids.

4.4.3 Gas hold-up

The hold-up for homogeneous bubble columns is determined by gas superficial velocity and bubble rise velocity according to:

$$\varepsilon = \frac{v_s}{v_{bs}} \qquad \text{(E - 4.19)}$$

In a bubble column, bubbles rise in a swarm. At low superficial gas velocities (< 0.04 m s^{-1}) for porous and perforated plates, the gas rise velocity decreases slightly with increasing gas superficial velocity. In contrast under heterogeneous flow conditions substantial liquid circulation velocities occur. Bubbles tend to be present mainly in the upward flow regions, thereby increasing the absolute bubble rise velocity, which is the sum of liquid velocity and v_{bs}. For high superficial gas velocities, gas rise velocities (relative to the wall) may reach up to 1.5 m s^{-1}. This is largely due to inhomogeneous effects like the forming of bubble clusters and coalescence to form some large bubbles with large rise velocities. If v_{bs} is calculated in a way similar to that given by Equation 4.19. It is found that v_{bs} is almost constant for different v_s values. The overall result is that the gas hold-up increases less than linearly with v_s for heterogeneous flow conditions.

For the simplified case in distilled water with $d_{vs} > 1$ mm we can assume that $v_{bs} = 0.25$ m s^{-1}. For the homogeneous flow regime, where the bubbles themselves contribute to mixing, it is assumed that $v_{lc} = 0$, hence applying Equation 4.19 gives:

$$\varepsilon = \frac{v_s}{0.25} \qquad \text{(E - 4.20)}$$

For fine bubble systems ($d_{vs} < 1$ mm), v_{bs} is smaller than 0.25 m s^{-1}, resulting in higher values than those given by Equation 4.20; yet linearity is maintained. For the heterogeneous flow regime, where a very different mixing process is accomplished through the action of large circulating flows, Equation 4.19 can be modified to include the sum of v_{bs} and the liquid circulation velocity v_{lc}, leading to:

$$\varepsilon = \frac{v_s}{0.25 + v_{lc}} \qquad \text{(E - 4.21)}$$

With v_{lc} derived from Equation 4.2 [$v_{lc} = 0.9\,(g * T * v_s)^{1/3}$].

The curves according to Equations 4.20 and 4.21 are given as curves a) and b) respectively in Figure 4.11. Also some data on air-water systems as measured and reviewed are presented in the same figure.

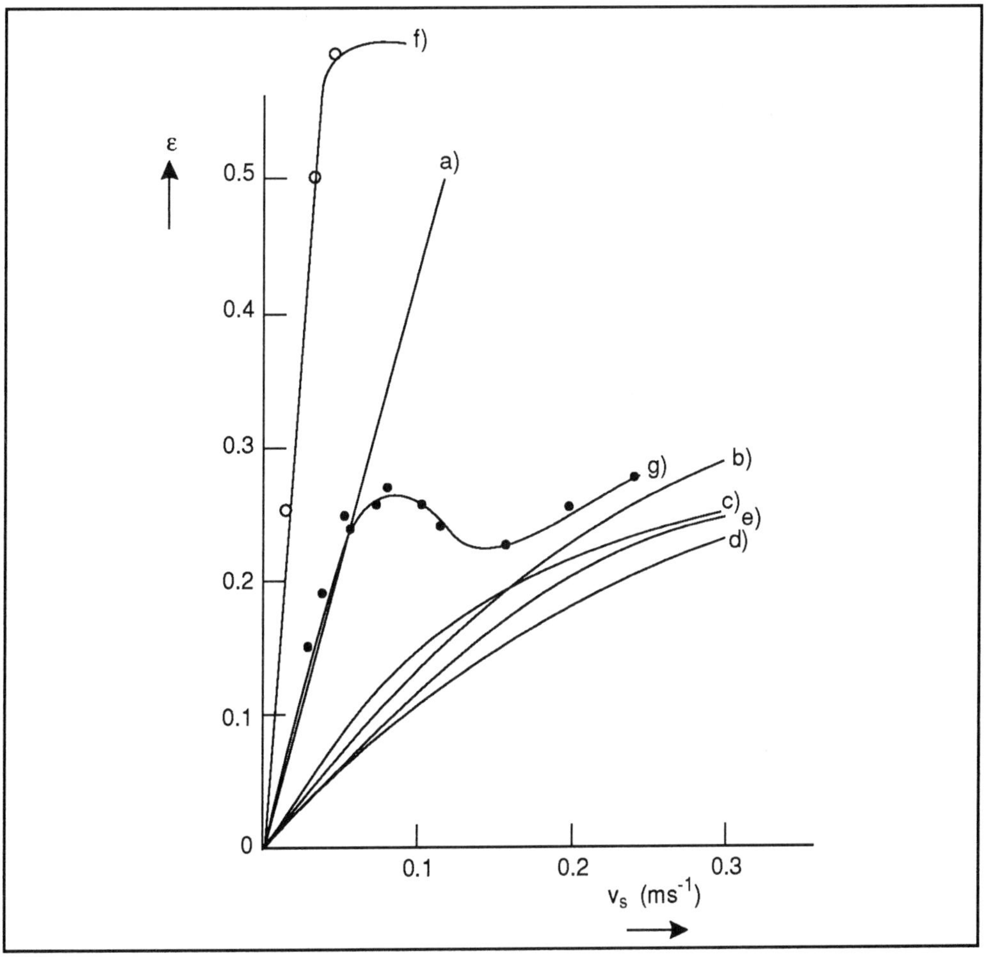

Figure 4.11 Gas hold-up in bubble columns (air-water systems). Line a) are theoretical values of ε calculated using Equation 4.20. Line b) are theoretical values of ε calculated using Equation 4.21. The other lines are from published data (see text for further details).

Π Examine Figure 4.11 carefully and decide which of Equations 4.20 and 4.21 are supported by experimental data.

It is clear that for homogeneous flow conditions (ie low v_s values), Equation 4.20 describes the hold-up rather accurately. A nearly linear relationship with superficial velocity is present, particularly for non-coalescing media with very small bubbles (curves f and g).

Note the localised maximum in curve g). This can be easily explained as a transition from homogeneous to heterogeneous flow. The theoretical curves (curve a) for homogeneous flow and curve b) for heterogeneous flow) lend support to this interpretation.

For the heterogeneous flow region hold-up values (curves c, d and e) are very similar to curve b).

experimental data support relationships between ε and v_s

We come to the conclusion that Equations 4.20 and 4.21 are, indeed, quite accurate for determining ε at different v_s values. But, we must consider other factors which may influence ε as not all of the curves c-g precisely follow the theoretical lines. From Figure 4.11 it is clear that the hold-up values are mainly determined by gas velocity. However other variables such as column diameter, sparger geometry and liquid properties are also important.

Column diameter

Because the liquid velocity increases for increasing column diameter (Equation 4.2), an increased column diameter may be expected to reduce hold-up values. Only at T < 0.1 m does a significant influence occurs. However, this is most likely due to wall effects.

Sparger type

The influence of the sparger type is rather complex, mainly depending on fluid characteristics. As already stated, the diameter of the bubbles in the column and thus the hold-up are determined by the coalescence behaviour of the liquid and initial bubble size at the sparger. For a non-coalescing system the bubble diameter is determined by the sparger.

Π Below is a list of three types of sparger, list them in the order you think that they will increase hold-up values.

- sintered discs;
- two phase injectors;
- perforated plates.

Unless you have had some experience of these spargers, there is actually no reason you should have been able to work this out, but experimentally the order of increase in hold-up values is perforated plates → sintered discs → two phase injectors under otherwise similar conditions. The difference may be as high as a factor 2. If you had worked out however that as under otherwise similar conditions, the order of decrease in bubble diameter is perforated plates → sintered discs → two phase injectors, the order of increase in hold-up is the same (as v_{bs} decreases with decreasing bubble diameter).

For a coalescing system the bubble diameter in the column is determined by the equilibrium value. Therefore, no large influence of the gas distributor can be expected. The influence will depend on the relative contribution of the transition area to the overall hold-up.

We can represent this in the following way.

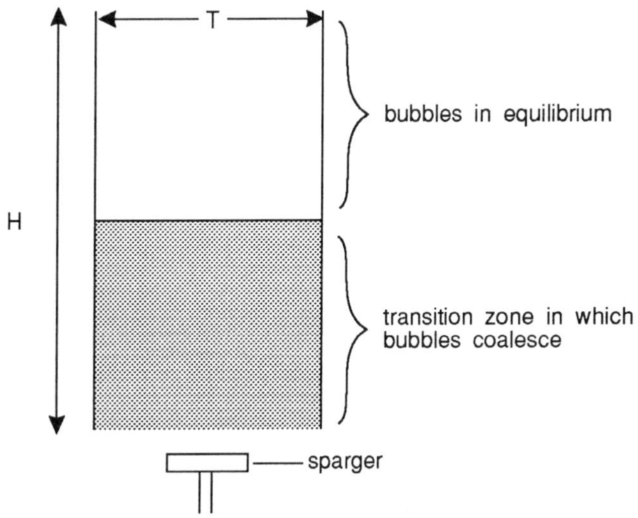

Experimentally there is no difference for perforated plates, sintered discs and two phase injectors. However, this is only true at H > 10T. At lower H/T values a transition region near the sparger where bubbles have not yet reached a final size becomes important, particularly at low gas rates because of a decreased collision frequency. At these lower H/T values there is a positive influence of porous plates or two phase fluid injectors compared to perforated plates or single orifices on hold-up values.

Liquid properties

The influence of liquid properties on gas hold-up is considerable and is strongly dependent on the sparger type. The most important liquid properties are the coalescence hindering, viscosity and surface tension of the medium. The influence of the properties of the gas on gas hold-up is relatively minor.

The coalescence properties of liquids can be manipulated by the addition of agents to the pure liquid.

Π Does salt increase or decrease the coalescence properties of aqueous media?

From what you learnt earlier you should know that salt increases surface tension and therefore addition of salt hinders coalescence. Its maximal effect is reached at a concentration corresponding with an ionic strength of 0.4 kmol m^{-3} (0.4 mol l^{-1}). For an injector nozzle, hold-up is increased by up to a factor of 3 by the addition of salt compared with the corresponding pure water value.

Π Will the effects of salt on hold-up be dependent upon sparger type? (Think what would happen if the sparger produced bubbles more-or-less equal to the equilibrium size).

addition of salt has little effect on hold-up in water

After a few moments thought, you should have realised that if the sparger produced bubbles of equilibrium size (about 6 mm diameter for water) then preventing coalescence by the addition of salt will have little influences on bubble size and, therefore, hold-up. It has been reported that an increase of only about 50% in hold-up occurs after salt addition using perforated plates which produce bubbles close to the equilibrium size. However using a single nozzle as sparger, other researchers achieved only a 10-20% increase in gas hold-up after salt addition to water.

surface active agents have limited effect on hold-up

Also addition of surface active agents, which decrease surface tension, change the coalescence behaviour. For example values have been reported in the literature of an increase in hold-up by a factor of 3 on the addition of 1% ethanol using porous plates or injector nozzles. For a perforated plate, which produced larger bubbles, only a 50% increase in hold-up on an addition of ethanol to water was achieved. Antifoam agents, are surface active agents and will hinder coalescence, but will, when added to a non-coalescing medium, promote coalescence. This interaction between the different additions is not well understood. For fermentation liquids in particular, where the micro-organisms can produce or consume both foaming and antifoam agents, these phenomena can lead to large variations in gas hold-up during the fermentation process.

Surface tension

The influence of surface tension, when pure liquids other than water are used, can be qualitatively described in that a lower surface tension gives a lower bubble rise velocity and therefore a higher hold-up. However, this effect is usually too small in comparison with the coalescence hindering effect to be measurable. With pure liquids the influence is very weak.

Viscosity of the medium

The hold-up is strongly dependent on the viscosity of the medium. The effect is brought about by two mechanisms:

- a high viscosity (above about 20 mPas) promotes coalescence, leading to large bubbles;
- a high viscosity decreases the bubble rise velocity.

In general it is found that an increased viscosity decreases hold-up, and thus the first mechanism seems to dominate. Sometimes, however, an increase in gas hold-up is found (eg by the addition of glycerin with viscosities below about 5 mPas), which is the result of a coalescence hindering effect of the (water-glycerin) mixture independent of viscosity.

Clearly the hold-up depends markedly on liquid properties, the dependency being governed by the sparger type. For the same manipulation of liquid properties (eg salt addition) the hold-up increase could be anywhere between 20% and 200%, varying with the sparger type. Consequently there will be no useful general relationships until research has been focused on the influence of these liquid properties in relation to sparger geometry. For engineering purposes predictions can be made for each individual case by careful estimation of the sparger and the liquid properties and by selection of literature relevant to those conditions. The starting point of such estimates is the hold-up value for coalescing media. This minimum value is more or less independent of sparger geometry.

The data we presented in Figure 4.11 shows a reasonable correlation with Equations 4.20 and 4.21 and permit a fairly simple expression for hold-up:

$$\varepsilon = 0.6 v_s^{0.7} \qquad \text{(E - 4.22)}$$

This holds for heterogeneous flow conditions in air-water systems.

For homogeneous flow conditions in air-water systems

$$\varepsilon = \frac{v_s}{0.25} \qquad \text{(see Equation 4.20)}$$

hold-up values lower in air lift than bubble column

The above discussion was concerned with batch columns without internal fillings and/or airlifts. In through-flow columns superficial liquid velocity is relatively unimportant under heterogeneous flow conditions as long as they are lower than gas superficial velocities. At higher liquid circulation velocities with concurrent flow, the hold-up decreases. This results in much lower hold-up values for an airlift reactor than for a batch bubble column. The magnitude of this decrease largely depends in the liquid recirculation velocity. These velocities can hardly be predicted from data reported in the literature. Therefore, at present, hold-up prediction in airlift reactors is not possible using general relationships.

Π Write down what you think the effect of having internal fillings, screens or sieve plates in a bubble column will be on hold-up.

You should have come to the conclusion that hold-up would be increased. Liquid circulation can be strongly reduced by internal fillings and screens and sieve plates. This results in a hold-up increase up to 100%, depending on liquid flow rate. Liquid flow rate is important for these applications because here liquid recirculation velocities are very low or absent.

Now that we have some knowledge of gas and liquid mixing, bubble size and hold-up values in bubble columns, we can move on to consider mass transfer. This we will do in the next section. Before doing so, attempt the following SAQ to test your knowledge about gas dispension in a bubble column.

SAQ 4.3

1) A bubble chamber under heterogeneous flow conditions produces bubbles of d_{vs} = 6 mm at the following gas superficial velocities 0.1 m s^{-1}, 0.2 m s^{-1} and 0.3 m s^{-1}. Give two circumstances under which this situation is likely to arise. (Assume that these refer to an air-water system).

2) The following data referring to d_{vs} and P/V input have been produced for a bubble column. Is the liquid in the system of the coalescing or of the non-coalescing type?

P/V (W m^{-3})	d_{vs}
1000	7 mm
2000	4.8 mm
3000	4.2 mm

3) The gas superficial flow rate is about ¼ of the hold-up value; this is typical of what sort of flow conditions?

4) Calculate ε when v_s = 1 m s^{-1} for:

a) heterogeneous flow conditions.

b) for homogeneous flow conditions.

5) A bubble column is fitted with two sieve plates. Will they increase or decrease hold-up?

4.5 Mass transfer

The gas liquid mass transfer in bubble columns is mainly determined by the diameter of the bubbles in the column. With a gas hold-up e in the bubble column the following relation holds for the volumetric mass transfer coefficient k_La, based on dispersion volume,

$$k_La = \frac{6 \cdot \varepsilon}{d_{vs}} \qquad \text{(E - 4.23)}$$

With respect to d_{vs}, it is clear from the section on bubble size that a division can be made between systems with coarse bubbles (d_{vs} = 4-6 mm) and fine bubbles (d_{vs} = 0.5-1 mm). The coarse bubbles occur in systems with orifice spargers, where the liquid may be coalescing or non-coalescing. The fine bubbles occur in systems with porous plate spargers or ejectors/injector spargers in non-coalescing liquids. However, for porous spargers or ejector/injector spargers in coalescing liquids the situation is not so clearly defined. The fine bubbles which are formed at the sparger coalesce in a short time to produce larger ones. In columns this effect is not negligible. Nonetheless, it seems useful to distinguish the mass transfer discussion into two groups. Firstly, systems with coarse bubbles where $d_{vs} \approx 6$ mm, ie with coalescing medium properties or with coarse bubble spargers and secondly, systems with fine bubble where $d_{vs} << 6$ mm, ie with non-coalescing medium properties combined with fine bubble spargers.

4.5.1 Coarse bubble systems

In Figure 4.12 we show the k_La^{20} for various superficial gas velocities, as found by the various researchers for coarse bubble systems. The results are scattered but remain within acceptable limits. Although a linear relationship between k_La and v_s is possible the results are best fitted by:

$$k_La^{20} = 0.32.v_s^{0.7} \qquad \text{(E - 4.24)}$$

This correlation is valid for $0.08 < T < 11.6$ m, $0.3 < H < 21$ m and $0 < v_s < 0.3$ ms^{-1}. At low v_s values this correlation slightly overestimates k_La values. Moreover, the mass transfer in coalescing (water) or non-coalescing (water/salt) media is described by the same relation. This is in good agreement with the finding that the bubble diameter for orifice spargers is not influenced much by media properties.

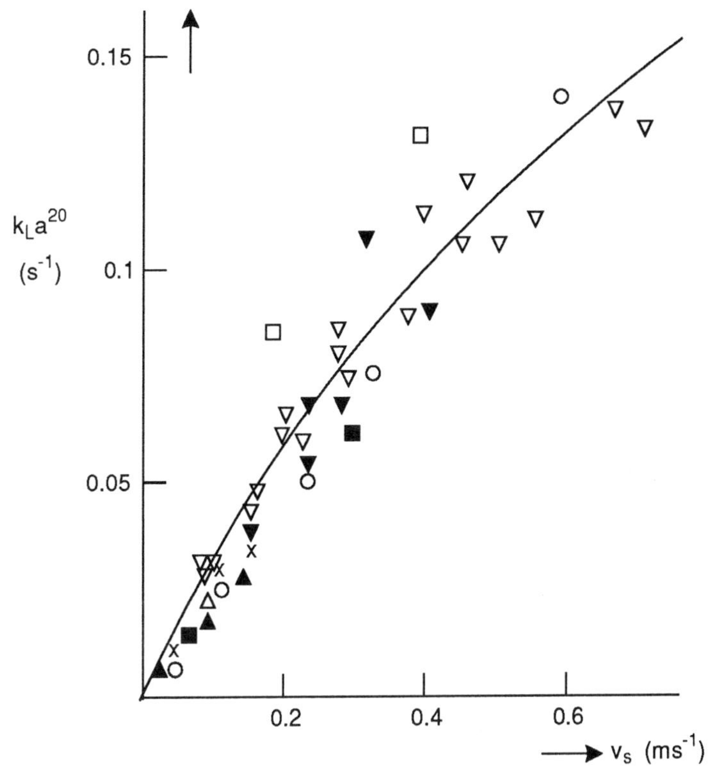

Figure 4.12 k_La^{20} values for air-water in coarse bubble systems. Adapted from TU Delft 1989 'Bioprocesstechnologie' Editor; Van der Lans, RGJM.(The data is plotted from the references sources in Table 4.1. We have not given full details of these references as we do not expect you to track them all down in the literature).

Table 4.1 gives the experimental details of how the k_La^{20} values plotted in Figure 4.12 were determined.

Reference	T (m)	H (m)	Gas sparger	Method	T(°C)	Medium
● Akita and Yoshida, 1974	0.08-0.3	2.5	Nozzle, perforated plate	Dynamic (O_2)	20	Coalescing
x Deckwer *et al*, 1978	0.2	7.2	Nozzle	CO_2 desorption, absorption steady state	14	Water; water-salt
■ Miller, 1974	0.15-0.68	0.3-1.37	Ring sparger	CO_2 desorption, steady state	20	Water
▼ Reith and Beek, 1968	0.14-0.29	1.6-3.7	Nozzle	Sulphite	30	Water-sulphite
❍ Towell et al, 1965	0.4	1.8-2.7	Nozzle, spider	CO_2 desorption, steady state	20	water
* Jackson and Shen, 1978	0.08-7.6	4-21	Orifice	Dynamic (O_2)	20	Water-sulphite
▲ Lin *et al*, 1976	0.08-0.15	3.0	Perforated plate	Dynamic (O_2)	20	Mineral medium
❐ Botton *et al*, 1980	0.1	0.2-0.8	Perforated plate	Sulphite (contr)	20	Non-coalescing
∇ Hikita *et al*, 1981	0.1-0.19	1.5-2.4	Nozzle	Dynamic O_2	22	Water
Δ Kataoka *et al*, 1974	5.5	5.9	Nozzle	CO_2 desorption	11	Water
+ Nakanoh and Yoshida, 1980	14.5	1.9	Orifice	Dynamic O_2	20	Water

Table 4.1 Details of experiments in which k_La^{20} values plotted in Figure 4.12 for air-water in coarse bubble columns were determined. (See text for details).

rule of thumb for oxygen transfer

Some researchers have expressed their mass transfer results (coarse bubble systems) in a simple rule of thumb. They have found that in the situation in which the dissolved oxygen concentration is zero, the oxygen transfer capacity can be calculated from the simple fact that, for each metre of unaerated liquid height in the bubble column, the % oxygen in the bubbles is reduced by 0.55. This is a useful fact to remember because it enables a quick calculation of oxygen removed from the sparged air.

Π If we have a liquid height of 15 m (unaerated) what will be the oxygen concentration in the off-gas? Assume air contains 21% O_2 and that the dissolved oxygen concentration is zero.

From our rule of thumb, the oxygen removal is 15 x 0.55 = 8.25%. Thus the oxygen left in the off-gas = 21-8.25% = 12.75%.

It can be demonstrated that the above, simple correlation is in good accordance with Equation 4.24. Because the saturated dissolved oxygen concentration decreases by about 2% °C^{-1} it can furthermore be shown that the value of 0.55% O_2/m is almost temperature independent.

4.5.2 Fine bubble systems

In Figure 4.13 , k_La^{20} is shown as a function of v_s, for fine bubble systems. In non-coalescencing media (water/salt) a very high k_La^{20} value is obtained at low v_s. Compared to coarse bubble systems, k_La^{20} increases by a factor of 6 at the same v_s due to the medium effect.

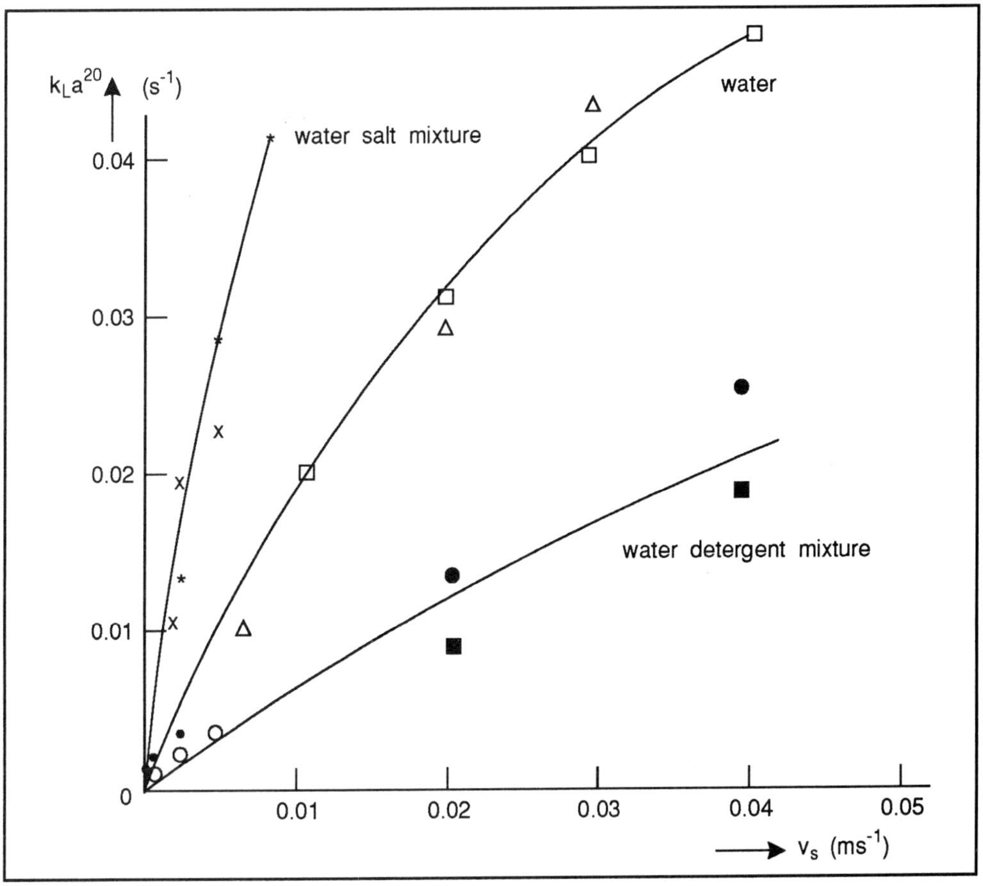

Figure 4.13 k_La^{20} values for air-water in fine bubble systems.

Table 4.2 gives details of the experiments in which the k_La^{20} values plotted in Figure 4.13 were determined.

Reference	T (m)	H (m)	Gas sparger	Method	T(°C)	Medium
. Lister and Boon, 1973	1.5	8.5	Dome diffuser	Dynamic (O_2)	20	Water
❍ Lister and Boon, 1973	1.5	8.5	Dome diffuser	Dynamic (O_2)	20	Water-detergent
❒ Sztatecsny *et al*, 1977	0.4	1.46	Sintered plate (5 μm)	Dynamic (O_2)	20	Water
■ Sztatecnsy *et al*, 1977	0.4	1.46	Sintered plate (25 μm)	Dynamic (O_2)	20	Water-detergent
Δ Deckwer *et al*, 1978	0.2	7.2	Sintered plate (170 μm)	CO_2 desorption, steady state	14	Water
x Jackson *et al*, 1975	0.076	4-7.3	Ceramic diffuser	Sulphite	20	Water-sulphite
● Akita and Yoshida, 1974	0.08-0.3	2.5	Porous plates (5, 50 and 120 μm)	Dynamic (O_2)	20	Coalescing medium
* Greenhalgh *et al*, 1975	0.09	0.5	Porous disc	Sulphite steady state	25-37	Water-sulphite; water-salt

Table 4.2 Details of experiments in which the k_La^{20} values plotted in Figure 4.13 for air-water in fine bubble systems were determined.

From these studies it can be shown that in non-coalescing media more than 3% O_2 m^{-1} can be transferred when two phase injectors are used, which is about six times the value for coarse bubble systems.

In strongly coalescing media (high viscosity or with antifoam) the k_La^{20} value of fine bubble spargers is about equal to the k_La^{20} values of the coarse bubble systems. This is due to the immediate coalescence of the fine bubbles at the sparger to coarse bubbles in the bulk of the bubble column. However, most media will occupy a position intermediate between the two extremes shown in Figure 4.13. For these media k_La is high in comparison with coarse bubble systems for low v_s values (less than 0.01 m s^{-1}) and at higher v_s values k_La levels off markedly. This influence of the low gas velocity in the air-water system with fine bubble spargers was also shown by some authors who found that, for $v_s > 0.02$ m s^{-1}, 0.6% O_2 m^{-1} is transferred and that for $v_s < 0.02$ m s^{-1}, this parameter can have a value up to 1% O_2 m^{-1}.

Π Write down the probable reason why superficial gas velocity has such an effect on oxygen transfer.

To answer this you really needed to remember the effect of gas velocity on bubble size. Look back at Figure 4.8. The reason that gas velocity has such an effect on oxygen transfer is the increasing d_{vs} value at higher v_s value. Another important parameter in the mass transfer of oxygen is the reactor height. The coalescence of the bubbles usually proceeds within a distance of 0.5-1 m from the sparger. However, the very small bubbles in this part of the column can bring about considerable mass transfer. It will be clear that the effect of this zone of small bubbles on the total mass transfer in the bubble column decreases rapidly when the column height is increased.

This is shown in Figure 4.14. The system used for sparging was a porous plate with a gas superficial velocity of 10^{-4} m s^{-1}

Π With a column height of 2 m and 6 m, how much oxygen was removed in this experiment? (Use Figure 4.14 to help you answer this).

From the graph, when the column height was 2 m, the rate of oxygen removal was about 1.8% m^{-1}. Thus over the 2 m column 2 x 1.8 = 3.6% of the oxygen would be removed. At a column height of 6m, the rate of oxygen removal was about 0.75% m^{-1}. Thus, over the whole column 6 x 0.75% of the oxygen will be removed = 4.5%. This calculation will have shown you the importance of producing and, if possible maintaining small bubbles. A course bubble column will have allowed the transfer of 6 x 0.55% = 3.3% oxygen over a 6 m column.

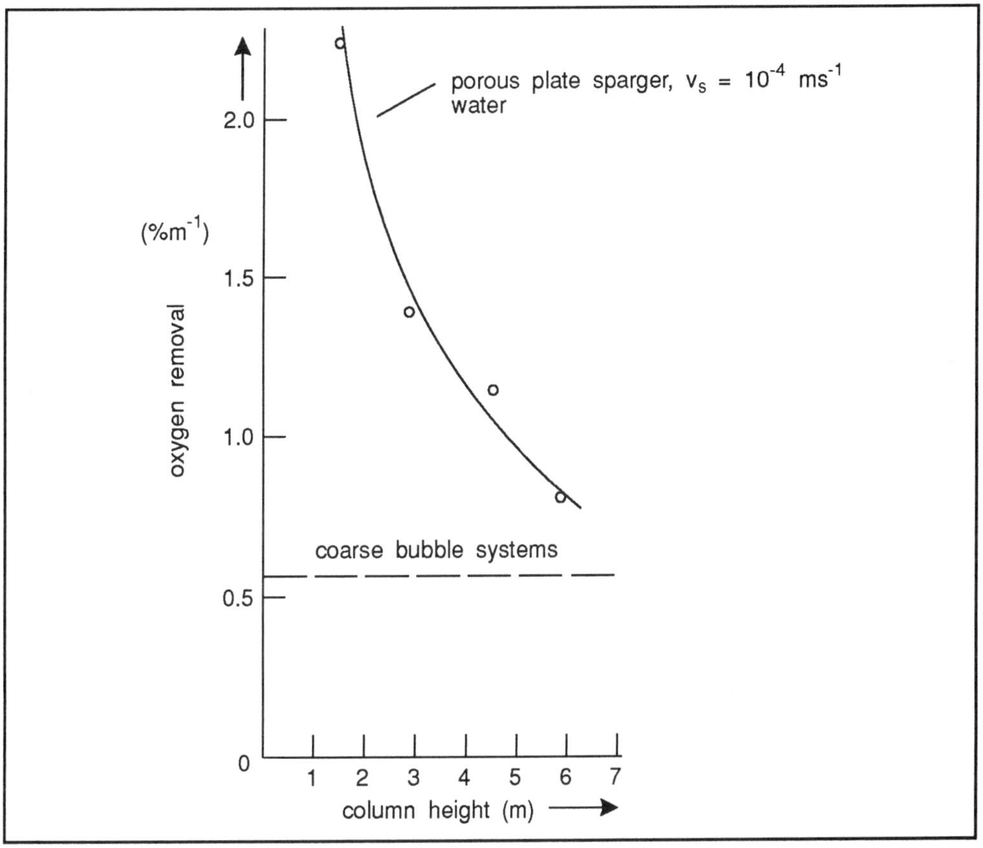

Figure 4.14 Influence of column height on mass transfer in a fine bubble system with a coalescing medium.

SAQ 4.4

1) Bubbles of d_{vs} of about 5-6 mm are produced in a bubble column of height 5 m. If air contains 21% oxygen and the temperature of the column is 20°C, what will be the oxygen concentration in the outflow gas?

2) The temperature of the column described in 1) is raised to 25°C. Will the concentration in the outflow gas be changed significantly?

3) The sparging device used in the column described in 1) was replaced by a porous plate sparger. If the medium is water and the gas superficial velocity is about 10^{-4} m s^{-1}, what will the concentration of oxygen in the gas outflow?

4.6 Heat transfer

heat transfer depends on v_s and η

Heat transfer from the broth to the wall or cooling coil is mainly determined by v_s. There is no influence of geometry (transfer to wall or coil). It has been found that high viscosity reduces the heat transfer coefficient considerably.

From measurements made on a variety of systems it can be shown that the following correlation may be used to calculate heat transfer coefficient α:

$$\alpha = 9391 \,.\, v_s^{0.25} \,.\, \left(\frac{\eta_w}{\eta}\right)^{0.35} \qquad \text{(E - 4.21)}$$

This correlation has proved applicable to $0.1 < T < 1$ m and $10^{-3} < \eta < 0.5.10^{-3}$ Pas. It has also been established that α does not increase further for $v_s > 0.1$ m s^{-1}.

4.7 New developments in bioreactor design

Various new biotechnological processes have been introduced in the last few years. Parallel to these developments in biotechnology, there is also an extensive development in the field of bioreactors. In fact, a number of trends can be seen. Firstly, there is a demand for bioreactors that can be used for many processes. This has led to the development of, for example, the Multi-Stage Stirred Bioreactor (three stirrers in one shaft). Secondly, there is also a demand for bioreactors that can be applied to specific biotechnological processes. Examples of this are the Liquid Impelled Loop Reactor and the Pumped Tower Loop Reactor (Surer^R reactor).

After the fermentation, the product has to be purified via various separations techniques from the broth. This process can be carried out in many different ways. Sometimes the fermentation broth (once the biomass has been removed) is the product itself (eg beer, soysauce); in this case separation is rather simple. However, in many processes the desired product is usually dissolved in the broth, and in such cases separation can be expensive. Furthermore, the re-use of biocatalyst (enzymes/micro-organisms/cells) may be economically necessary. In order to facilitate this re-use, the so-called fluidised and packed-bed reactors and membrane reactors have been developed. These new developments in the field of bioreactors are briefly discussed in the following sections.

Π It might be helpful to produce a summary table as you read the remainder of this chapter. We would suggest you use the following headings.

Type of reactor	Key feature	Applications
Cell-lift	Cells are gently kept in suspension by lifting them up into a draft tube and then expelling them through a discharge part	Shear sensitive cells

4.7.1 The cell-lift reactor

The large scale culture of plant and animal cells has increased greatly in importance during the past few years. It is now widely recognised that many valuable biochemicals can be mass-produced by cell cultures. However, animal cells, for example, are more fragile and shear-sensitive than micro-organisms. They cannot be stirred intensely or exposed to turbulence. Moreover, cells frequently require an anchorage, and are therefore grown on tiny glass or plastic beads known as microcarriers. Hence, the conventional CSTR is not suitable for the cultivation of shear-sensitive cells and an alternative CSTR has been developed.

microcarriers

This alternative CSTR (Figure 4.15) uses the so-called cell-lift principle. It contains a draft tube in which a low differential pressure is created by low-shear rotation of three-swept back impeller discharge ports. This action lifts microcarriers or suspended cells up into the draft tube and then expels them through the discharge ports in a continuous recirculation loop. Such movement keeps the cells uniformly dispersed throughout the culture vessel. Oxygen transfer to the medium is achieved by sparging large volumes of gas in a cell-free cavity. This cavity (shown in figure as a shaded area) is obtained by surrounding the draft tube with a very fine stainless steel screen which allows the medium to pass through, but not the microcarriers or cells.

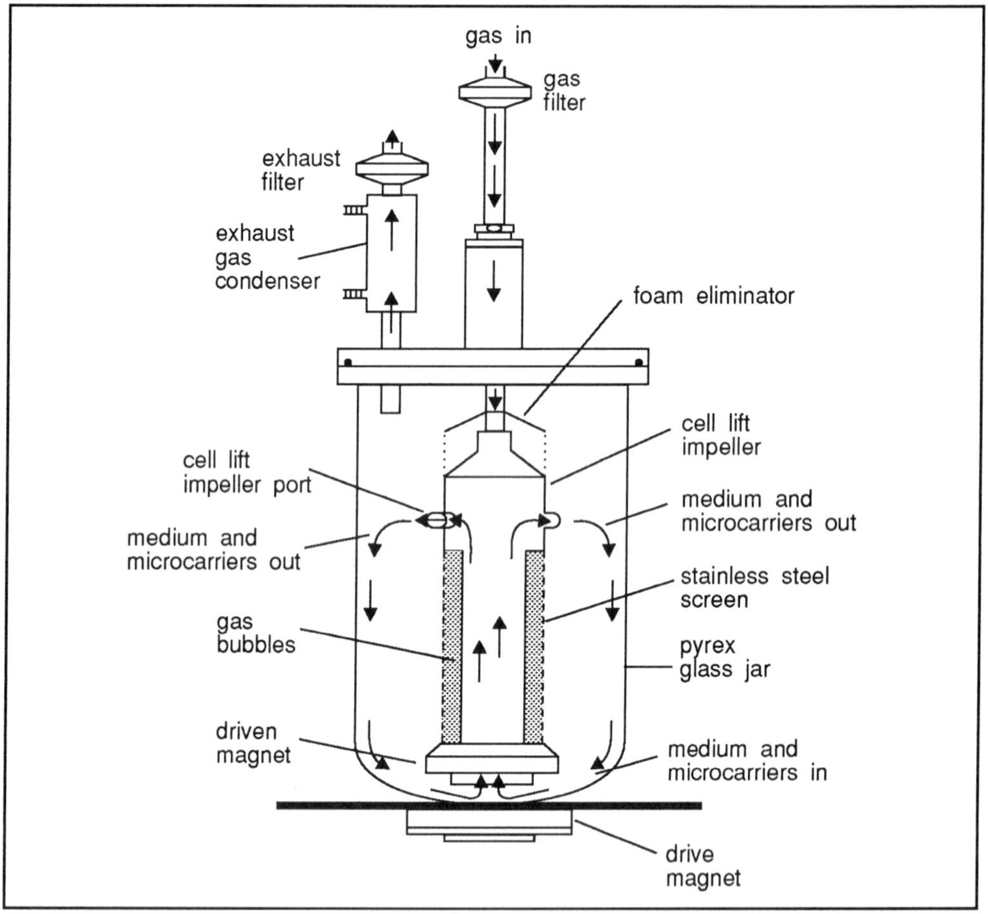

Figure 4.15 Bioreactor vessel equipped with a cell lift impeller.

4.7.2 The multi-purpose tower bioreactor

three modes of operation

The multi-purpose tower bioreactor (MTB) uses the airlift principle which we have already mentioned in Section 4.1. The advantage of the MTB over the conventional bubble column or airlift reactor is that the modular design allows one to assemble a system according to the demands of the bioprocess. The reactor can be used in three different modes: as a bubble column (mode a), as an airlift inner loop reactor with the draft tube as a riser (mode b) and finally as an airlift inner loop reactor with the draft tube as a downcomer (mode c). These three modes are illustrated in Figure 4.16. Due to high oxygen transfer, minimal shear forces and low foaming properties, the MTB can be used for shear sensitive cells/organisms (mode a, b or c), flocculating organisms (mode b or c) or foaming cultures (mode c). The MTB also contains a settler to prevent the biomass from being pumped out of the bioreactor.

Currently, results are available for the operation of MTBs on a small scale. However, from these experiments this reactor show promising advantages for cultivation of fragile cells, eg animal cells and plant cells.

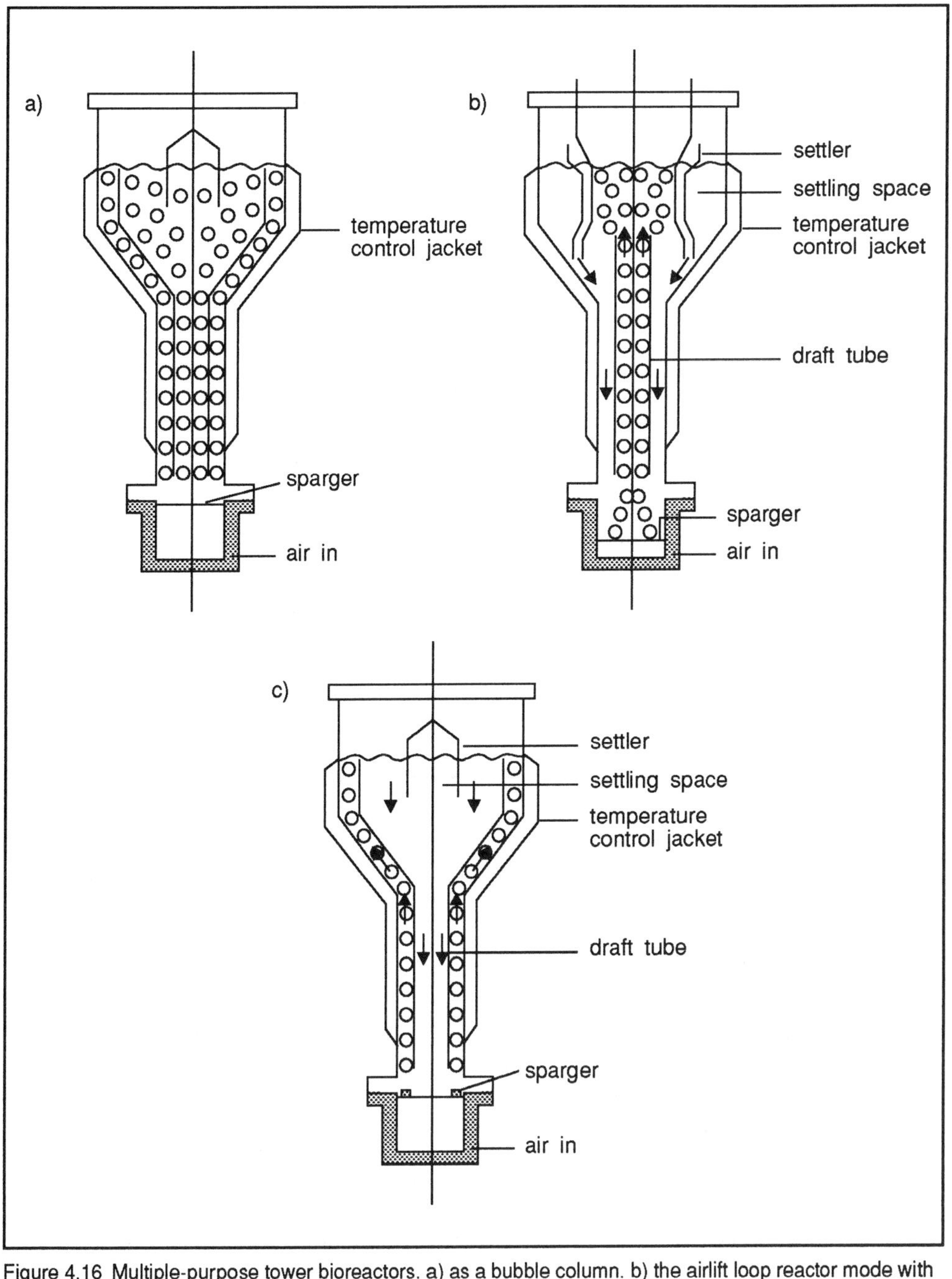

Figure 4.16 Multiple-purpose tower bioreactors. a) as a bubble column. b) the airlift loop reactor mode with the draft tube as riser. c) the airlift loop reactor mode with the draft tube as a downcomer. (b and c redrawn from Applikon 1989 Biotekknowledge, ApplikonDept Instr Schiedam.2 3-6.

4.7.3 The liquid impelled loop reactor

The liquid impelled loop reactor (LILR) is a new type of bioreactor, developed for among others application of biocatalysis in organic solvents. Although different configurations are possible, only one configuration (Figure 4.17) will be discussed. The reactor consists of two parallel longitudinal tubes which are connected at the top as well as at the bottom. Organic solvent is dispersed via the liquid sparger, at the bottom in to a continuous non-organic liquid phase. The continuous and the dispersed phase do not mix and have different densities (in this case the organic phase has a lower density). Due to the differences in densities the organic solvent droplets rise to the surface. Two distinct liquid phases are present at the top. Since the densities of the liquid in the two longitudinal tubes of the reactor are different, the continuous phase liquid circulates.

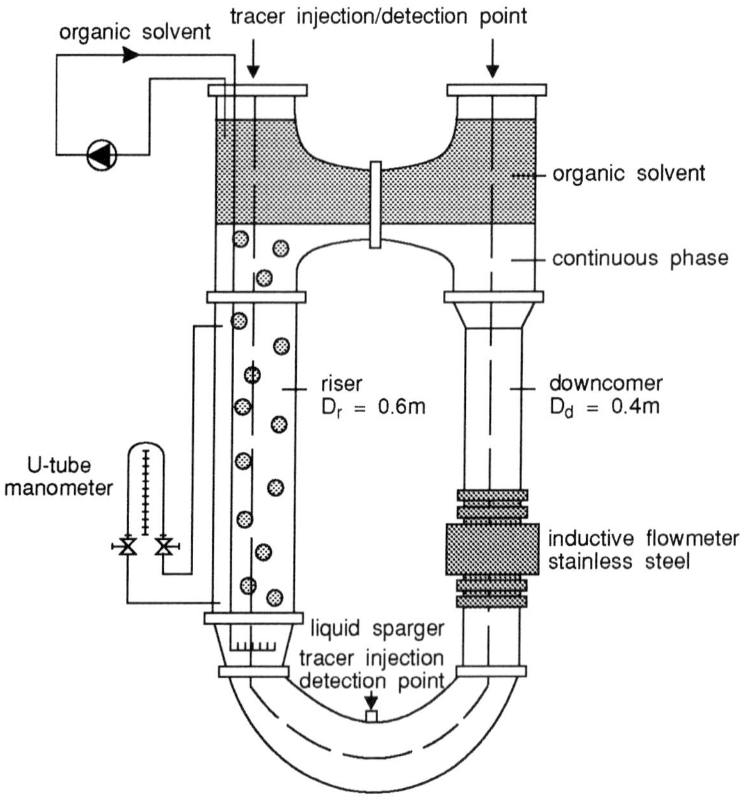

Figure 4.17 The liquid impelled loop reactor (see text for details). Redrawn from Sonsbeek H M and Verlaan P 'Characterisation of a liquid impelled loop reactor' Dept Food Science Lu Wageningen, The Netherlands.

4.7.4 The pumped tower loop reactor

high viscosity products

The Pumped Tower Loop Reactor (PTLR, Figure 4.18) has been developed especially for the batch production of the polysaccharide xanthan. Xanthan has particular specific functional properties such as high viscosity and pseudoplastic rheological behaviour. Furthermore, viscosity increases during fermentation. These properties make the production of xanthan expensive. High energy input is required for mixing and for heat and oxygen transfer during fermentation in a CSTR. This results in local limitations which have a negative effect on the fermentation-time and productivity of the organism. Furthermore, because of these properties, scale-up of the xanthan process is rather difficult.

If the fermentation conditions are improved, it becomes important that the reactor design is based upon the flow pattern of the liquid. Due to the strong pseudoplastic behaviour, the viscosity of xanthan depends on the shear rate. Increase in shear rate leads to a lower viscosity. If the liquid is allowed to flow under defined conditions (eg plug flow) this results in uniform flow patterns. This is in contrast to the tank reactor in which large variations in shear rate occurs.

In principle, an airlift reactor can be applied for the production of xanthan. However, the driving force of the gas, on which the airlift principle depends, is usually not enough to keep the viscous fluid in motion. Furthermore, due to the stratification of the gas to the middle of the column of the airlift reactor canal formation occurs and the fluid in the vicinity of the walls will remain stagnant.

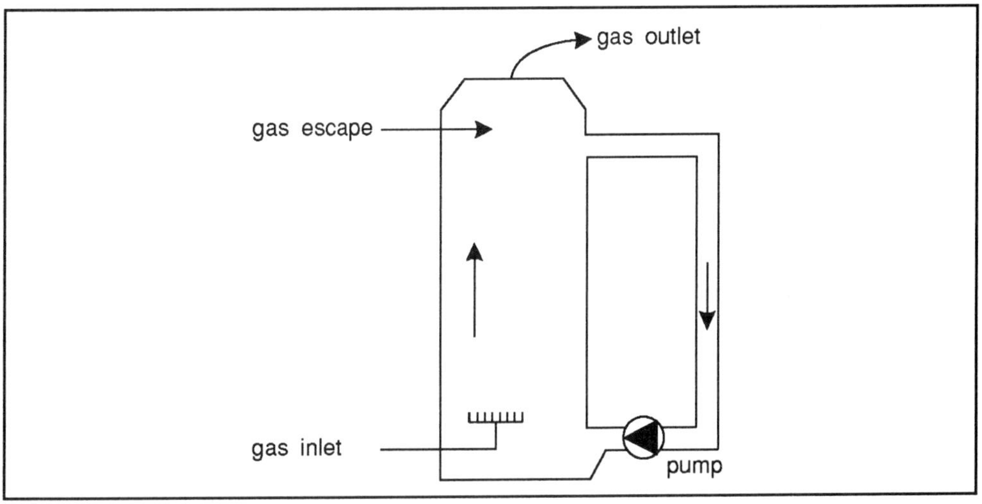

Figure 4.18 Stylised pumped tower loop reactor.

A tube reactor, called the Pumped Tower Loop Reactor, was developed to solve the above problem. In this PTLR, the liquid is kept in constant motion with the help of a specially designed pump which is placed in the connection between the riser and the downcomer. The organism is confronted only to the varying conditions in the height of the column. This leads to shorter fermentation time and higher product concentration. Furthermore, the energy-input is lower and the reactor design is easier than that for a tank reactor, because the liquid flow pattern can be described optimally in a simple scale-up procedure.

4.7.5 The fluidised and packed-bed reactors

To reduce the processing costs and to prevent the loss of biocatalyst, fluidised and packed-bed reactors were developed. These are illustrated in Figure 4.19. In both configurations, glass or plastic beads are used on which the biocatalyst is immobilised. Immobilisation occurs, for example, via gel entrapment, covalent binding or encapsulation. In the fluidised-bed reactor, the immobilised catalyst floats around freely due to the airlift principle which moves the liquid. After the fermentation period, the beads can be easily separated from the broth and can be reused. However, in the packed-bed the biocatalyst is stationary. In the packed-bed, the medium is saturated with oxygen in a special tank before it is trickled along the packed-beads. Fluidised and packed-bed bioreactors are used in the glucose-isomerisation process and for the hydrolysis of lactose.

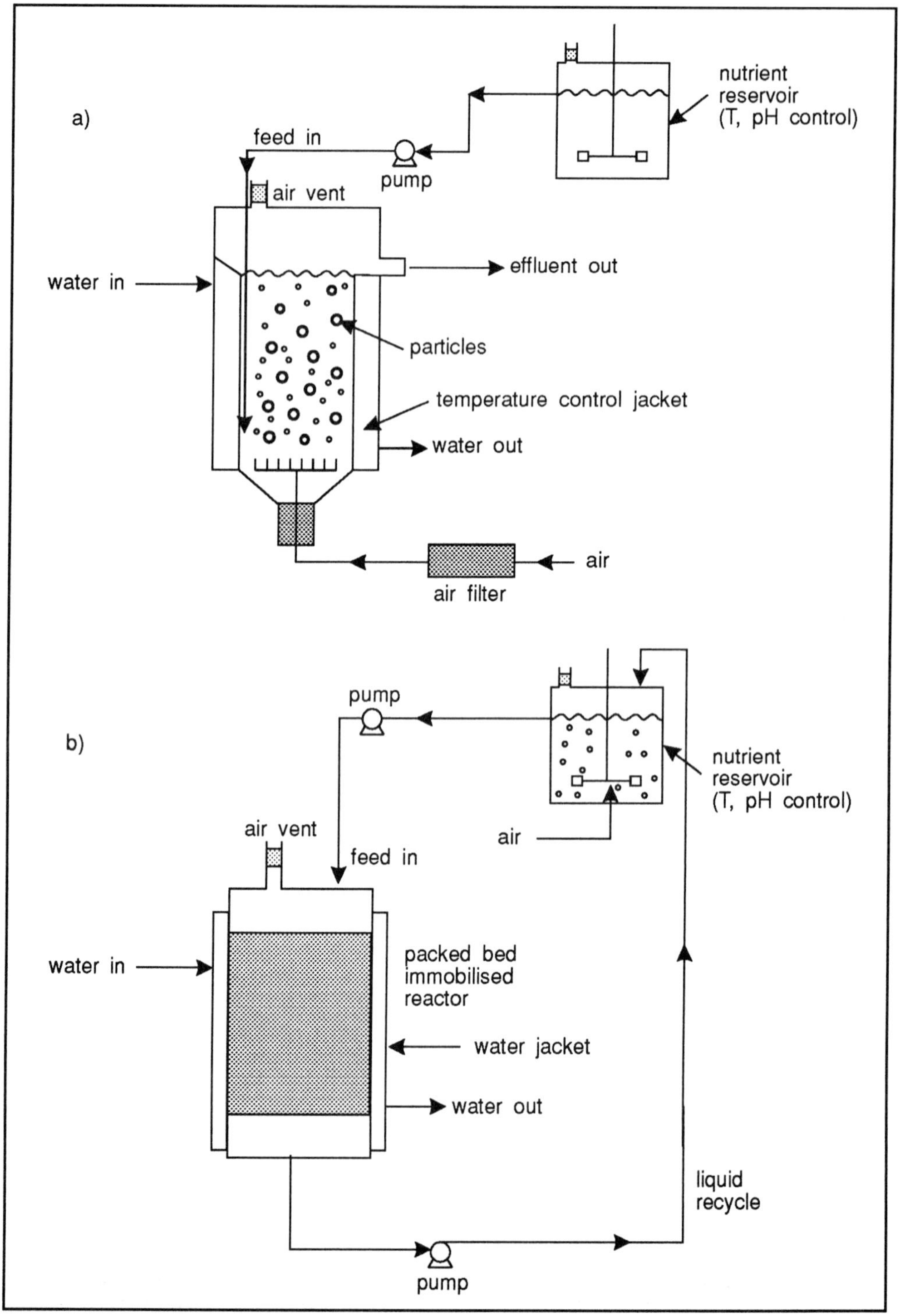

Figure 4.19 Fluidised a) and packed-bed b) reactors (see text for details).

Due to the adhesion capacity of micro-organisms or cells, biofilms can be formed on the beads. If these beads are used in a reactor one speaks not of a fluidised or packed-bed, but of a biofilm reactor. There are various forms of biofilm reactors, such as the sloping plane biofilm reactors and the rotating drum biofilm reactors. Film bioreactors with a fluidised or packed-bed configuration are, for example, used in waste-water treatment.

4.7.6 The membrane reactor

Membranes are often used in biotechnology in the separation process. Proteins, for example, can be separated from low molecular weight substances via ultrafiltration. This principle is applied, for example, in the preparation cheese.

The membrane acts as a selective barrier to proteins whereas low-molecular weight substances can freely permeate through the membrane. In addition, membranes can be used to separate the biocatalyst from the fermentation broth (microfiltration).

continuous removal of low molecular weight products

In a membrane reactor continuous processing is possible if production and separation can be incorporated in one system. There are various configurations possible to achieve this. Figure 4.20a shows the circulation of the liquid broth from a conventional stirred tank reactor along a series of membranes. In Figure 4.20b a similar system without the stirred tank reactor is shown. In both cases the biocatalyst is present in the liquid. These reactors are useful if auto-catalysis occurs in the kinetic process. An example of the system as shown in Figure 4.20a is the production of lactic acid by bacteria. Lactic acid bacteria produce a limited amount of the product in a batch process. This is because the produced lactic acid reduces the pH to such an extent that this inhibits the growth of the organisms. However, by making the process continuous, lactic acid is separated via the membrane and pH can then be regulated to the optimum level. Another example of the membrane reactor is the continuous removal of ethanol and /or butanol via pervaporation. In this process vapour is sucked from the external side of the membrane. The vapour contains a relatively high concentration of alcohol. The cultivation of animal cells for the production of monoclonal antibodies is an example of the membrane system shown in Figure 4.20b. For this process membranes are arranged in the form of hollow fibres. The cells are allowed to grow outside the fibres in the reactor while the medium is pumped through the hollow fibres. Another advantage of the membrane reactor is that shear sensitive cells remain productive for a long time.

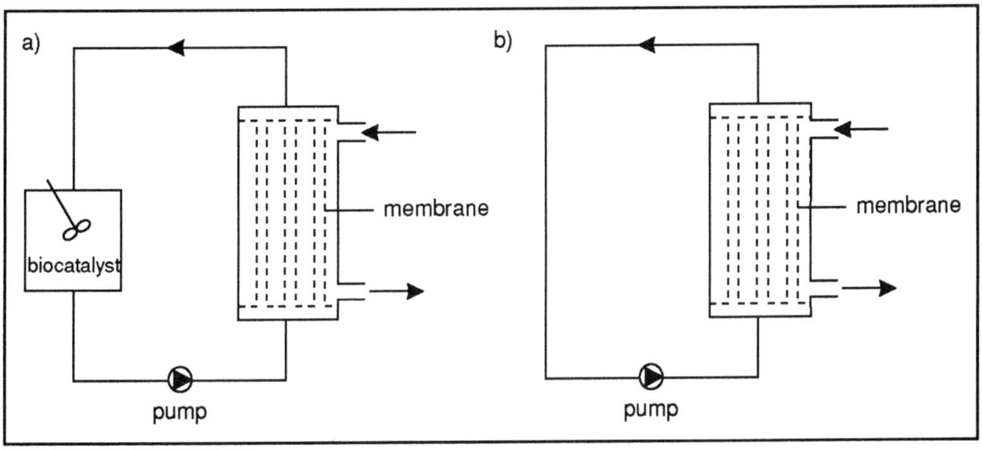

Figure 4.20 Membrane reactors a) separate fermenter and separator. b) integrated fermenter and separator. In each case, low molecular weight components diffuse through the membrane and are removed from the system.

It is also possible to immobilise enzymes on the membrane as shown in Figure 4.21. Cells or enzymes can be incorporated during the preparation of the membrane. Such membranes can be used for separation and can also function as a reaction surface between two different liquid phases. This avoids, in contrast to a conventional two phase system, the formation of emulsions. This technique is very useful when two non-mixable liquids, such as oil and water, are present. For example, the enzyme lipase converts oil to glycerol and fatty acids in the presence of water.

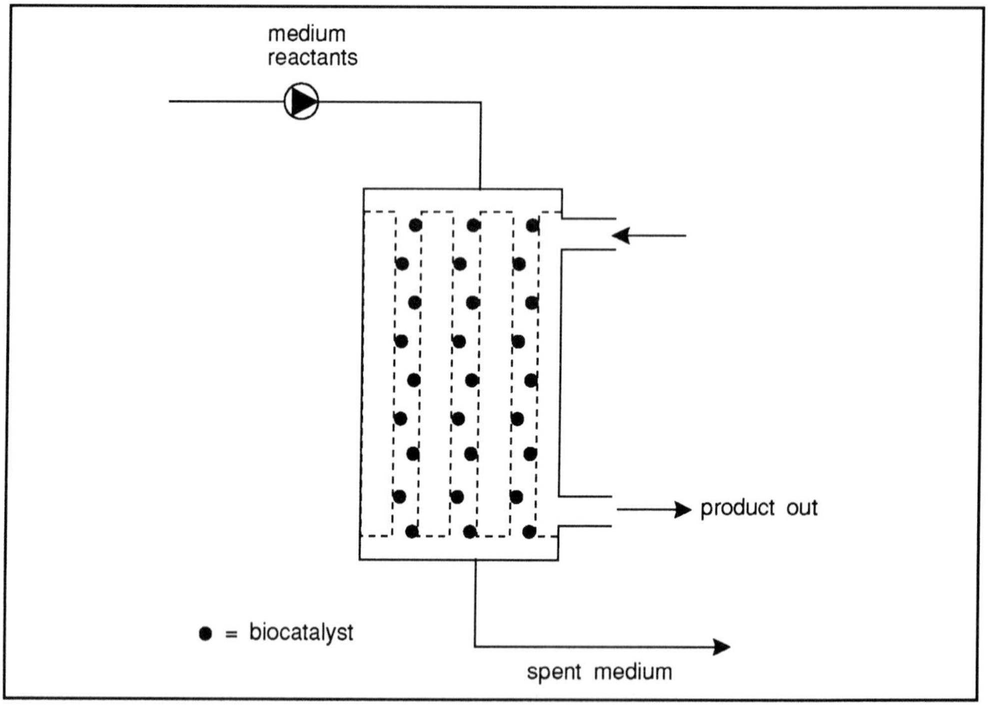

Figure 4.21 Membrane reactor incorporating an immobilised biocatalyst.

The commercial applications of the membrane reactors are, however, limited. This is due to unpredictability of the membranes. Membranes tend to clog quite often and have a limited life span, varying between six months to two years. Furthermore, membranes are relatively expensive.

SAQ 4.5

Here is a list of newly developed bioreactors:

Cell lift reactor;

Multi-purpose tower reactor;

Liquid impelled loop reactor;

Pumped tower loop reactor;

Fluidised-bed reactor;

Membrane reactor.

Choose the appropriate type(s) of reactor to match each of the following statements.

1) Particularly useful for the production of antibodies by appropriate animal cells.
2) Can be used to continuously remove lactic acid from a culture of bacteria.
3) Uses glass or plastic beads to provide anchorage for animal cells and to act as microcarriers.
4) Suitable for processes in which high viscosity products such as xanthan are made.
5) Can be used in several different modes.
6) Useful when two immiscible solvents (eg water and an organic solvent) have to be used. For example. when the reactants (or products) of a bioprocess are only sparingly soluble in water.

Summary and objectives

In this chapter we have discussed the main features of bubble column and airlift reactors. Central to this discussion has been an examination of bubble characteristics and mass transfer processes especially relating to oxygen transfer. Included in this discussion has been a description of flow patterns, sparger design and the influence of liquid properties on bubble sizes. We concluded the chapter by briefly describing some recent developments in reactor design needed to meet the demands of shear-sensitive cells, high viscosity products and the use of immiscible liquids.

Now that you have completed this chapter you should be able to:

- select and use relationships between gas flow rates, gas superficial velocities and column dimension to calculate values of liquid circulation velocities and liquid circulation times;
- calculate the number of ideally mixed compartments in bubble chambers from supplied data;
- give mean bubble sizes for different spargers in coalescing and non-coalescing systems;
- identify non-coalescing and coalescing systems from data relating to sparger type and bubble size;
- identify factors which influence gas hold-up values;
- use data relating to gas superficial velocity and hold-up values to identify homogenous and heterogeneous flow conditions;
- use 'rules of thumb' to calculate oxygen concentrations in outflow gas from aerated bubble columns;
- identify suitable types of reactors for use in a range of particular circumstances.

An introduction to immobilised systems

An introduction to immobilised systems

5.1 Introduction

In Chapters 3 and 4 we examined Conventional Stirred Tank Reactors, Bubble Columns and Air-lift reactors. In each of these, the biocatalyst used for the biotransformation is held in suspension. This is not, however, the only way in which biocatalysts can be used. The use of agglomerated systems is becoming increasingly an important strategy. Agglomerated or immobilised systems are those in which the biocatalyst is attached to (absorbed onto), or entrapped in, a solid or suspended substratum. Attachment of the biocatalyst to a substratum within the reactor provides a means of retaining the catalyst in the reactor. Other reaction species (eg substrates) may be present in both suspension and in the attached phase.

In this chapter, we will outline the principles of immobilised systems and introduce some of the terminology that has been applied to these systems. We will also discuss the advantages and disadvantages of immobilised systems. In the next chapter (Chapter 6) we will explain the physical background to the modelling of immobilised systems. This discussion is underpinned by an understanding of the principles of transport phenomena and reaction kinetics. Although it is assumed the reader has some previous knowledge of these processes, some revisionary help is given. (Note transport phenomena and reaction kinetics are examined in depth in the partner BIOTOL text 'Bioprocess Technology: Modelling and Transport Phenomena').

Using this knowledge, we will examine the modelling of immobilised systems in Chapter 7 and consider the configuration of immobilised reactors in Chapter 8. Further discussion of immobilised reactors is provided in the BIOTOL text 'Bioreactor Design and Product Yield'. Let us begin with some of the terminology used in describing immobilised systems.

5.2 Introduction to some terms used in describing immobilised systems

In biotechnology, enzymes or cells (usually micro-organisms) are used to convert reactants (substrates) into products. Whether single enzymes or whole cells are used, essentially they catalyse the desired reactions and thus either system may be regarded as a biocatalyst. The reaction catalysed by such systems may also be referred to as bioconversion or biotransformation.

Although we may consider both single enzymes and whole cells as biocatalysts we should, however, remember that there are some quite fundamental differences between the two.

Π Make a list of the differences between single enzymes and whole cells you may consider as being important in the discussion of immobilised reactors.

You were probably able to generate quite an extensive list. The major differences are that a single enzyme (or a small number of enzymes) will catalyse a specific reaction converting a substrate into a quite closely related chemical product. Whole cells represents a very complex network of enzymes and may carry out a whole sequence of reactions and the products may be, chemically, quite distantly related to the substrate. Furthermore, a single substrate may give rise to many different products. For example a bacterium using glucose as a substrate may produce CO_2, as well as the many biochemicals which make up biomass. Of course, often in biotechnology a single (or limited number of) product(s) from such a complex series of reactions may be of interest. Thus a fundamental difference between systems in which the biocatalyst is a single enzyme and systems using whole cells is in the simplicity of the mixture of products made. There are, however, other important differences. Single enzymes are incapable of self-replication. Thus if we begin with a certain amount of an enzyme in a bioreactor then this cannot replicate itself. In contrast, whole cells are capable of replication and can replenish themselves. This of course has important consequences in terms of using substrate for maintenance and growth. It also has important consequences in terms of the need to reload the reactor with the biocatalyst.

In this and the following three chapters, we will use the generalised term biocatalyst inferring both single enzyme and whole cell systems. It is, however, important to keep in mind the difference between the two systems. Immobilised systems based on using single enzymes (or a small number of enzymes) are specifically discussed in the BIOTOL text 'Technological Applications of Biocatalysts'.

The biocatalyst used in biotechnology may be distributed evenly throughout the bulk fluid or can be agglomerated. The first case is referred to as a free biocatalyst with kinetic behaviour according to free kinetics and the second is referred to as an immobilised biocatalyst with kinetic behaviour according to apparent kinetics. For the immobilised biocatalysts the apparent (or observed) kinetics are due to the combination of the intrinsic (or true) kinetics of the biocatalyst and mass transport (mass transfer). The intrinsic kinetics can differ from the free kinetics due to, for example, the different configuration of the biocatalyst in an immobilised state compared to a free state.

Immobilised systems using whole cells are also described by a variety of other terms including, attached growth, fixed film, biofilm or immobilised cell systems. When the cells grow in the form of a layer on a substratum, the terms used often reflect the nature of the business in which the system is being used. Thus 'immobilised cell layers' is commonly used in contemporary biotechnology and 'fixed films' in public health engineering. The growth of unwanted organisms as layers in pipes and vessels is sometimes referred to as 'slime layers'. All, of course, represent immobilised systems and there are no simple rules which govern which term should be used. Perhaps the most widely accepted term is to describe such systems as 'biofilms'!

5.3 Kinetic considerations of immobilised and free biocatalysts

As long as the biocatalyst is distributed evenly over the culture fluid, the immediate environment of the biocatalyst is identical to the average (or bulk) environment and the kinetic behaviour of the system can be predicted from the true (or micro) kinetics and bulk properties such as substrate concentration, pH and temperature. These bulk properties are the only properties which can be readily measured (by means of taking samples). Therefore, in a well distributed system a relationship between measurable quantities and the behaviour of the system can be obtained (homogeneous catalysis). It hardly needs to be mentioned that this facilitates the design and control of these systems.

In agglomerated systems the situation becomes more complex. Conversion of substrate (or biomass growth accompanied with substrate conversion) occurs simultaneously with diffusion of substrate into the immobilised biocatalyst and diffusion of product towards the surrounding bulk environment (heterogeneous catalysis). An extra complication in bioreactors may be a low liquid velocity such that low mass transfer causes a substantial difference between the bulk concentration and the concentration at the interface between the immobilised biocatalyst and the liquid. This, of course, makes these systems more difficult to analyse.

SAQ 5.1

Put a tick in the appropriate column for each of the following features.

Feature	Immobilised system using an enzyme as the biocatalyst	Immobilised system using whole cells as the biocatalyst
Products are usually chemically very similar to substrate		
The reaction rate is likely to slow down as time passes even though the substrate concentration may be maintained at a constant level.		
Energy is expended by the biocatalyst to maintain itself		
The products of the process are likely to contain a complex mixture of chemicals		

5.4 Advantages and disadvantages of immobilised systems

First we will briefly list the advantages and disadvantages of biocatalysts compared to chemical catalysts. Compared to chemical catalysts, biocatalysts:

- are highly specific;
- have very high turnover numbers at relatively low temperatures;
- are relatively sensitive to external parameters (pH, temperature);
- are readily inhibited or deactivated;
- are often require cofactors and/or metal ions.

Despite the disadvantages, the high specificity of biocatalysts and the ability to bring about sequential chemical changes (especially in whole cells), makes them industrially attractive. They have, however, one other feature which can be considered as a negative factor. Biocatalysts, especially purified enzymes, are expensive to produce. Thus it is important to make maximal use of the biocatalyst and not to waste it. One such strategy is to fix (immobilise) the biocatalyst in the reactor to prevent it being washed out an lost to the system.

Π Below we list the advantages and disadvantages of immobilising biocatalysts compared to using free (suspended) biocatalysts. We would suggest that as you read through this list you make a summary table. We would suggest you use the following format.

Feature	Free system	Immobilised system
Biocatalyst retained and used over and over again	-	+

By constructing such a table, you will have produced for yourself an easy to learn revision sheet.

Advantages of immobilisation are:

- the biocatalyst can be kept in a reactor relatively easily. If a free biocatalyst (enzyme solution or suspension of cells) is added to a continuous flow reactor, the biocatalysts would leave the reactor as fast as the outflow;
- high flow rates in a reactor are possible in combination with a high biocatalyst retention. Due to the fact that the biocatalyst retention time can be operated independently from the material flow through the reactor, high flow rates can be used without loss of biocatalysts. Lower loses, lower costs;
- no contamination of the processed solution with biocatalysts. In the case of a system which employs free cells or enzymes, the solution leaving the reactor containing the product, also contains the biocatalysts. With immobilisation, the biocatalysts perform the conversion of substrate in the reactor and stay separate from the product. This simplifies the down stream processing leading to reduced costs;
- improved physical and chemical stability. By immobilisation the biocatalysts are protected from shear forces due to aeration or mixing and from the negative (inhibitory) influence of certain compounds outside the immobilisation matrix;
- possible regeneration of cofactors (in the case of immobilisation of living cells). If living cells are immobilised the complex biochemical network can be left intact and regeneration of co-factors takes place as they would in free cells;
- possible processing of multi-enzyme reactions in the case of immobilised cells and/or a combination of enzymes. Combinations of reactions performed by different enzymes, in which the intermediate products are converted to new

products, can be done in a one process step by the simultaneous immobilisation of the different enzymes.

Possible disadvantages of immobilisation are:

- loss of activity by the immobilisation process. The chemical and physical steps necessary for immobilisation can damage the biocatalyst resulting in loss of activity;
- heterogeneous catalysis. Due to immobilisation mass transport problems are introduced resulting in different catalytic behaviour compared to the homogeneous catalysis using the free biocatalysts;
- extra cost due to an extra step in the process;
- danger of parallel or consecutive reactions in case of living cells;
- change of kinetics due to immobilisation. Due to the immobilisation process and due to the different environment of the immobilised biocatalyst compared to the free biocatalyst, a change of kinetics may occur. Possible causes are denaturation and effects of the carrier material (charge on the carrier, steric hinderance etc).

5.5 Immobilisation methods

Now that we have established that immobilisation offers many advantages, we will turn our attention to the general methods available for immobilisation. In this description we will not deal with the practical details of the immobilisation process. We can however identify two main sub-groups of immobilisation. One group depends upon making use of physical forces. For example we can adsorb the biocatalyst onto a solid support using, for example, ionic interactions. Alternatively the biocatalyst may be physically entrapped in a gel or encapsulated. The second group of processes involves using chemical attachment. Thus the biocatalyst may be covalently attached to a solid support or cross-linked with each other to form an agglomerated mass. These different procedures are illustrated in Figure 5.1.

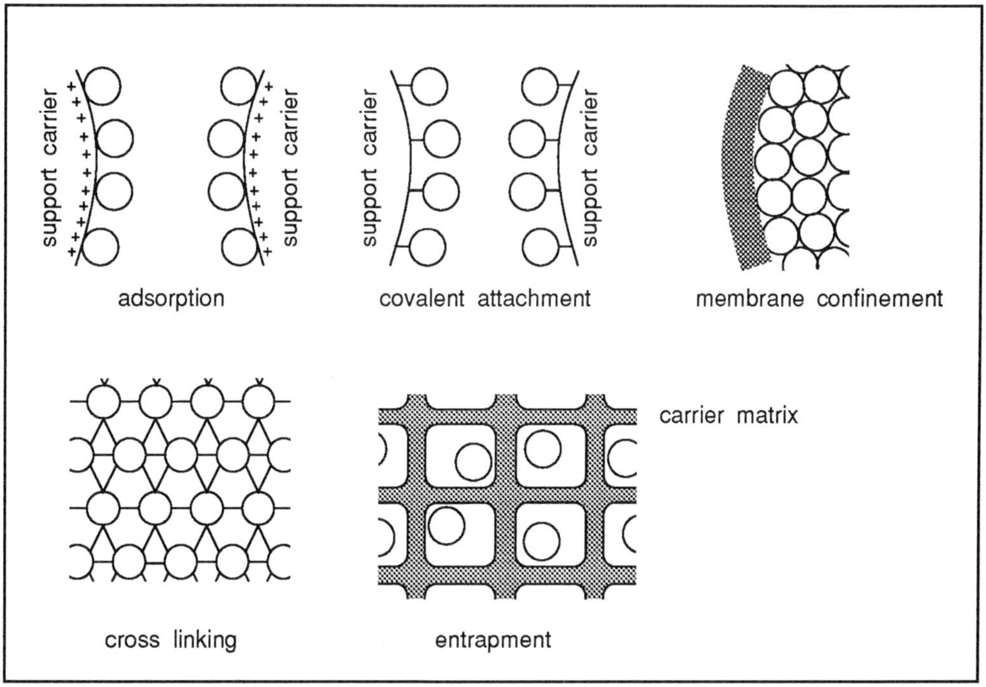

Figure 5.1 Various methods of immobilisation of biocatalysts.

Attachment to or adsorption on a solid support can, in the case of micro-organisms, form layers of cells (biolayers). Sometimes micro-organisms can agglomerate spontaneously and pellets, flocs or beads can develop. It is also possible to immobilise pure enzymes or whole cells. In the case of intracellular enzymes the high costs of isolation and purification can sometimes be avoided by immobilisation of whole permeable dead cells. On a large scale, immobilised biocatalysts may consist of a core particle which is coated with a layer of biologically active material, or of a bead of an immobilisation matrix of polymeric gels (agarose alginate). These techniques are sometimes applied in fluidised-bed reactor or in a packed-bed reactor configurations to retain the biocatalyst in the reactor. Further details on the agents and processes used to immobilise biocatalysts are given in the BIOTOL text 'Technological Application of Biocatalyst'.

SAQ 5.2

Answer true or false to the following:

1) Immobilisation may simplify downstream processing aimed at purifying the product.

2) When using an immobilised, purified enzyme, the regeneration of cofactors needed for enzymatic activity may present a problem.

3) Immobilisation usually leads to a decrease in enzyme activity.

4) Using immobilised cells, we can employ high dilution rates.

5) Immobilisation is always cheaper than using free biocatalysts.

5.6 Practical use of immobilised biocatalysts

Only a limited number of industrially used biocatalysts are immobilised. Some are so cheap that there are no economic reasons for immobilisation or recovery. Some examples of industrially used immobilised biocatalysts are:

- amino-acylases from *Pseudomonas putida,* by which L-amino acids can be produced;
- glucose-isomerase, in which whole cells of *Actinoplanes missouriensis* are immobilised. Using this enzyme, high-fructose syrup can be produced;
- lactase; lactose from milk or whey can be hydrolysed to galactose and glucose resulting in lactose-free dairy products;
- antibiotic production eg penicillin;
- in anaerobic waste water treatment by an UASB (upflow anaerobic sludge blanket) reactor, waste water is treated by naturally developed agglomerates (pellets, beads) of methanogenic and acidogenic micro-organisms. These organisms can also be found in the biolayer on sand grains found in fluidised-bed reactors for anaerobic waste water treatment;
- a biolayer of aerobic organisms can be found on sand grains in air lift loop reactors used for aerobic waste water treatment.

The applications of immobilised systems are rapidly increasing and are expected to grow with the growth of biotechnology.

Summary and objectives

In this chapter, we have briefly introduced the immobilised systems used in biotechnology. The discussion has centred mainly around some of the terms used in describing immobilised systems and the advantages and disadvantages of immobilisation.

In the next chapters we will build upon this description to develop models of immobilised systems before further examining some of the applications of immobilised systems.

Now that you have completed this chapter you should be able to:

- list differences between immobilised whole cells and immobilised enzyme systems;
- list advantages and disadvantages of using immobilised systems;
- describes in outline the physical and chemical devices which may be used to immobilise biocatalysts.
- use a variety of terms applied to immobilised systems such as biofilms, attached growth, fixed films, slime layers and free biocatalyst.

Physical background to modelling immobilised systems

Physical background to modelling immobilised systems

6.1 Introduction

To understand the conversion rates of substrates which can occur in an immobilised biocatalyst system it is necessary to describe the physical and chemical processes which take place. For this, the basic aspects of mass transport and reaction kinetics will be briefly discussed. This is done in the form of a brief revision of mass transport and reaction kinetics phenomena. If you have difficulty in comprehending this brief chapter we would recommend you to read the appropriate chapters of the BIOTOL text 'Bioprocess Technology: Modelling and Transport Phenomena'. In subsequent chapters we will use this description to establish substrate balances in immobilised systems.

We begin by discussing mass transport in a geometrically simple system and then move on to a reminder of the kinetics of biocatalyst processes.

6.2 Mass transport

6.2.1 Within the biocatalyst matrix

We will use a flat plate or slab as our example.

This has the geometry which is infinitely long with small sides compared to the length (Figure 3.1). The slab represents a matrix containing biocatalysts.

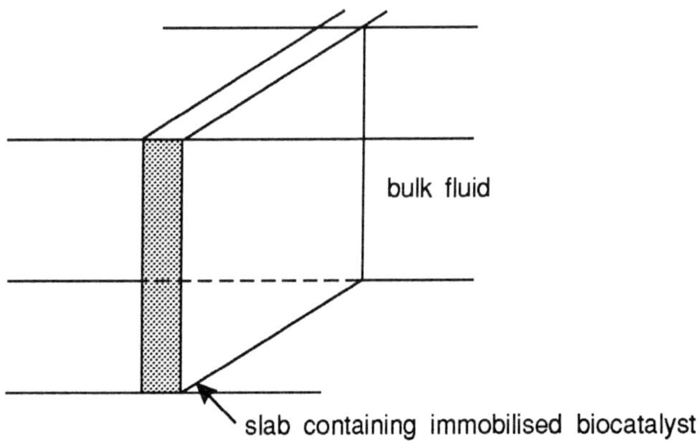

Figure 6.1 Schematic representation of an immobilised biocatalyst in slab geometry.

Outside of the matrix is evenly distributed bulk fluid which contains the substrate and product.

It is assumed that:

- the biocatalyst is homogeneously distributed in the matrix;
- there is no interaction between the substrate and/or the product and the immobilisation matrix.

Ficks law

Mass transport within the slab is described by Ficks law.

$$\phi''_{M,A} = -D \frac{dC_A}{dx} \tag{E - 6.1}$$

where $\phi''_{M,A}$ is the flow of mass (M) of component A per unit area. (Note the convention, the superscript " represents per unit area); D is a transport coefficient called the diffusion coefficient; C_A is the concentration of component A; x is the distance in the slab.

Essentially what Ficks law states is that the rate of transport of a component (ie the flow) is proportional to the diffusion coefficient and to the concentration gradient.

Thus: | flow | = - | diffusion coefficient | * | concentration gradient |

The symbols we have used in the equation above are the ones we have used throughout the BIOTOL series of technology orientated texts. Unfortunately many other conventions have been used in other literature. For example, biochemists often write S or [S] to represent substrate concentration rather than C_A. Likewise some authors use J to represent flow rather than ϕ and r to represent a distance in the matrix.

So we could re-write Equation 6.1 as $J'' = -D \frac{dS}{dr}$; its meaning is exactly the same. For whatever symbols we use the meaning of Ficks law of course remains the same. Thus mass transport by diffusion is proportional to the concentration gradient of substrate or product. The proportionality factor is the diffusion coefficient. It is assumed that:

- internal mass transport is described by diffusion with an effective diffusion coefficient;
- the effective diffusion coefficient is independent of the substrate or product concentration.

By introduction of an effective diffusion coefficient, in contrary to a (free) diffusion coefficient in the bulk solution, the effects of the matrix are taken into account.

Thus we could replace D in Equation 6.2 by D_e to signify that we mean the effective diffusion coefficient within the matrix rather than the D diffusion coefficient of the component free in the bulk solution.

Now let us turn our attention to the bulk liquid and boundary layer.

6.2.2 Bulk liquid and boundary layer

External mass transport from the bulk solution to the outside of the slab, the interface between immobilisation matrix and bulk solution, is described by the film theory. It is assumed that there is no partition, so the solubility of a particular compound in the bulk equals the solubility in the matrix. This means that the concentration on both sides of the interface is the same. The concentration gradient outside the biocatalyst matrix can be regarded as linear (film theory), the external mass transport is concentrated in a stagnant boundary layer. We have represented this in Figure 6.2. Let us work through this figure by considering the concentration of a component in the bulk liquid, then in the stagnant film close to the surface of the biocatalyst matrix and finally inside of the biocatylic matrix. We have chosen to use a biochemical-like nomenclative to describe the concentration of a component. Since we are usually concerned with a substrate, we have denoted the substrate concentration as S (rather than C_s). We have also used the subscripts b to represents the bulk fluid, i to represent the interface between the fluid and the biocatalyst matrix and m to indicate the matrix.

film theory

Now let us examine Figure 6.2.

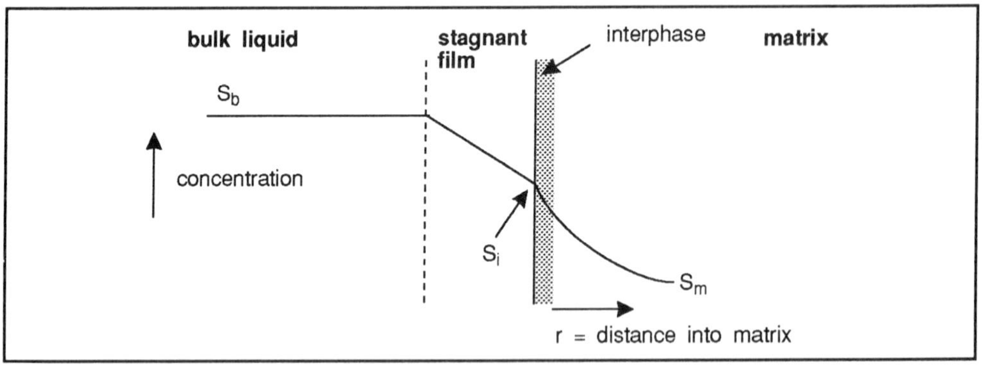

Figure 6.2 Concentration profile of substrate according to the film theory (see text for details).

In a well mixed system the concentration of substrate in the bulk liquid (S_b) will be the same throughout the liquid except for a thin film at the surface. This thin film is unmixed (stagnant) and thus substrate can only pass through it by diffusion. Within the matrix, the substrate is being used by the catalyst. Thus the substrate concentration in the matrix is low. Note that the concentration profile for the substrate in the matrix will depend upon its rate of transfer across the interphase and the rate of its bioconversion within the matrix. Note also that the concentration of the substrate S on either side of the interphase (S_i) is the same. At this stage we will not consider the concentration of substrate within the matrix further. We will concentrate on the mass transfer of substrate from the bulk liquid to the matrix.

The flux at the surface is assumed to be, according to the film theory, proportional to the difference between the bulk substrate concentration S_b and the substrate concentration at the interface S_i and with the external mass transport coefficient k_L. When no accumulation of substrate occurs at the interface between slab and liquid, the transport towards the biocatalyst must equal the transport of substrate into the particle:

$$\mathcal{D}_e \cdot \frac{dS}{dr} = k_L (S_b - S_i) \qquad \text{(E - 6.2)}$$

external mass transport is often not rate limiting

In many immobilised systems the stagnant film is very small and diffusion across this layer is fast. Thus external mass transport is often not rate limiting and for most cases we can assume that the interface concentration of the substrate (S_i) is more-or-less equal the bulk concentration (S_b). This approximation allows for greater simplification. Remember, however, that this can only be applied to systems in which the bioconversion by an immobilised biocatalyst is not limited by external mass transport. Fortunately in most cases this is true.

SAQ 6.1

1) Using Figure 6.2 as a model, draw a concentration profile for a product made within a biocatalytic matrix. Assume that their is no partitioning of the product between matrix and bulk liquid.

2) Repeat 1) but in this case assume there is partitioning between the bulk liquid and the matrix, such that at equilibrium the concentration of the product is higher in the matrix than in the bulk liquid.

6.3 Reaction kinetics

In this section some kinetics equations will be described. The basic properties of zeroth and first order kinetics are discussed. We begin by reminding you of the general form of the relationship between substrate concentration and reaction rate (Figure 6.3).

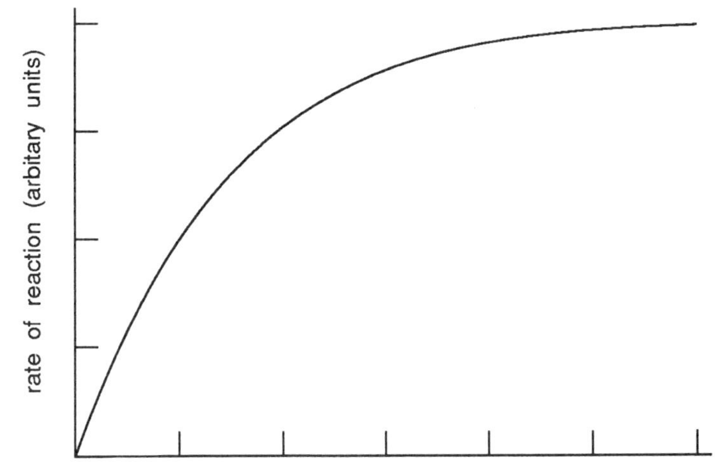

Figure 6.3 General form of the relationship between reaction rate and substrate concentration for an enzyme catalysed reaction.

Michaelis Menten (Monod) kinetics

Enzymes which display the kinetics shown in Figure 6.3 are said to exhibit Michaelis Menten kinetics. A similar relationship is also found for specific growth rate and substrate concentration for whole cells, (the so called Monod relationship).

Thus we could draw Figure 6.3 again by replacing rate of reaction by specific growth rate (μ) for whole cells. The main point to realise is that at high substrate concentration, the process exhibits zeroth order kinetics. At low substrate concentrations, the rate of reaction (growth rate) is proportional to substrate concentration, thus first order kinetics are found in this concentration range.

In truly zeroth order reactions the reaction rate or substrate utilisation (v) is independent of the substrate concentration and has a constant value. The constant is called the zeroth order reaction constant k_0:

$$v = -k_0 \quad \text{(E - 6.3)}$$

Figure 6.4 shows the time course of a zeroth order reaction as a first order reaction and a reaction following Michaelis Menten kientics.

For the zeroth order reaction, we can write:

$$S_b(t) = S_b(0) - k_0 \cdot t \quad \text{(E - 6.4)}$$

The reaction rate is maximal until the substrate concentration is zero.

In first order reactions the reaction rate is proportional to the substrate concentration. The proportionality factor is called the first order reaction rate constant k_1:

$$v(S_b) = -k_1 \cdot S_b \quad \text{(E - 6.5)}$$

Figure 6.4 shows the time course of a first order reaction rate as described by:

$$S_b(t) = S_b(0) \cdot \exp(-k_1 \cdot t) \quad \text{(E - 6.6)}$$

The reaction rate approaches zero as the substrate concentration approaches zero.

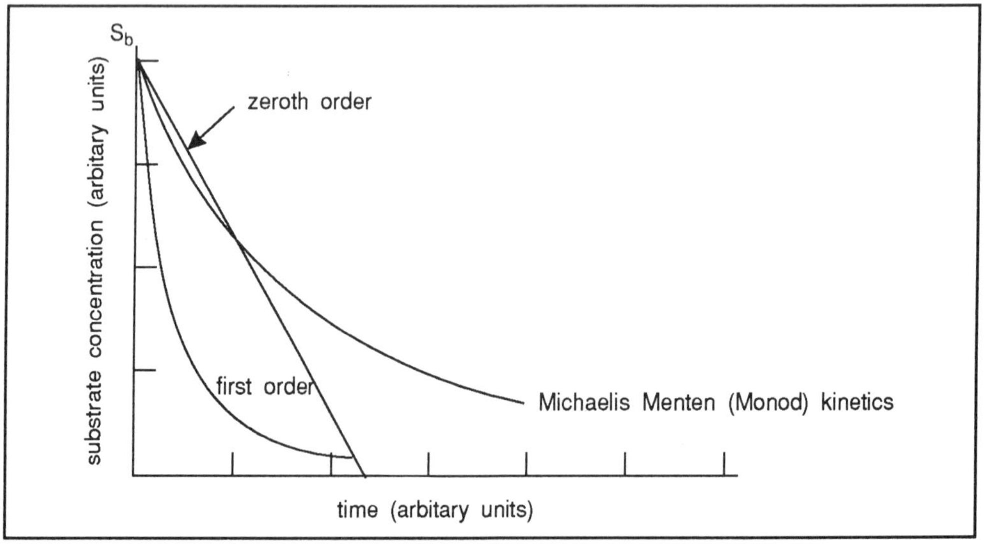

Figure 6.4 Comparison of the time course of zeroth, first order reactions and reactions following Michaelis Menten kinetic.

Michaelis Menten and Monod kinetics

Enzyme reactions often show the kinetic behaviour depicted by the third curve in Figure 6.4. Such enzyme systems are said to obey Michaelis Menten kinetics. A similar type of kinetic behaviour is often observed in microbial growth and in that situation the system is described by the Monod equation. In these situation, the reaction rate (or specific growth rate) is non-linear with the substrate concentration. It can be described by the equation:

$$v(S) = V_{max} \cdot \frac{S_b}{K_M + S_b} \qquad (E - 6.7)$$

In which V_{max} is the maximal conversion on rate and K_M is the Michaelis Menten constant (or Monod constant (K_s) in case of microbial growth). K_M is defined as the substrate concentration where v(S) = 0.5. V_{max}, Note in some text V_{max} is written V^{max} or V_m. The conversion rate reaches for high substrate concentrations ($S>K_M$) a maximum and is independent of the concentration (zeroth order). For low values of the substrate concentrations ($S<K_M$) the reaction rate is first order. Between these extremes values the reaction rate is a mixture of first and zeroth order reaction. Due to the non-linearity of Equation 6.7 the time course of a reaction can not be given analytically.

Π Redraw Figure 6.4, but this time plot it as velocity (rate of reaction against substrate concentration for the three types of kinetics. Do this before looking at the next figure.

The figure you should have drawn should have looked like this:

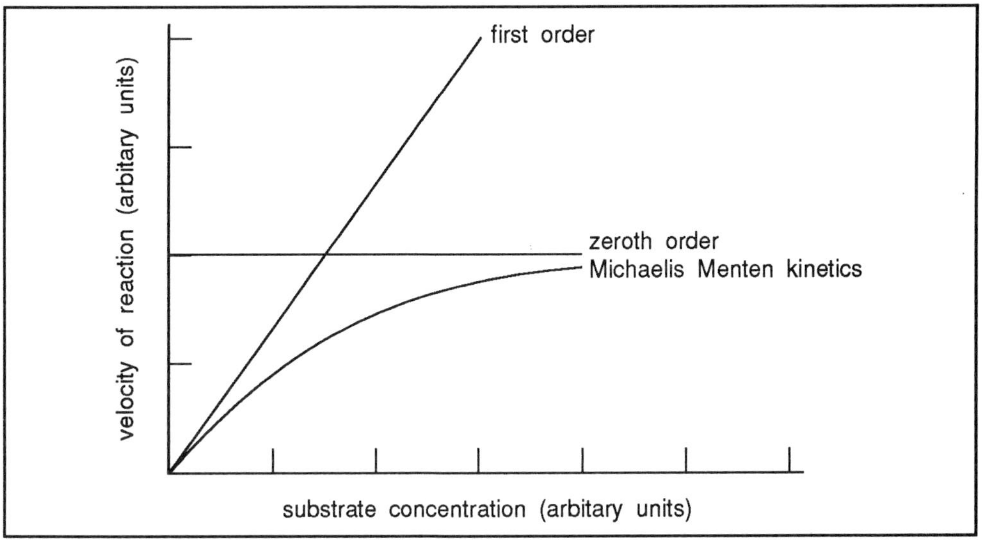

Figure 6.5 Comparison of the effect of substrate concentration on the velocity of reaction for zeroth and first order reactions and for reactions displaying Michaelis Menten kinetics.

The Michaelis Menten constant (K_M) and Monod constant (K_s) are useful values in describing the kinetics of a system. Remember K_M = substrate concentrations at ½ V_{max} and K_s = substrate at ½ μ_{max} (½ maximum specific growth rate). In practice it is not easy to determine V_{max} or μ_{max} directly from a plot of v (μ) against substrate concentration. It is more usual to use a double reciprocal plot since:

$$v(S) = \frac{V_{max}\, S_b}{K_M + S_b} \text{ then } \frac{1}{v(S)} = \frac{K_M}{V_{max}}\left(\frac{1}{S_b}\right) + \frac{1}{V_{max}}$$

Thus a graph of $\frac{1}{v(S)}$ against $\frac{1}{S_b}$ should give a straight line with a slope of $\frac{K_M}{V_{max}}$, with an intercept on the $\frac{1}{v(S)}$ axis at $\frac{1}{V_{max}}$. Such a plot is called a Lineweaver-Burk plot and is represented in Figure 6.6.

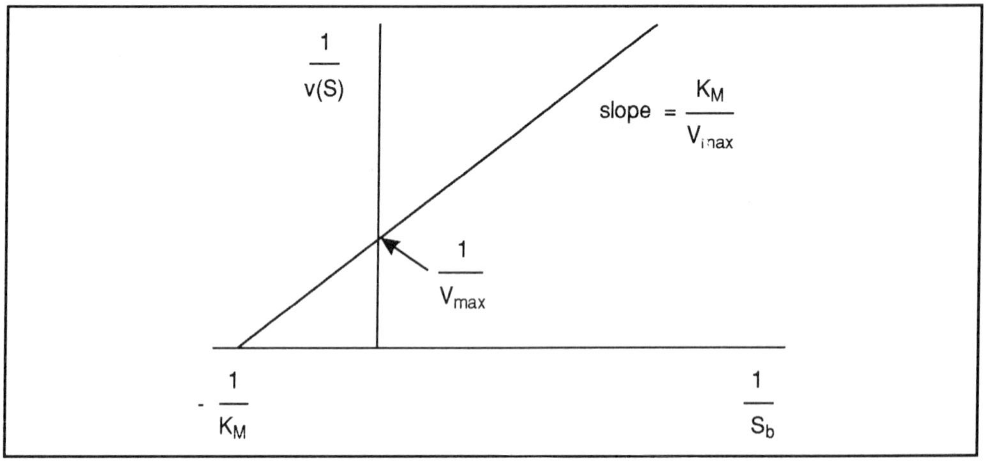

Figure 6.6 A Lineweaver-Burk plot (see text for details).

The Monod relationship for growth rate and substrate concentration is:

$\mu = \frac{\mu_{max} S_b}{K_S + S_b}$ where μ = specific growth rate.

This type of kinetic behaviour can be handled in exactly the same way as the Michaelis Menten kinetics of an enzyme. Thus a plot of $\frac{1}{\mu}$ against $\frac{1}{S_b}$ gives a slope of $\frac{K_s}{\mu_{max}}$ and intercepts of $\frac{1}{\mu_{max}}$ and $-\frac{1}{K_s}$

SAQ 6.2

The following set of velocity-concentration data has been obtained for an enzyme catalysed reaction.

S_b (mol l^{-1})	velocity (mol min^{-1})
2×10^{-4}	2.12×10^{-8}
3×10^{-4}	2.7×10^{-8}
4×10^{-4}	2.94×10^{-8}
6×10^{-4}	3.33×10^{-8}
10×10^{-4}	3.85×10^{-8}

Use the data to:

1) prove the reaction follows Michaelis Menten kinetics;

2) determine K_M and V_{max}.

Summary and objectives

This has been a very brief chapter in which we have attempted to remind you of two basic principles underpinning the modelling and analysis of immobilised systems. In the first we have refreshed your knowledge of Ficks law relating the rate of mass transfer to a concentration gradient. In the second we reminded you that biocatalysed processes (reactions and growth) follow a mixture of first and zeroth order kinetics depending upon the concentration of reactants (substrates). Thus mixed type of kinetics is described by Michaelis Menten kinetics when applied to enzymes or by Monod kinetics when applied to whole cells. We are now sufficiently prepared to model immobilised systems.

Now that you have completed this chapter you should be able to:

- predict the general forms of concentration gradients of substances diffusing into and out of an immobilised biocatalyst matrix;
- calculate various parameters of biocatalysed processes from supplied data.

Modelling of immobilised systems

Modelling of immobilised systems

7.1 Introduction

The two previous chapters have introduced you to the overall features of immobilised systems especially relating to their advantages and disadvantages. You have also been reminded of some important principles of mass transfer and the reaction kinetics associated with biocatalysed processes. We now are ready to examine the core issues in modelling immobilised systems.

If we think about immobilised systems we will realise that two basic geometrics are encountered in these systems. Biofilms coating a surface can be regarded as 'slabs' rather like those described in the previous chapter. This is not, however, the only configuration. The biocatalyst can be produced in the form of pellets. In other words they can have a spherical geometry. As you might anticipate these are significantly different systems. Think for a moment about the ratio of surface to volume of spheres and slabs. Clearly the surface:volume ratio has an influence on the availability of substrate. Thus it is appropriate for us to consider these two basic geometries. For simplicity we have omitted discussion of biocatalysts in other configurations (eg small, solid cylinders). Furthermore, we might anticipate that since the kinetics of bioconversions may be first order (at low substrate concentration), zeroth order (at high substrate concentration) or a mixture of both (at intermediate values of substrate concentration), we ought to discuss immobilised systems operating in different substrate, and therefore different kinetic regimes.

This chapter is divided into a number of stages. In the first, we will model systems in which the reaction is coupled to diffusion into a slab. This will be followed by an analogous modelling in which the reaction is coupled to diffusion into a sphere. Both of these will be based on zeroth order kinetics. Subsequently we will then extend the discussion to examine spherical and slab systems coupled to reactions showing first order kinetics. We will then consider these systems in relation to external mass transport and Michaelis Menten kinetics within immobilised systems and examine internal diffusion and external mass transport as likely rate limiting factors.

In parts of this chapter some quite difficult mathematical manipulations are used. Therefore we have provided two options. In the main text, we describe the principle relationships and explain their meaning. The mathematical derivation of these relationships are developed within panels. Thus, for those without the appropriate mathematical expertise, the relationships described can still be understood. Of course, understanding the mathematical derivations provides a firmer foundation upon which to use the relationships.

This is quite a long chapter, so we suggest you do not attempt it all at one sitting.

7.2 Modelling of reactions of zeroth order kinetics coupled to diffusion in a slab

In this section an immobilised biocatalyst with slab geometry in which the reaction takes place with zeroth order kinetics is modelled. A solution of the model is presented giving the substrate concentration in the slab and the conversion rates.

7.2.1 The substrate balance for a slab

In Figure 7.1 we once again use an infinitely long flat plate or slab (the thickness is small compared to the length, so the side effects at the edges are negligible). The thickness of the slab is Rp and the location in the slab is represented by the coordinate r.

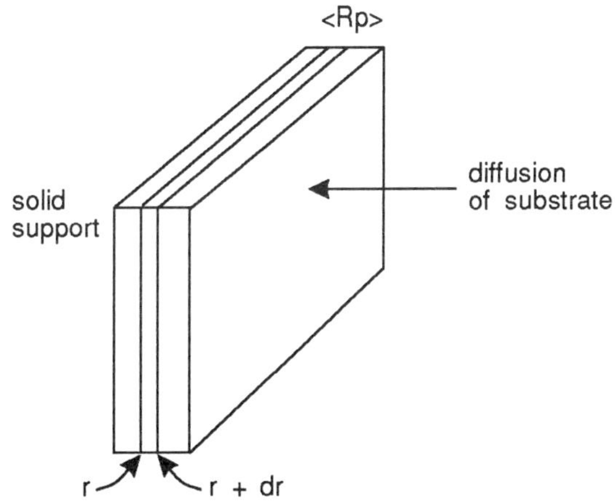

Figure 7.1 Slab geometry (see text for description).

The slab represents a matrix containing enzymes or cells (immobilised biocatalyst) and is supported by a solid support. Substrate is transported into the slab by diffusion and is consumed according to a specified reaction rate. To model these processes an infinitely small part of the slab with a thickness of dr is considered (Figure 7.1). Substrate is transported into this slab, consumed within it and transported out. In other words, in steady-state conditions:

Flow in - Flow out - Amount consumed in reaction = 0

Flux at location r + dr - Flux at location r - consumption = 0.

Let us convert this into a mathematical notation.

Diffusion transport takes place according to Fick's law. Thus:

$$[A \,.\, D_e \,.\, \frac{dS}{dr}]_{r+dr} - [A \,.\, D_e \,.\, \frac{dS}{dr}]_r - v(S) = 0 \qquad \text{(E - 7.1)}$$

In which A is the area of the slab, v(S) is the rate of consumption of S at concentration S. Note that we have used the effective diffusion coefficient (D_e) of the substrate in the matrix, not its diffusion coefficient in free solution (D).

For those with a more mathematical background, we can take this a little further.

Taylors law

Equation 7.1 can be evaluated by using Taylors law which says that the function value of a function f(x) at x + dx is approximately equal to the function value at x plus the derivative of the function at x multiplied by dx:

$$f(x + dx) = f(x) + f'(x) \cdot dx \qquad \text{(E - 7.2)}$$

Or in this case:

$$[A \cdot D_e \frac{dS}{dr}]_{r+dr} = [A \cdot D_e \frac{dS}{dr}]_r + \frac{d}{dr}[A \cdot D_e \frac{dS}{dr}]_r \cdot dr \qquad \text{(E - 7.3)}$$

Combined with 7.1 this gives:

$$v(S) = \frac{d}{dr}[A \cdot D_e \frac{dS}{dr}]_r \cdot dr$$

$$v(S) = D_e \cdot \frac{d^2S}{dr^2} \qquad \text{(E - 7.4)}$$

What is the meaning of this equation?

Equation 7.4 is the basic second order differential equation for diffusion coupled with reaction. It says that the amount of substrate which is transported by diffusion under the influence of a concentration gradient equals the amount of substrate converted. The solution of Equation 7.4 gives the substrates concentration S as a function of the location r. A reaction rate term for v(S) must be introduced. It is assumed that only one substrate is limiting so only one substrate concentration profile is relevant.

7.2.2 Solution of the transport/reaction equation, zeroth order kinetics

substrate not limited by external mass transport

For zeroth order reactions Equation 7.4 is solved fairly easy, the term v(S) is replaced by k_0. We remind you that k_0 = zeroth order reaction constant (see Chapter 6). For solution of this second order differential equation, however, two boundary conditions have to be formulated. To start with, the situation is considered without accounting for external mass transport limitation which means that substrate concentration on the surface of the slab equals the concentration in the bulk. The situation in which substrate limitation arises due to external mass transport through a stagnant layer can be accounted for by film theory and will be considered later.

For now we will focus on a slab in which the concentration at the surface of the slab is equal to the concentration in the bulk. What is the concentration of substrate at the other surface of the slab?

Intuitively one would probably draw a concentration profile in the slab which looks like that in Figure 7.2.

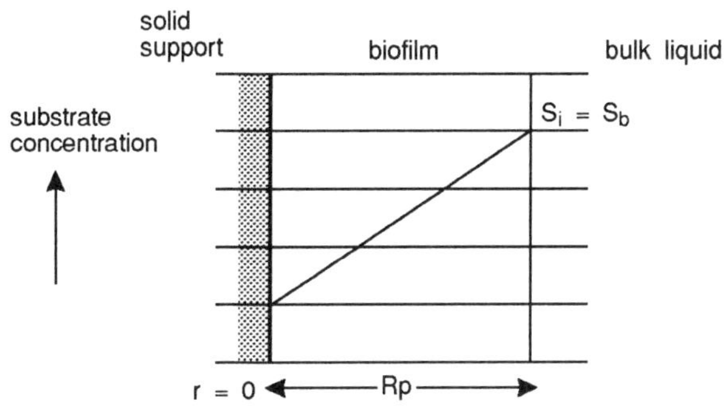

Figure 7.2 Concentration profile in the slab. (see text).

The substrate concentration on the surface equals the bulk substrate concentration and further towards the other edge the concentration decreases as transport by diffusion can not cope up with consumption of substrate by reaction. But if the profile at the face next to the solid support of the slab is considered it can be seen that there is a slope in the profile. In other words $dS/dr \neq 0$. According to Ficks law this means that there is transport of substrate, which of course is not the case since the compound will not diffuse into the solid support. So obviously if there is no transport in the solid support and $dS/dr = 0$ next to the solid support. Actually dS/dr at $r = 0$ is not defined, but it can be shown that the limit value for dS/dr for r approaching zero equals zero. So most probably the substrate concentration in the slab schematically looks like that shown in Figure 7.3.

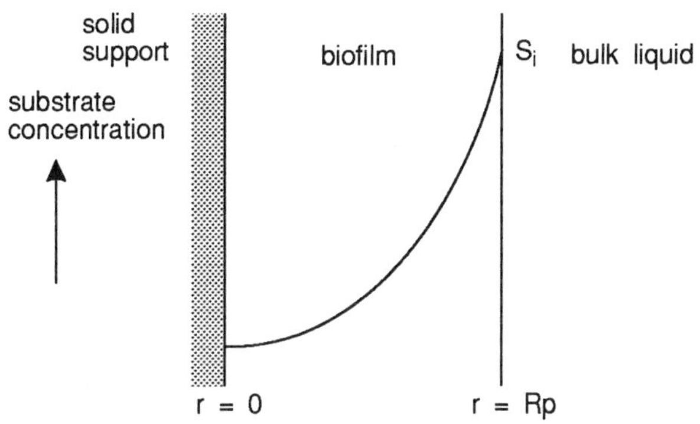

Figure 7.3 Stylised concentration profile in the slab (biofilm). (see text).

Π Use Figure 7.3 and see if you can formulate two boundary conditions for the substrate in the biofilm. Write down your ideas and then read on.

We can write down the following boundary conditions.

$\frac{dS}{dr} = 0$ for $r = 0$; no transport of substrate at the solid support: slab interphase.

$S_i = S_b$ for $r = Rp$; the substrate concentration at the surface equals the bulk substrate concentration.

Now Equation 7.4 can be solved by integrating and using the two boundary conditions:

$$\frac{d^2S}{dr^2} = \frac{k_0}{D_e} \qquad \text{(E - 7.5)}$$

The first integration yields:

$$\frac{dS}{dr} = \frac{k_0}{D_e} . r + IC_1 \qquad \text{(E - 7.6)}$$

The second integration yields:

$$S(r) = 0.5\frac{k_0}{D_e} . r^2 + IC_1 . r + IC_2 \qquad \text{(E - 7.7)}$$

By substitution of the two boundary conditions the integration constants IC_1 and IC_2 can be found. It follows that:

$$IC_1 = 0 \text{ (because } \frac{dS}{dr} = 0 \text{ for } r = 0\text{) and } IC_2 = S_b - 0.5\frac{k_0}{D_e} . Rp^2 \text{ (because } S_i = S_b \text{ for } r = Rp\text{)} \qquad \text{(E - 7.8)}$$

Consequently the substrate concentration in the slab is given by:

$$S(r) = S_b - 0.5\frac{k_0}{D_e}(Rp^2 - r^2) \qquad \text{(E - 7.9)}$$

Let us stop at this point and interpret what this mathematics is telling us. In essence it is saying that we can predict (calculate) the substrate concentration for any point in the slab if we know the effective diffusion coefficient of the substrate in the slab, the reaction constant, the thickness of the slab and the concentration of substrate in the bulk solution. Prove that you can use relation Equation 7.9 by attempting SAQ 7.1.

SAQ 7.1

1) The glucose concentration in the mid point in a biocatalytic matrix of slab geometry is 1×10^{-2} mol l^{-1} whilst the glucose concentration in the bulk liquid is 4×10^{-2} mol l^{-1}. If the bulk liquid is replaced by a solution containing 6×10^{-2} mol glucose l^{-1}, what will the new steady state glucose concentration be at the mid point of the slab? Assume that the consumption of the glucose follows zeroth order kinetics and that the new glucose regime does not influence the amount of biocatalyst.

2) In the new regime (ie glucose solution containing 6×10^{-2} mol glucose l^{-1} in the bulk liquid) described in 1, what would happen to the glucose concentration at the mid point in the slab if we halved the amount of catalyst present?

We illustrated the profile of substrate concentration in a slab in Figure 7.3. If in fact we use Equation 7.9 to calculate the substrate concentration ($S_{(r)}$) for a whole series of r values we will produce a profile like that illustrated in Figure 7.3. The actual values will of course depend upon the thickness of the biofilm (Rp), the effective diffusion coefficient, the reaction rate and the bulk liquid concentration.

SAQ 7.2

Here are three profiles of substrate concentration in three biofilms of the same thickness (Rp).

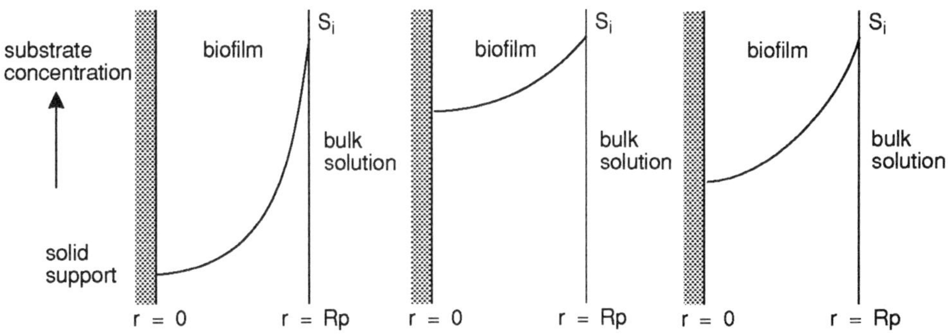

1) Which of the three profiles most likely represents a substrate with a high effective diffusion coefficient?

2) Which of the three profiles probably represents the fastest consumption of substrate in a reaction within the biofilm.

3) Which of the three profiles most likely represents a substrate with a low effective diffusion coefficient?

SAQ 7.3

Draw a schematic profile for substrate concentration for a biofilm in which the substrate can be transported from both sides of the film (that is, the solid support in Figure 7.1 is replaced by bulk solution).

7.2.3 Net conversion rates, the effectiveness factor

In the profiles of substrate concentration we have examined so far, the concentration of substrate diminishes in the biofilm the further away from the bulk solution. If the substrate concentration diminishes to zero (for example, in a thick film or in one with a high k_O value), then we can describe such a situation as substrate depletion. Although substrate concentrations are of interest especially if we wish to ensure substrate is not being depleted, of greater interest is the net conversion rate of substrate into product. In this section we will consider the conversion rates for complete and incomplete (ie substrate depletion) substrate penetration into the slab (biofilm). We will relate these conversion rates to a so called 'effectiveness' factor.

effectiveness factor

Conversion rate with and without substrate depletion in the slab

In the previous part, only the situation is shown in which the substrate concentration has a value different from zero in all parts of the biofilm. For this case it is easy to calculate the net conversion rate, if it is kept in mind that the reaction rate for zeroth order reactions is maximal if the substrate concentration differs from zero, and equals zero if the substrate concentration is zero. For the situation in Figure 7.3 (see also Figure

7.4) the reaction rate is maximal in the whole slab and has the value k_0. Suppose the volume of the slab is V_s, than the total conversion rate v is given by:

$$v = V_s \cdot k_0 \qquad (E - 7.10)$$

Thus for a slab with volume V_s, containing for example an oxygen consuming enzyme with reaction rate k_0, then $V_s \cdot k_0$ mol of oxygen is consumed per unit time would imply that the same amount of oxygen must be transported from the gas phase to the bulk liquid adjacent to the slab.

Now let us consider the situation in which the reaction rate k_0 is so high that the substrate is depleted before the end of the slab, for example at a specific location r = rd (see Figure 7.4).

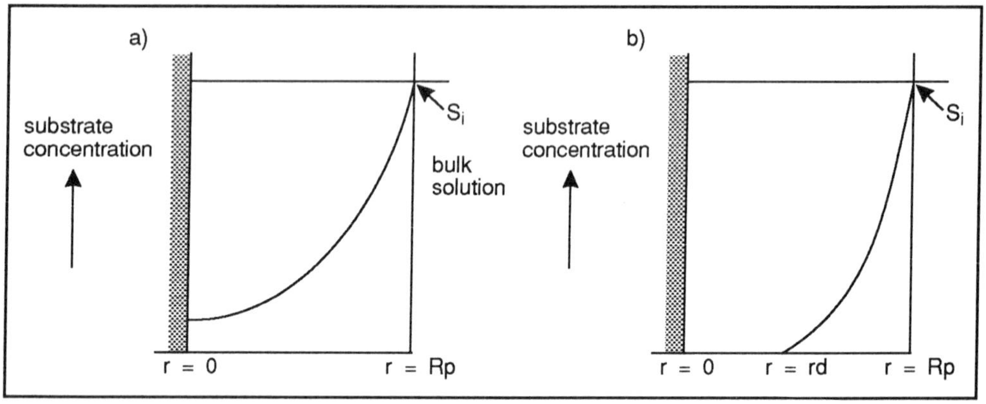

Figure 7.4 a) Substrate concentration profile of a slab with complete substrate penetration. b) Substrate concentration profile of a slab with substrate depletion at r = rd.

One may ask, what is the conversion rate in this case? If again the properties of zeroth order reactions are kept in mind the volume of the slab in which the reaction takes place is A_s (Rp-rd), in which A_s is the outer surface area of the slab. So the total conversion rate v is given by:

$$v = k_0 A_s (Rp - rd) \qquad (E - 7.11)$$

It can be concluded that the full reaction potential of the slab is not effectively used.

SAQ 7.4

The thickness of a biofilm is 1 mm. The zeroth order reaction constant for substrate consumption of the biofilm is 1 mol m^{-3} h^{-1} and its surface area is $100 m^2$. The rate of consumption of substrate by this biofilm has been measured as 0.08 mol h^{-1}.

1) Does the substrate penetrate the whole biofilm?

2) If not, how far does the substrate penetrate the membrane?

The effectiveness factor

The situation in which the slab is not completely used for reaction can be expressed by means of an effectiveness factor. Various effectiveness factors can be defined as we will discuss later on. For now the effectiveness factor is represented by the symbol η. How effectively is the slab from Figure 7.4b used? The maximal conversion can be obtained if the substrate reaches the end of the slab and the whole volume is used.

Let us define the effectiveness factor as the fraction of the thickness of the slab that is used.

In other words $\eta = \frac{Rp - rd}{Rp}$ (E - 7.12a)

or $\eta = 1 - \frac{rd}{Rp}$

Note that we have followed the general convention of representing the effectiveness factor by η. Do not confuse its use in this context with the widespread use of the same symbol to represent viscosity.

The net conversion rate can now simply be expressed by:

$v = \eta \cdot Vs \cdot k_0$ (E - 7.13)

The next problem is to calculate η because the value of rd is not usually known and so η is not known. The value for rd can be calculated if the substrate concentration profile is known for the situation when substrate is depleted. For this the second order differential Equation 7.5 has to be solved again with one different boundary condition. Beyond the location rd there is no transport of substrate, (in the previous solution with no substrate depletion this location was at r = 0). The two boundary conditions to solve Equation 7.5 for the case with substrate depletion are as follows:

$\frac{dS}{dr} = 0$ at $r = rd$; (no transport of substrate at $r \leq rd$)

and

$S_i = S_b$ when $r = Rp$; (the substrate concentration at the surface equals the bulk substrate concentration).

After the second integration of Equation 7.5 the result was:

$S(r) = 0.5 \frac{k_0}{D_e} \cdot r^2 + IC_1 \cdot r + IC_2$ (see Equation 7.7)

By substitution of the two boundary conditions, the integration constants IC_1 and IC_2 can be found. It follows that:

$S(r) = S_b - 0.5 \frac{k_0}{D_e} (Rp^2 - r^2) + \frac{k_0}{D_e} (rd \cdot Rp - rd \cdot r)$ (E - 7.14)

Π It might be a good exercise to derive Equation 7.14 for yourself by solving Equation 7.7 using the appropriate boundary conditions. (Note that IC_1 follows from $\frac{dS}{dr} = 0$ for r = rd and IC_2 follows from $S_i = S_b$ for r = Rp).

Here is our solution

From Equation 7.6 $\frac{dS}{dr} = \frac{k_0}{D_e} r + IC_1$

Therefore when $\frac{dS}{dr} = 0$ (at r = rd) then,

$0 = \frac{k_0}{D_e} r + IC_1$ and $IC_1 = -\frac{k_0}{D_e} . rd.$

From Equation 7.7

$$S_b = 0.5 \frac{k_0}{D_e} Rp^2 - \frac{k_0}{D_e} . rd\ Rp. + IC_2$$

$$IC_2 = S_b - 0.5 \frac{k_0}{D_e} Rp^2 + \frac{k_0}{D_e} rd. Rp.$$

Therefore we obtain

$$S(r) = 0.5 \frac{k_0}{D_e} r^2 - \frac{k_0}{D_e} rdr + S_b - 0.5 \frac{k_0}{D_e} Rp^2 + \frac{k_0}{D_e} rd. Rp.$$

$$S(r) = S_b - 0.5 \frac{k_0}{D_e} (Rp^2 - r^2) + \frac{k_0}{D_e} (rd. Rp - rd. r)$$

You will see by comparing Equation 7.9 (for complete substrate penetration) with Equation 7.14 for substrate limitation, that the two equations differ by a factor:

$$\frac{k_0}{D_e} (rd . Rp - rd . r)$$

$$S(r) = S_b - 0.5 \frac{k_0}{D_e} (Rp^2 - r^2) \qquad \text{(E - 7.9)}$$

$$S(r) = S_b - 0.5 \frac{k_0}{D_e} (Rp^2 - r^2) + \frac{k_0}{D_e} (rd\ Rp - rd . r) \qquad \text{(E - 7.14)}$$

It is obvious that this equation is only valid if the substrate does not reach the end of the slab. Otherwise the slab is fully penetrated and the value of rd is zero. Thus Equation 7.14 becomes Equation 7.9 for full penetration. With Equation 7.14, rd can be calculated because for r = rd the substrate concentration equals, by definition, zero. For r = rd and S(rd) = 0 we can calculate rd.

Π Using r = rd and S(rd) = 0 and Equation 7.14 show that:

$rd = Rp - \sqrt{2S_b D_e/k_0}$

It might be helpful to try this for yourself before reading on.

Here is our solution.

$$\text{Since } S(r) = S_b - 0.5\frac{k_0}{D_e}(Rp^2 - r^2) + \frac{k_0}{D_e}(rd\,Rp - rd\,r) \qquad \text{(E - 7.14)}$$

Then for r = rd

$$S(rd) = S_b - 0.5\frac{k_0}{D_e}(Rp^2 - rd^2) + \frac{k_0}{D_e}(rd\,Rp - rd^2)$$

but S(rd) = 0

$$\text{thus } S_b = 0.5\frac{k_0}{D_e}(Rp^2 - rd^2) - \frac{k_0}{D_e}(rd\,Rp - rd^2)$$

multiply by 2

$$2S_b = \frac{k_0}{D_e}(Rp^2 - rd^2) - \frac{2k_0}{D_e}(rd\,Rp - rd^2)$$

Multiply by $\frac{D_e}{k_0}$

$$\text{Thus } \frac{2\,S_b\,D_e}{k_0} = (Rp^2 - rd^2) - 2\,(rd\,Rp - rd^2)$$

$$\frac{2\,S_b\,D_e}{k_0} = Rp^2 - rd^2 - 2rd\,Rp + 2rd^2$$

$$\frac{2\,S_b\,D_e}{k_0} = Rp^2 - 2\,rd\,Rp + rd^2$$

$$\frac{2\,S_b\,D_e}{k_0} = (Rp - rd)^2$$

$$\sqrt{2S_b\,D_e/k_0} = Rp - rd$$

$$\text{Thus } rd = Rp - \sqrt{2S_b\,D_e/k_0} \qquad \text{(E - 7.15)}$$

What Equation 7.15 is showing us is that the penetration depth is related to the thickness of the slab (Rp), the substrate concentration in the bulk liquid (S_b), the effective diffusion coefficient (D_e) and the zeroth order rate constant k_0.

Before we attempt to use this equation in a numerical example, let us take the discussion on stage further.

With Equation 7.14, the location (rd) where substrate depletion starts (the penetration depth) can be calculated. Remember however that the effectiveness factor $\eta = 1 - \frac{rd}{Rp}$ (E -7.12).

We can combine Equations 7.12 and 7.15.

Thus $\sqrt{(2S_b\, D_e)/(k_0)} = Rp - rd$

divide through by Rp

$$\sqrt{(2S_b\, D_e)/(Rp^2\, k_0)} = \frac{Rp - rd}{Rp} = \eta \qquad \text{(E - 7.16)}$$

With Equation 7.16 the effectiveness factor can be calculated if the values for the bulk substrate concentration, the slab thickness, the zeroth order reaction rate constant and the effective diffusion coefficient are known. If the effectiveness factor is known, the conversion rate given by Equation 7.13 can be calculated.

Let us now apply these relationships by attempting the following SAQ.

SAQ 7.5

We provide the following data.

The biocatalyst is a 2 mm thick biofilm supported by a packing material. Oxygen is consumed by the biolayer according to zeroth order kinetics with a zeroth order reaction rate constant of 1×10^{-3} mol s^{-1} kg^{-1} . 100 kg of the biocatalyst is immobilised with a concentration of 100 kg m^3. The effective diffusion coefficient of oxygen is 2.2×10^{-9} m^2 s^{-1}. Oxygen is consumed at a rate of 0.06 mol s^{-1}.

1) Calculate the volume of the biofilm and its area.

2) Determine whether oxygen penetrates the whole depth of the biofilm and, if not to what depth does it penetrate.

3) Calculate the effectiveness factor for the biofilm.

4) Calculate the concentration of oxygen in the bulk liquid.

effectiveness factor and Ficks law

As we have seen the effectiveness factor can be calculated directly from the penetration depth rd. Another way to calculate the effectiveness factor is to compare the amount of substrate which is transported into the slab, and which is converted in the slab, with the maximal amount of substrate that can be converted. The flux of substrate is given, according to Ficks law, by the derivative dS/dr at the interface of the slab multiplied by the effective diffusion coefficient (we dealt with this at the beginning of Chapter 6 - see Equation 6.1). The flux multiplied by the area of the slab gives the total amount of substrate transported into the slab and consequently in the slab. The effectiveness factor now is given by:

$$\text{Effectiveness} = \frac{\text{Total flux}}{\text{Maximal substrate conversion in the slab}}$$

$$\eta = \frac{A_S \,.\, D_e \,.\, \left.\frac{dS}{dr}\right|_{Rp}}{V_s \,.\, k_0} \qquad \text{(E - 7.17)}$$

NB $\left.\frac{dS}{dr}\right|_{Rp}$ is a short hand notation meaning the substrate concentration gradient at the bulk liquid, slab interface (ie r = Rp).

The value for dS/dr at the interface can be obtained from the solution of Equation 7.5:

$$\frac{d^2S}{dr^2} = \frac{k_0}{D_e}$$

Thus by integration we get:

$$\frac{dS}{dr} = \frac{k_0}{D_e} \,.\, r + IC_1 \qquad \text{(E - 7.6)}$$

With dS/ dr = 0 at r = rd it follows that:

$$\left.\frac{dS}{dr}\right|_{Rp} = \frac{k_0}{D_e}(Rp - rd) \qquad \text{(E - 7.18)}$$

If you think carefully for a moment about Equation 7.18 you will realise that Rp - rd is the penetration depth of the substrate. Thus Equation 7.18 tells us that the penetration of the substrate is governed by the concentration gradient at the interface ($\left.\frac{dS}{dr}\right|_{Rp}$), the zeroth order reaction rate constant (k_0) and the effective diffusion coefficient.

Π Use a piece of paper to substitute Equation 7.18 into Equation 7.17 and see what you can produce. You should be able to show that:

$$\eta = \frac{Rp - rd}{Rp}$$

Since $\eta = \frac{A_S \,.\, D_e \,.\, \left.\frac{dS}{dr}\right|_{Rp}}{V_s \,.\, k_0}$ (E - 7.17) and $\left.\frac{dS}{dr}\right|_{Rp} = \frac{k_0}{D_e}(Rp - rd)$

Then $\eta = \frac{A_s D_e k_0}{V_s k_0 D_e}(Rp - rd)$

$$\eta = \frac{A_s}{V_s}(Rp - rd)$$

But A_s = area of slab V_s = volume of slab. Thus $\frac{V_s}{A_s}$ = depth of slab = Rp.

Thus $\eta = \frac{Rp - rd}{Rp}$ which is the relationship we derived for η in Equation 7.16.

The practical significance of the effectiveness factor

We have spent some time considering the flux of substrate into a slab biofilm. We have learnt that, from basic principles we can model the substrate concentration profile within the slab in the situation where: a) the substrate does not become depleted. b) the substrate becomes depleted.

We remind you that for the situation in which the substrate does not depleted.

$$S(r) = S_b - 0.5 \frac{k_0}{D_e} (Rp^2 - r^2) \qquad \text{(E - 7.9)}$$

and for the situation in which substrate becomes depleted:

$$S(r) = S_b - 0.5 \frac{k_0}{D_e} (Rp^2 - r^2) + \frac{k_0}{D_e} (rd\, Rp - rd\, r) \qquad \text{(E - 7.14)}$$

We went on to develop the discussion to consider the effectiveness factor (η). This factor is really a measure of the proportion of the biofilm which is converting substrate and is given by $\frac{Rp - rd}{Rp}$ where (Rp - rd) is the penetration depth of the substrate and Rp is the thickness of the biofilm.

In most instances, biotechnologists would like to operate with an effectiveness factor of 1 (ie all of the biofilm is converting the substrate). In other words, they would like to operate under situations where the substrate completely penetrates the biofilm (rd = 0). A key calculation is therefore to determine the circumstances in which the substrate concentrations at the support edge of the film just reaches zero.

We can use Equation 7.14 to help us solve this since under the conditions described rd = 0, r = 0 and S(r) = 0. Thus $0 = S_b - 0.5 \frac{k_0}{D_e} (Rp^2 - 0) + \frac{k_0}{D_e} (0.Rp - 0.0)$

$$\frac{2\, D_e\, S_b}{Rp^2\, k_0} = 1 \qquad \text{(E - 7.19)}$$

or, of course $\frac{Rp^2\, k_0}{2\, D_e\, S_b} = 1$

Since we normally wish to operate such a process without substrate limitation, we have the option to modify some of the operational parameters to ensure that this does occur.

Π Examine Equation 7.19 and decide which of the parameters are most easily modified to ensure that substrate limitation does not occur in a biofilm?

Equation 7.19 contains four parameters, (D_e, S_b, Rp and k_0). Let us examine each in turn.

In most instances D_e is not easy to modify. It is an intrinsic property of the compound and the milieu in which the compound is diffusing. Thus D_e is not usually a parameter we can readily modify.

S_b is the substrate concentration in the bulk liquid. In principle if we have substrate limitation within the biofilm, we can overcome this by increasing S_b. There are, of course, some limits to this approach including the solubility and cost of substrate. Also we must bear in mind that very high substrate concentration may have a deleterious effect on the organisms present in the outer layer of the biofilm.

Another approach to ensuring that substrate limitation does not occur is to produce a thinner film (reduce Rp). If the volume of the biofilm is kept constant, it is better to use a thinner film (that is, one with a large surface area) thereby reducing the probability of substrate limitation arising.

We could also alter (decrease) the density of the biocatalyst within the biofilm. This, in effect would reduce k_0. Thus the substrate penetration depth would be increased. Again, however, we would not increase the net conversion of substrate using this strategy.

Thus in response to the previous question, we would suggest that increasing S_b generally offers the easiest approach to ensuring that substrate limitation does not occur and also maximise substrate conversion rates.

Before we attempt a numerical example, let us look at a stylised representation of what happens to the substrate concentration profile when the value of $2D_e\,S_b/Rp^2\,k_0$ is altered.

Π On a piece of paper, draw concentration profiles for a biofilm when:

1) $2D_e\,S_b/Rp^2\,k_0 < 1$

2) $2D_e\,S_b/Rp^2\,k_0 = 1$

3) $2D_e\,S_b/Rp^2\,k_0 > 1$

(Try this without looking at our figure below). What is the η value for situations 1), 2) and 3) described above. The sort of figure we anticipates you would draw are:

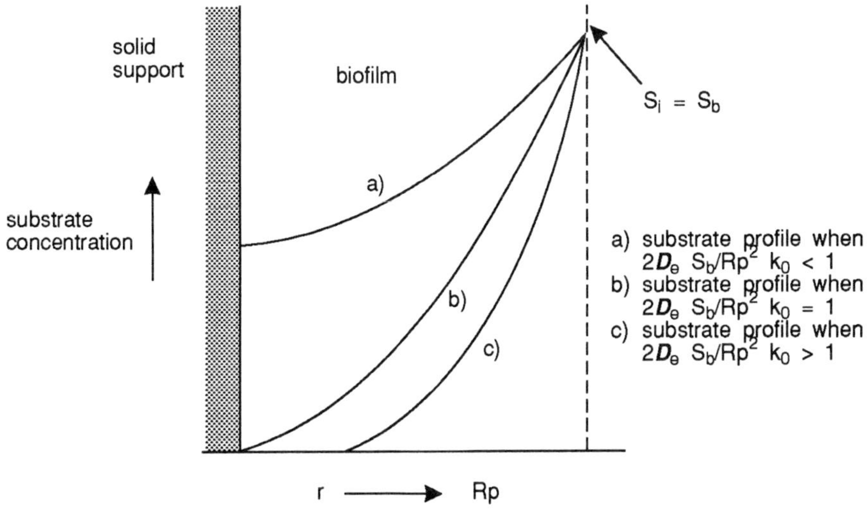

Line a) represents a system which has an effectiveness factor $\eta = 1$. Likewise line b) has an η value = 1. On the other hand line c) represents a systems which has an η value < 1.

From the previous figure and Equation 7.19, we can see that the absolute values of the parameters are not critical, it is the ratio of these parameters values that is important if we are to get full penetration of the biofilm by the substrate.

Graphically we can represent the relationship between η and $2D_e\ S_b/Rp^2\ k_0$ in the following way.

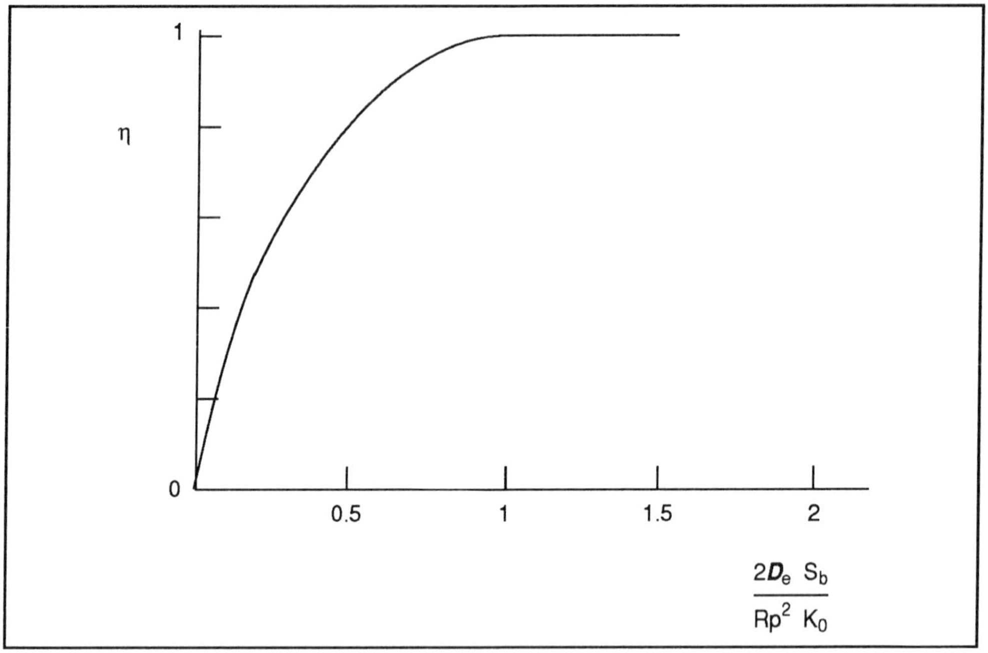

Figure 7.5 Relationship between η and $2D_e\ S_b/Rp^2\ k_0$ (see text for discussion).

SAQ 7.6

The biocatalyst is a 2mm thick biofilm supported by a packing material. Oxygen is consumed by the biolayer according to zeroth order kinetics with a zeroth order reaction rate constant of 1×10^{-3} mol s^{-1} m^{-3}. The effective diffusion coefficient of oxygen is 2.2×10^{-9} m^2 s^{-1}. What concentration of oxygen should we use in the bulk fluid to ensure oxygen does not become limiting in the biofilm?

7.3 Modelling of reaction coupled to diffusion in a sphere, zeroth order kinetics

In the previous part slab geometry with a zeroth order reaction was discussed. Now a zeroth order reaction in combination with spherical geometry will be considered.

In the previous sections a few assumptions were made concerning the development of a model for slab geometry. For example we assumed that the substrate concentration at the interface (S_i) was the same as the substrate concentration of the bulk liquid. These assumptions are also valid here. There is a strong analogy between treatment of a slab and a sphere. As most aspects are extensively treated in the previous section here we will only cover the main points.

The complication in examining a system with spherical geometry is merely a mathematical consequence of the more complex geometry. In all other respects the treatment will be similar to that described for a slab geometry in Section 7.2.

7.3.1 The substrate balance for a sphere

We have shown a stylised system in Figure 7.6 in which an immobilised biocatalyst is held in a sphere. The figure also shows the dimensions we will use in this discussion.

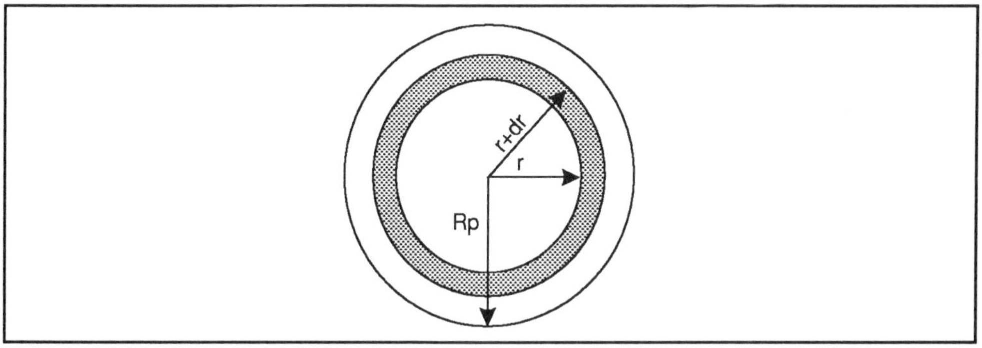

Figure 7.6 Spherical biocatalyst particle with radius Rp and a shell with thickness dr (see text for details).

The substrate balance over a very small shell with thickness dr can be made. Under steady state conditions the flux of substrate through the area at the location r + dr minus the flux of substrate through the area at the location r must equal the conversion rate in the corresponding volume:

flux in at r + dr	-	flux out a r	=	conversion in shell

$$[4\,\pi\,r^2 \cdot \frac{dS}{dr}]_{r+dr} - [4\,\pi\,r^2 \cdot \frac{dS}{dr}]_r = 4\,\pi\,r^2\,dr \cdot \frac{k_0}{D_e} \qquad \text{(E - 7.20)}$$

Here we have made the assumption that the surface area at r = surface area at r + dr. This is valid providing dr is small compared to r.

7.3.2 Substrate concentration profile

What we are really interested in is to be able to construct a substrate concentration profile within the particle.

We can do this from the substrate balance described above, but it requires some mathematical manipulation and definition of the boundary conditions. Since the first step demands some mathematical expertise we have made it 'optional' by including it in a separate panel (panel 7.1) entitled 'Manipulation of Equation 7.20'.

The outcome of this manipulation is that we can describe the substrate concentration at position r by the equation.

$$S(r) = Sb - \frac{1}{6}\,\frac{k_0}{D_e}\,(Rp^2 - r^2) - \frac{1}{3}\,\frac{k_0}{D_e}\,(\frac{rd^3}{Rp} - \frac{rd^3}{r}) \qquad \text{(E - 7.21)}$$

Panel 7.1

Manipulation of Equation 7.20

Since $[4 \pi r^2 \frac{dS}{dr}]_{r+dr} - [4 \pi r^2 \frac{dS}{dr}]_r = 4 \pi r^2 dr \frac{k_0}{D_e}$ (E - 7.20)

We can carry out a Taylor expansion on Equation 7.20 which on neglection of the higher order terms gives:

$$\frac{1}{r^2} \cdot \frac{d}{dr} [r^2 \cdot \frac{dS}{dr}] = \frac{k_0}{D_e} \qquad \text{(E - 7.20a)}$$

Carrying out the differentiation gives:

$\frac{1}{r^2} [2r \frac{dS}{dr} + r^2 \frac{d^2S}{dr^2}] = \frac{k_0}{D_e}$ which can be written as:

$$\frac{d^2S}{dr^2} + \frac{2}{r} \cdot \frac{dS}{dr} = \frac{k_0}{D_e} \qquad \text{(E - 7.20b)}$$

Equation 7.20b describes the substrate concentration profile in the spherical catalyst. The second order differential equation can be solved by integration if the appropriate boundary conditions are formulated. Integrating Equation 7.20a twice gives:

$$\frac{dS}{dr} = \frac{1}{3}\frac{k_0}{D_e} r + \frac{IC1}{r^2} \text{ and } S(r) = \frac{1}{6}\frac{k_0}{D_e} r^2 - \frac{IC1}{r} + IC2 \qquad \text{(E - 7.20c)}$$

where IC1 and IC2 are integration constants.

Now let us define the boundary conditions. The first boundary condition is formulated for the surface of the particle:

$S(r) = S_b$ at $r = Rp$; the substrate concentration at the interface equals the bulk substrate concentration, external mass transport limitation is neglected.

For the second boundary condition the situation is considered when substrate depletion occurs, yielding:

$\frac{dS}{dr} = 0$ at $r = rd$; there is no transport of substrate beyond the location $r = rd$.

For the two integration constants from Equation 7.20c, it follows that:

$$IC1 = -\frac{1}{3}\frac{k_0}{D_e} rd^3 \quad IC2 = S_b - -\frac{1}{6}\frac{k_0}{D_e} Rp^2 - \frac{1}{3}\frac{k_0}{D_e} \cdot \frac{rd^3}{Rp}$$

Substitution in 7.20c gives:

$$S(r) = S_b - \frac{1}{6}\frac{k_0}{D_e}(Rp^2 - r^2) - \frac{1}{3}\frac{k_0}{D_e}\left(\frac{rd^3}{Rp} - \frac{rd^3}{r}\right) \qquad \text{(E - 7.21)}$$

Equation 7.21 is quite straight forward, although it is a little long to remember. The main point to bear in mind is that Equation 7.21 gives the substrate concentration as a function of the location r. If there is no substrate depletion (rd = 0) Equation 7.21 can be modified to Equation 7.22 which can also be found if for the second boundary condition dS/dr = 0 at r = 0 is used.

$$S(r) = S_b - \frac{1}{6}\frac{k_0}{D_e}(Rp^2 - r^2) \qquad \text{(E - 7.22)}$$

If rd differs from zero, the value for rd has to be calculated if the substrate concentration profile is wanted. For this, Equation 7.21 is solved for rd with S(rd) = 0. By definition the substrate concentration at r = rd is zero, yielding:

$$S_b - \frac{1}{6}\frac{k_0}{D_e}(Rp^2 - rd^2) - \frac{1}{3}\frac{k_0}{D_e}\left(\frac{rd^3}{Rp} - rd^2\right) = 0 \qquad \text{(E - 7.23)}$$

Equation 7.23 can be solved for rd as a function of Rp, D_e and k_0 by means of the standard solution technique for cubic equations. The result is as follows:

$$rd = \left(\cos\left(\frac{\alpha + 4\pi}{3}\right) + 0.5\right) Rp$$

$$\alpha = \text{invcos}\left(\frac{12\, S_b\, D_e}{Rp^2\, k_0} - 1\right) \qquad \text{(E - 7.24)}$$

Equation 7.23 describes the substrate concentration profile under conditions of substrate depletion. The situation where substrate depletion just occurs is characterised by the fact that rd = 0 and , by definition, S(0) = 0. In other words if the substrate concentration in the centre just starts to deviate from zero substrate, depletion occurs. If rd = 0 is substituted in Equation 7.24 it follows that for:

$$\frac{Rp^2\, k_0}{S_b\, D_e} = 6 \qquad \text{(E - 7.25)}$$

substrate depletion starts to play a role.

You may not have been familiar with all of the mathematical manipulations done in this section so let us make sure you have understood what the results of these manipulation have told us.

First we can determine the substrate concentrate profile within the biocatalyst.

Under conditions in which substrate is limiting

$$S(r) = S_b - \frac{1}{6}\frac{k_0}{D_e}(Rp^2 - rd^2) - \frac{1}{3}\frac{k_0}{D_e}\left(\frac{rd^3}{Rp} - rd^2\right) \qquad \text{(E - 7.21)}$$

If substrate is not limiting, then

$$S(r) = S_b - \frac{1}{6}\frac{k_0}{D_e}(Rp^2 - r^2) \qquad \text{(E - 7.22)}$$

Secondly, we have learnt that if the value of $\frac{Rp^2\ k_0}{S_b\ D_e}$ is greater than 6, then substrate limitation begins to occur.

Π You may find it useful to construct a summary sheet of the key relationships established in Sections 7.2 and 7.3. You could use the following format.

Zeroth order reaction

Slab geometry, substrate concentration profile, substrate limitation

$$S(r) = S_b - 0.5\ \frac{k_0}{D_e}\ (Rp^2 - r^2) + \frac{k_0}{D_e}\ (rd\ .\ Rp - rd.r)$$

Slab geometry, substrate concentration profile, substrate limitation.

S(r) =

Value of $\frac{Rp^2\ k_0}{2D_e\ S_b}$ when substrate becomes limiting in a slab = 1

Spherical geometry, substrate concentration profile, substrate limitation

S(r) =

Spherical geometry, substrate concentration profile, no substrate limitation.

S(r) =

Value of $\frac{Rp^2\ k_0}{2D_e\ S_b}$ when substrate just becomes limiting in a sphere.

In the following sections we will be able to add further relationships as we examine systems in which the reaction does not follow zeroth order kinetics.

SAQ 7.7

Plot a generalised substrate concentration profile for a substrate in a spherical immobilised system when:

1) $Rp^2\ k_0/S_bD_e < 6$.

2) $Rp^2\ k_0/S_bD_e = 6$.

3) $Rp^2\ k_0/S_bD_e > 6$.

7.3.3 Effectiveness factor

As we described in Section 7.2, for zeroth order reactions, when substrate penetrates the whole biocatalyst then the effectiveness factor (η) has a value of 1. In situations in which substrate limitation occurs, η is less than 1.

The effectiveness factor, (the ratio of the actual conversion rate and the maximal possible conversion rate with no substrate depletion) can be calculated from the volume of the sphere where substrate is present.

Thus:

$$\eta = \frac{\text{volume of the sphere where substrate is present}}{\text{volume of the whole sphere}}$$

$$\eta = \frac{(4/3\,\pi\,Rp^3 - 4/3\,\pi\,rd^3)}{4/3\,\pi\,Rp^3} = 1 - \left(\frac{rd}{Rp}\right)^3 \qquad \text{(E - 7.26)}$$

The effectiveness factor is one for $Rp^2k_0/S_b\mathbf{D}_e$ = < 6 (rd = 0) and smaller than one for $Rp^2k_0/S_b\mathbf{D}_e > 6$ (rd > 0).

Π Plot a graph of η against $\frac{Rp^2\,k_0}{S_b\,\mathbf{D}_e}$ using the following axis.

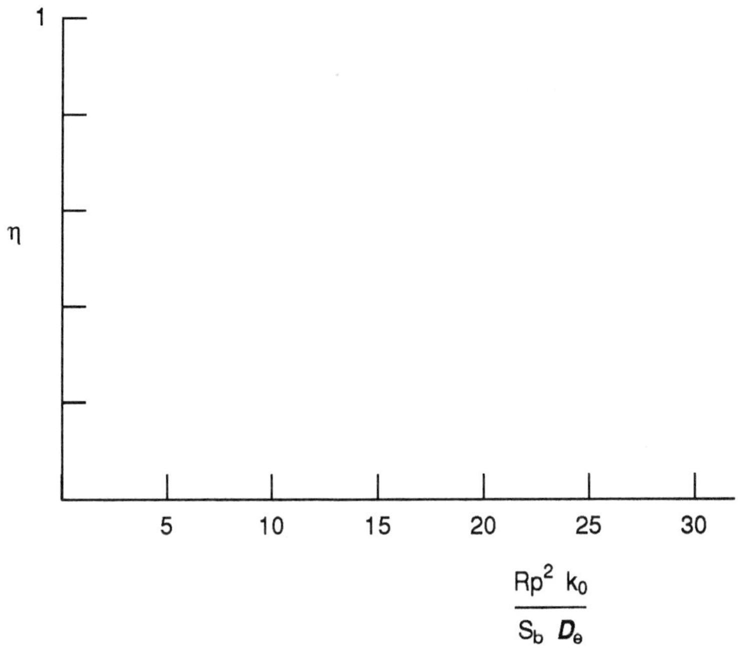

To calculate η you need to know rd. This can be determined from Equation 7.24. In fact your graph should look like this. The value of η = 1 from $\frac{Rp^2\,k_0}{S_b\,\mathbf{D}_e} = 0$ to $\frac{Rp^2\,k_0}{S_b\,\mathbf{D}_e} = 6$. As $\frac{Rp^2\,k_0}{S_b\,\mathbf{D}_e}$ increases above 6, then η declines.

At $\frac{Rp^2 k_0}{S_b D_e}$ = 10, η is approximately 0.80, at $\frac{Rp^2 k_0}{S_b D_e}$ = 20 η is approximately 0.65 and at $\frac{Rp^2 k_0}{S_b D_e}$ = 30, η is approximately 0.52.

If you did not calculate these values, plot them on the graph provided.

SAQ 7.8

A biocatalyst is produced in the form of small spherical pellets with a radius of 4mm. Glucose is consumed by the biocatalyst according to zeroth order kinetics with a zeroth order reaction rate constant of 2 x 10^{-3} mol s^{-1} m^{-3}. The effective diffusion coefficient of glucose in the system is 1.2 x 10^{-9} m^2 s^{-1}. Calculate the concentration of glucose that would be required to ensure that glucose limitation did not occur on this system.

SAQ 7.9

It is proposed to treat an industrial effluent using an immobilised biocatalyst in the form of spherical beads. The effluent contains 1 mol m^{-3} of the pollutant that is to be removed by the biocatalyst.

Trials using a small scale model indicate that 1 cm^3 of biocatalyst will degrade 0.01 mmol pollutant min^{-1} and exhibits zeroth order kinetics with respect to pollutant removal. The effective diffusion coefficient of the pollutant in the immobilised system is 0.8 x 10^{-9} m^2 s^{-1}.

Using this data, determine the maximum size the beads should be made to avoid using more catalyst than is necessary.

7.4 Modelling of immobilised biocatalyst, spherical and slab geometry with first order reaction rate

In section 7.2 and 7.3 the heterogeneous catalysis of spherical and slab geometry with a zeroth order reaction rate was discussed. Due to the specific property of a zeroth order reaction (a maximal reaction rate if substrate is present) the substrate profile and the effectiveness factor could be calculated relatively easy. The most important aspects of this approach is to calculate the position where substrate depletion occurs (also called the penetration depth), as beyond that location no reaction takes place. However, when the reaction rate depends on the substrate concentration, theoretically the substrate never gets depleted because the reaction rate decreases as the substrate concentration decreases. In an immobilised biocatalyst this means that there is no region where no reaction occurs. This also implies that the effectiveness factor never equals exactly one. In the next part, first order reaction rate coupled with diffusion in spherical and flat plate geometry is discussed. Because this involves some quite complex mathematical manipulation, we have adopted a dual approach. In the main text we provide a straight forward account of the relationships describing the concentration profiles and the effectiveness factor. The mathematical derivation of these relationships is given in the form of separate panels. In this ways the reader who has difficulties with the mathematical concepts can, nevertheless, relate to the conclusions drawn from the various relationships described in the main text.

7.4.1 Substrate concentration for a slab

The derivation of the balance equation is exactly the same as for the zeroth order reaction except that the reaction rate term depends on the concentration according to the first order reaction constant k_1.

Thus consider our slab (see Figure 7.1), we can write the following balance:

flow of substrate in	-	flow of substrate out	-	amount consumed in reaction	= 0

This we could write as an approximation as:

$$\left[A\, D_e \frac{dS}{dr} \right]_{r+dr} - \left[A\, .\, D_e \frac{dS}{dr} \right]_r - A\, dr\, k_1\, Sr = 0$$

Note A. dr. k_1 S_r is only an approximation of the amount of substrate consumed in reaction in the volume of the slab described by A.dr since the substrate concentration will not be constant throughout this slab. S will be dependant upon r. A more specific solution to this is provided in Panel 7.2.

From using such a balance equation, it can be shown that the substrate concentration as a function of location is given by:

$$S(r) = S_b - \frac{\exp(rK) + \exp(-rK)}{\exp(RpK) + \exp(-RpK)} \qquad \text{(E - 7.27)}$$

where $K = \sqrt{(k_1/D_e)}$

Equation 7.27 can be written in terms of hyperbolic functions. Mathematically, Equation 7.27 can also be written as:

$$S(r) = S_b \frac{\cosh(\sqrt{(k_1/D_e)}\, .\, r)}{\cosh(\sqrt{(k_1/D_e)}\, .\, Rp)} \qquad \text{(E - 7.28)}$$

We will perhaps need to provide a little bit of an explanation here.

Exponential functions (for example e^x and e^{-x}) are related to, so called, hyperbolic functions. These hyperbolic functions are cosh, sinh and tanh. The more important of these are cosh and sinh. These are related in the following way:

$$\cosh x = \frac{1}{2}(e^x + e^{-x})$$

and $\sinh x = \frac{1}{2}(e^x - e^{-x})$ and $\tanh x = \frac{\sinh x}{\cosh x}$

Π Let us do a calculation. What is cosh x when x = 1? (e = 2.7183). Try it before reading on.

$e^x = 2.7183$ when $x = 1$

$e^{-x} = 0.3679$ when $x = 1$

Thus $\cosh = \frac{1}{2}(2.7183 + 0.3679) = 1.5431$.

Π What is the value of sinh x when x = 1?

Since $\sinh = \frac{1}{2}(e^x - e^{-x})$ then $\sinh = 1.1752$.

Although the values of cosh x, sinh x, tanh x can be calculated from first principles, tables of data are available and many modern calculators allow calculations of these values quite simply. In fact if you have a programmable calculator you may well be able to program in the calculation of these hyperbolic functions.

Equations 7.27 and 7.28 describe the substrate concentration at position r. Clearly the ratio of k_1/D_e as very important. Essentially what we can conclude is that the substrate concentration becomes very low near the solid support of the slab (at r = 0) if the value of $\sqrt{(k_1/D_e)}$ is high: S(r) however never becomes zero because of the first order reaction rate. Graphically we can represent this as shown in Figure 7.7.

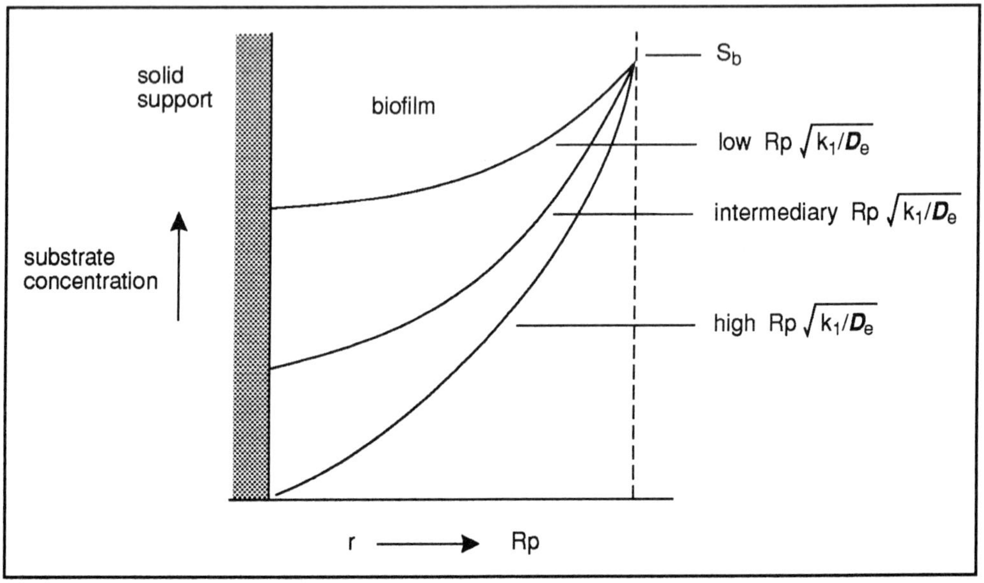

Figure 7.7 Substrate concentration profiles in a slab as a function of; Rp $\sqrt{(k_1/D_e)}$. Conditions include the boundary condition that the substrate concentration at Rp = the substrate concentration in the bulk liquid (S_b) and that the substrate is consumed by a first order reaction.

If you think carefully about Figure 7.7 you will see that in a general way the relationships are predictable.

Even if you found the mathematics of this analysis difficult, the conclusions are quite straight forward. Equation 7.27 and 7.28 enables us to calculate the substrate concentration at any point (r) within the biofilm, providing we know k_1, D_e, Rp and the concentration of the substrate within the bulk liquid (S_b).

Thus we would expect the thicker the film (large Rp), the lower the concentration of substrate would be. We would also anticipate the higher the reaction rate (k_1 = consumption of substrate), the lower the substrate concentration would be. Likewise the higher the diffusion rate (D_e), the higher the substrate concentration would be. It is, however, only by conducting a mathematical analysis the exact relationships between S(r) and these three parameters described in Equation 7.27 and 7.28 can be established.

Panel 7.2

Derivation of the substrate concentration profile for a flat biofilm in which substrate is consumed by a first order reaction.

The derivation of the balance equation is exactly the same as for the zeroth order reaction. The only difference is the reaction rate term which depends on the concentration according to the first order reaction constant k_1. The differential equation describing the substrate concentration profile is as follows:

$$\frac{d^2S}{dr^2} = \frac{k_1}{D_e} \cdot S \qquad \text{(E - P1)}$$

This second order differential equation can be solved by means of solution of the characteristic equation: $\lambda^2 - k_1/D_e = 0$, yielding $\lambda_{1,2} = +/- \sqrt{(k_1/D_e)}$. The standard solution is as follows:

$S(r) = IC1 . \exp(\lambda_1 . r) + IC2 . \exp(\lambda_2 . r_2)$

Introducing $K = \sqrt{(k_1/D_e)}$ yields:

$$S(r) = IC1 . \exp(Kr) + IC2 . \exp(-Kr) \qquad \text{(E - P2)}$$

The integration constants can be calculated if the boundary conditions are given. The two boundary conditions, at the centre and at the surface of the particle, are:

$S = S_b$ at $r = Rp$; no mass transport limitation at the surface, the substrate concentration at the surface equals the bulk substrate concentration.

$\frac{dS}{dr} = 0$ at $r = 0$; no mass transport beyond the centre.

Substitution of the first boundary condition in Equation P2 gives:

$$S_b = IC1 . \exp(RpK) + IC2 . \exp(-RpK) \qquad \text{(E - P3)}$$

The derivative of Equation P2 is:

$$dS/dr = IC1 . K . \exp(rK) - IC2 . K . \exp(-rK) \qquad \text{(E - P4)}$$

Substitution of the second boundary condition in Equation P4 results in IC1 = IC2 or:

$$dS/dr = IC1 \cdot K \cdot (\exp(rK) - \exp(-rK)) \qquad \text{(E - P5)}$$

Equation P3 now can be rewritten as:

$$S_b = IC1 \cdot (\exp(RpK) + \exp(-RpK)) \qquad \text{(E - P6)}$$

Resulting in:

$$IC1 = S_b/(\exp(RpK) + \exp(-RpK)) \qquad \text{(E - P7)}$$

Because of the fact that IC1 = IC2 Equation P2, the substrate concentration profile, can be rewritten as:

$$S(r) = IC1 \cdot (\exp(rK) + \exp(-rK)) \qquad \text{(E - P8)}$$

Substitution of IC1 from Equation P7 gives the substrate concentration as a function of the location:

$$S(r) = S_b \cdot \frac{\exp(rK) + \exp(-rK)}{\exp(RpK) + \exp(-RpK)} \qquad \text{(E - P9)}$$

In which $K = \sqrt{(k_1/D_e)}$.

7.4.2 The effectiveness factor for a slab

As discussed previously the effectiveness factor can be calculated from the flux of substrate at the surface of the slab and the reaction rate at bulk conditions. The first can be calculated from Equation P5 (see Panel 7.2) the latter now is a function of the bulk substrate concentration due to the first order kinetics. The expression for η is:

$$\eta = \frac{\text{flux of substrate at the surface}}{\text{reaction rate at bulk conditions}}$$

$$\eta = \frac{A_s \cdot D_e \cdot \left.\frac{dS}{dr}\right|_{Rp}}{V_s \cdot k_1 \cdot S_b} \qquad \text{(E - 7.29)}$$

dS/dr at r = Rp can be found by substitution of IC1 in Equation P5 and substitution of r = Rp yielding:

$$\left.\frac{dS}{dr}\right|_{Rp} = S_b \cdot \frac{K \cdot (\exp(RpK) - \exp(-RpK))}{\exp(RpK) + \exp(-RpK)} \qquad \text{(E - 7.30)}$$

Making use of the hyperbolic functions results in:

$$\left.\frac{dS}{dr}\right|_{Rp} = S_b \cdot K \cdot \frac{\sinh(RpK)}{\cosh(RpK)} = S_b \cdot K \cdot \tanh(RpK) \qquad \text{(E - 7.31)}$$

Substitution of Equation 7.31 in Equation 7.29 with $A_s/V_s = 1/Rp$ gives for the effectiveness factor:

$$\eta = \frac{\tanh (Rp . \sqrt{(k_1/D_e)})}{Rp . \sqrt{(k_1/D_e)}} \qquad \text{(E - 7.32)}$$

This is a key relationship for it relates η to Rp, k_1 and D_e.

Thus the equations to be able to use are 7.27 and 7.28 which relate substrate concentration to Rp, k_1 S_b D_e and 7.32 which relates effectiveness η to Rp, k_1, D_e. A plot of η against $1/Rp \sqrt{(k_1/D_e)}$ is provided in Figure 7.8.

Let us now work out the effectiveness of a biofilm from some data by attempting SAQ 7.10.

SAQ 7.10

1) A biofilm is 2mm thick and it consumes oxygen according to first order kinetics in which the rate constant = 6 min^{-1} and in which the effective diffusion coefficient of oxygen is 1×10^{-9} m^2 s^{-1}. Calculate the effectiveness factor (η) of this biofilm. (If you do not have access to hyperbolic function tables, leave your answer in terms of a hyperbolic function).

2) If the rate constant (k_1) for oxygen consumption was decreased to 0.06 min^{-1}, what would be the effectiveness factor (η) of the biofilm. (If you do not have access to hyperbolic function tables, leave your answer in terms of a hyperbolic function).

7.4.3 Substrate concentration for a sphere

The approach used is exactly the same as for a biocatalyst in a slab geometry.

Note we will use the same nomenclature that we used in Figure 7.6. It would be a good idea to refresh your memory of this figure before reading on.

flow of substrate into a shell within the sphere	-	flow of substrate out of a shell within the sphere	-	amount consumed by reaction within the shell	= 0

Again the mathematical derivation for the substrate concentration at position r is a little complex but using the derivation described in Panel 7.3 we can establish that the substrate concentration is given by the relationship:

$$S(r) = \frac{S_b . Rp}{r} \; \frac{\exp (Kr) - \exp (- Kr)}{\exp (KRp) - \exp (- KRp)} \qquad \text{(E - 7.34)}$$

or by introducing hyperbolic functions as

$$S(r) = \frac{S_b . Rp}{r} \; \frac{\sinh (Kr)}{\sinh (KRp)} = \frac{S_b . Rp}{r} \; \frac{\sinh (\sqrt{(k_1/D_e)} . r)}{\sinh \sqrt{(k_1/D_e)} . Rp)} \qquad \text{(E - 7.35)}$$

Where $K = \sqrt{(k_1/D_e)}$

The essential message is that again we can predict the concentration profile within the sphere providing we known k_1, D_e, the radius (Rp) of the sphere and the substrate concentration (S_b) in the bulk liquid.

Panel 7.3

Derivation of the substrate concentration profile for a spherical immobilised biocatalyst in which the substrate is consumed by a first order reaction.

The derivation of the balance equation is exactly the same as for the zeroth order reaction, the only difference being the first order reaction rate term. The differential equation describing the substrate concentration profile is as follows:

$$\frac{1}{r^2}\frac{d}{dr}[r^2\frac{dS}{dr}] = \frac{k_1}{D_e} . S$$

Carrying out the differentiation gives:

$$\frac{1}{r^2}[2r\frac{dS}{dr}+r^2\frac{d^2S}{dr^2}] = \frac{k_1}{D_e} . S$$

$$\frac{d^2S}{dr^2}+\frac{2}{r}\frac{dS}{dr}=\frac{k_1S}{D_e} \qquad \text{(E - R1)}$$

With the same boundary conditions as in the previous section concerning slab geometry:

$\frac{dS}{dr} = 0$, r = 0; (ie the centre of the sphere)

$S = S_b$, r = Rp;

The second order differential equation can be solved by introducing a new function s(r) for which S(r) = s(r)/r is valid. Equation R1 can be rewritten as function of s(r). The first and second derivative of S(r) as a function of s(r) are as follows:

$$\frac{dS}{dr}=\frac{1}{r}\frac{ds}{dr}-\frac{s}{r^2}$$

$$\frac{d^2S}{dr^2} = -\frac{1}{r^2}\frac{ds}{dr}+\frac{1}{r}\frac{d^2s}{dr^2}-\frac{1}{r^2}\frac{ds}{dr}+\frac{2s}{r^3} \qquad \text{(E - R2)}$$

Substitution of R2 in R1 yields:

$$\frac{d^2s}{dr^2} = \frac{k_1}{D_e} . s \qquad \text{(E - R3)}$$

Which is an analog of Equation P1 (see Panel 7.2). The solution of R3 is already discussed and can be obtained by means of the characteristic equation:

$$s(r) = IC1 . \exp(Kr) + IC2 . \exp(-Kr) = S(r) . r \qquad \text{(E - R4)}$$

With $K = \sqrt{(k_1 / D_e)}$

The boundary conditions are the same as in the previous case of the flat plate:

$S = S_b$ at r = Rp;

ds/dr = 0 at r = 0;

From the first boundary condition it follows by substitutions in Equation R4.

$$S_b = IC1/Rp . \exp(KRp) + IC2/Rp . \exp(-KRp) \qquad \text{(E - R5)}$$

For the substitution of the second boundary condition the derivative of Equation R4 has to be calculated yielding:

$$\frac{dS}{dr} = \frac{IC1}{r^2} \cdot (-\exp(Kr) + Kr \cdot \exp(Kr) + \exp(-Kr) + Kr(\exp(-Kr))$$

(E - R6)

By substitution in Equation R6 of the second boundary condition it follows that:

$$IC1 = -IC2 \qquad \text{(E - R7)}$$

Consequently Equation R4, describing the substrate concentration profile, changes in:

$$S(r) = IC1/r \cdot (\exp(Kr) - \exp(-Kr)) \qquad \text{(E - R8)}$$

From R5 the integration constant can be calculated by substitution of IC1 = -IC2:

$$IC1 = \frac{S_b \cdot Rp}{\exp(KRp) - \exp(-KRp)} \qquad \text{(E - R9)}$$

The initial condition is known and now the substrate concentration profile can be calculated by substitution of R9 in R8:

$$S(r) = \frac{S_b \cdot Rp}{r} \cdot \frac{\exp(Kr) - \exp(-Kr)}{\exp(KRp) - \exp(-KRp)} \qquad \text{(E - R10)}$$

By introduction of hyperbolic functions Equation R10 can be written as:

$$S(r) = \frac{S_b \cdot Rp}{r} \cdot \frac{\sinh(Kr)}{\sinh(KRp)} = \frac{S_b \cdot Rp}{r} \cdot \frac{\sinh(\sqrt{(k_1/D_e)} \cdot r)}{\sinh(\sqrt{(k_1/D_e)} \cdot Rp)}$$

(E - R11)

SAQ 7.11

Draw generalised figures of substrate concentration profiles for a spherical immobilised biocatalyst with

1) a high $Rp\sqrt{(k_1/D_e)}$

2) a low $Rp\sqrt{(k_1/D_e)}$

7.4.4 The effectiveness factor for a sphere

The mathematical derivation for the effectiveness factor for a sphere is given in Panel 7.4.

The actual relationship is

$$\eta = \frac{3}{\sqrt{(k_1/D_e)}\,.\,Rp} \cdot \left\{ \frac{\cosh(\sqrt{(k_1/D_e)}\,.\,Rp)}{\sinh(\sqrt{(k_1/D_e)}\,Rp)} - \frac{1}{\sqrt{(k_1/D_e)}\,.\,Rp} \right\} \qquad \text{(E - 7.36)}$$

See Panel 7.4 for derivation.

This perhaps looks a little daunting but again the message is quite simple: η is related to k_1, D_e and Rp.

As no region can be pointed out where substrate is depleted, due to the first order reaction rate, the value of η will theoretically always be smaller than one.

Panel 7.4

Derivation of the effectiveness factor for a sphere in which the substrate is consumed by a first order reaction.

The effectiveness factor can be calculated from the flux of substrate at the surface of the slab and the reaction rate at bulk conditions. The first can be calculated from Equation R6 (see Panel 7.3) the latter is a function of the bulk substrate concentration due to the first order kinetics. The expression for η for spheres is (see Equation 7.29):

$$\eta = \frac{3\,.\,D_e\,.\,\left.\frac{dS}{dr}\right|_{Rp}}{Rp\,.\,k_1\,.\,S_b} \qquad \text{(E - S1)}$$

The derivative at the surface can be obtained from Equation R6 and the integration constant as given by Equation R9 (see Panel 7.3).

$$\left.\frac{dS}{dr}\right|_{Rp} = \frac{S_b}{Rp} \cdot \frac{-\exp(KRp) + KRp.\exp(KRp) + \exp(-KRp) + KRp.\exp(-KRp)}{\exp(KRp) - \exp(-KRp)} \qquad \text{(E - S2)}$$

Substitution of S2 in S1 yields the effectiveness factor:

$$\eta = \frac{3}{KRp} \cdot \left\{ \frac{\exp(KRp) + \exp(-KRp)}{\exp(KRp) - \exp(-KRp)} - \frac{1}{KRp} \right\} \qquad \text{(E - S3)}$$

When hyperbolic functions are introduced Equation S3 becomes:

$$\eta = \frac{3}{KRp} \cdot \left\{ \frac{\cosh(K\,.\,Rp)}{\sinh(K\,.\,RP)} - \frac{1}{KRp} \right\};$$

$$\eta = \frac{3}{\sqrt{(k_1/D_e)}\,.\,Rp} \cdot \left\{ \frac{\cosh(\sqrt{(k_1/D_e)}\,.\,Rp)}{\sinh(\sqrt{(k_1/D_e)}\,.\,Rp)} - \frac{1}{\sqrt{(k_1/D_e)}\,.\,Rp} \right\} \qquad \text{(E - S4)}$$

$$\eta = \frac{3}{\sqrt{(k_1/D_e)}\,.\,Rp} \cdot \left\{ \frac{1}{\tanh(\sqrt{(k_1/D_e)}\,.\,Rp)} - \frac{1}{\sqrt{(k_1/D_e)}\,.\,Rp} \right\}$$

Equation S4 describes the effectiveness factor as a function of k_1/D_e.

Figure 7.8 shows the effectiveness factor as a function of Rp $\sqrt{(k_1/D_e)}$ according to Equation 7.36.

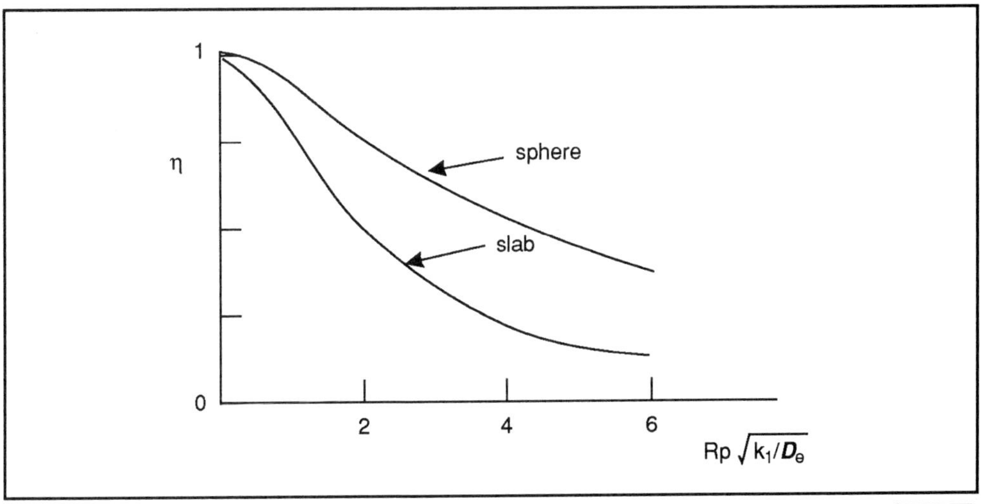

Figure 7.8 Effectiveness factor for a first order reaction in a sphere and a slab as a function of Rp $\sqrt{(k_1/D_e)}$.

7.4.5 Summary

Let us take a break from learning something new and review what we have covered so far in this chapter and re-emphasis its importance.

In the previous sections the substrate concentration profiles and the effectiveness factors was derived for zeroth and first order reactions in flat plate and spherical geometry. For all the cases it was pointed out that the ratio of the reaction rate constant and the effective diffusion coefficient determines the performance of the biocatalysts.

For engineering purposes the effectiveness factor is important because the conversion rate can be calculated if the effectivity of the biocatalyst is known. In Table 7.1 the expressions for the effectiveness factor as derived in the previous section are summarised. We suggested earlier that you made your own summary table of the important equations - here we focus on the effectiveness factor.

Table 7.1 Expressions for the effectiveness factor for zeroth and first order, flat plate and spherical geometry.

Biocatalyst geometry	Reaction order	Effectiveness factor and remarks	
Slab	0th	$\eta_o = \sqrt{(2\, S_b\, D_e)/Rp^2\, k_0} = \dfrac{Rp - rd}{Rp}$	$\eta_o = 1$ providing Rp $\sqrt{(k_0/D_e\, S_b)} =< \sqrt{2}$
Sphere	0th	$\eta_o = 1 - \left(\dfrac{rd}{Rp}\right)^3$	$\eta_o = 1$ providing Rp $\sqrt{(k_0/D_e\, S_b)} =< \sqrt{6}$
Slab	1st	$\eta_1 = \dfrac{\tanh(Rp\sqrt{(k_1/D_e)})}{Rp\sqrt{(k_1/D_e)}}$	
Sphere	1st	$\eta_1 = \dfrac{3}{\sqrt{(k_1/D_e)}\,.\,Rp} \cdot \left\{\dfrac{\cosh((k_1/D_e)\,.\,Rp)}{\sinh((k_1/D_e)\,.\,Rp)} - \dfrac{1}{\sqrt{(k_1/D_e)}\,.\,Rp}\right\}$	

Π Examine Table 7.1 carefully and see if you can spot a common feature in each of the equations.

We hoped you would spot that the expression $\sqrt{(k/D_e)}$ is found throughout the formulas. This is not surprising as we discussed before, the performance of a catalyst depends of the ratio of the reaction constant and the effective diffusion coefficient. A high reaction constant combined with a high effective diffusion coefficient results in the same performance as a low reaction rate constant and low effective diffusion coefficient. It can also be seen from Table 7.1 that the radius or thickness Rp is connected with the term $\sqrt{(k/D_e)}$. It is obvious that the performance of a catalyst also depends on the size (or surface area). At a constant value of $\sqrt{(k/D_e)}$ large particles will be less effectively used than small particles, depending, of course, on the exact values of the parameters. This observation has led to the introduction of a dimensionless quantity known as the *Thiele modulus* ϕ:

7.4.6 Thiele modulus

We anticipate that you have already met with the concept of dimensionless quantities (or numbers). If not we would recommend the BIOTOL text 'Bioprocess Technology: Modelling and Transport Phenomena' where dimensionless numbers are discussed more fully. Essentially a dimensionless numbers is a ratio of two process, each of which has the same units. For example, the ratio of mass transport by convection to mass transfer by diffusion produces a dimensionless number (Reynolds number).

Here we are interested in another ratio, the ratio of the rate of removal of substrate by reaction to the rate of diffusion of the substrate. The Thiele modulus ϕ effectively gives us such a ratio. It is defined as:

$$\phi = Rp\sqrt{(k_n/D_e)} \qquad \text{(E - 7.37)}$$

where n is the zeroth or first order reaction rate.

Π See if you can prove that ϕ is, in fact, dimensionless. (Try this using first order reaction rate before reading on).

The units for Rp = m.

The units for k_n for a first order reaction are (quantity)(quantity)$^{-1}$(time)$^{-1}$ eg kg kg^{-1} s^{-1}.

The units for D_e = m^2 s^{-1}.

Thus $\phi = m\sqrt{(kg\ kg^{-1}\ s^{-1}/m^2\ s^{-1}}$

$= m\sqrt{m^{-2}} = m \, . \, m^{-1}$

The whole point about using the Thiele modulus is that it enables us to produce generalised expressions for the effectiveness factor and the substrate concentration profile more-or-less independent of geometry and kinetics.

Previously we have plotted graphs of the effectiveness factor η against $2D_eS_b/Rp^2k_0$ (Figure 7.5) or $Rp^2\,k_0/S_bD_e$ (in text activity in Section 7.3.3) or $Rp\sqrt{k_1/D_e}$ (Figure 7.8).

Since the effectiveness factor is related to Rp, k_n and D_e for all systems we have studied (see Table 7.1), we can, in principle, express the effectiveness factors as a function of the Thiele modulus. For first order kinetics this is shown in Figure 7.9.

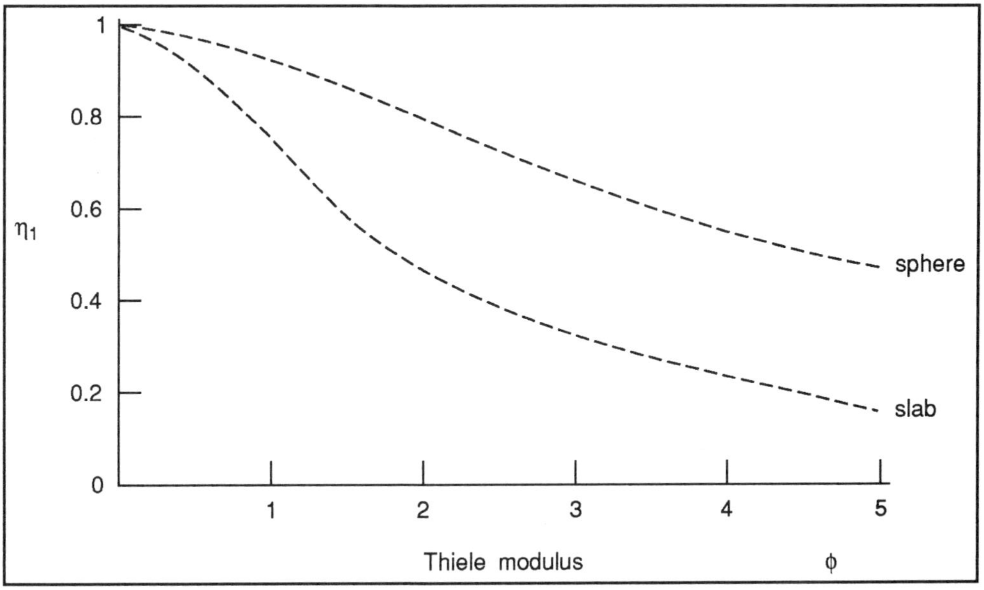

Figure 7.9 The relationship between the effectiveness factor for a slab and sphere in which substrate is consumed by first order kinetics and Thiele modulus. For slabs η_1, is related to $\frac{1}{\phi}$ at high ϕ values, whilst for spheres η_1 is related to $\frac{3}{\phi}$ for high ϕ values (see text below).

Π Examine Figure 7.9 carefully and describe the relationship between η and ϕ.

It can be seen from Figure 7.9 that at low values of ϕ the effectiveness factor is close to unity and at high values the effectiveness factor is approximately inversally proportional to Thiele modulus.

We can rewrite Equations 7.32 (slabs) and 7.36 (spheres) as a function of the Thiele modulus ϕ. Thus:

$$\eta_1 = \frac{\tanh(\phi)}{\phi} \text{ (slab)}$$

$$\text{and } \eta_1 = \frac{3}{\phi} \cdot \left\{ \frac{1}{\tanh(\phi)} - \frac{1}{\phi} \right\} \text{ (sphere)}$$

Mathematically, when $\phi < 0.3 : \phi \approx \tanh(\phi)$ and for $\phi > 2 : \tanh(\phi) = 1$. From this it can be shown that for high values of the Thiele modulus that η is related to $\frac{1}{\phi}$ for slabs and to $\frac{3}{\phi}$ for spheres.

For both slab and sphere geometrics, the value of the effectiveness factor is close to unity at low values of the Thiele modulus. Notice, however, that the lines for η against ϕ for the two geometrics do not coincide at high values of ϕ. Can we in fact produce a formula that can be used for any kind of geometry? The answer is that we can approach this ideal by using a modified Thiele modulus. We will explain this approach in the next section.

7.4.7 A generalised Thiele modulus

We have already explained that the lines relating η to ϕ do not coincide at high values of ϕ (Figure 7.9). We may describe such lines as asymptotic (note an asymptote is a mathematical term which applies to a straight line that approaches a curve so that the perpendicular distance from a moving point to the line approaches zero as the point moves an infinite distance from the origin. Thus lines A and B below are asymptotic).

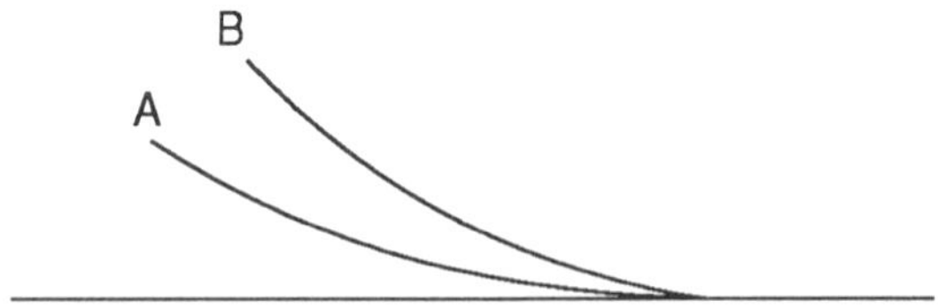

What we are attempting to do here is to find a way in which plots of η against some function produce coincident (symptotic) lines. One approach to this is to redefine the Thiele modulus.

Previously we have defined ϕ as $Rp\sqrt{(k_n/D_e)}$ (see Equation 7.37).

Let us now define ϕ in a different way.

$$\Phi = \frac{Vp}{Ap}\sqrt{(k_n/D_e)} \qquad \text{(E - 7.38)}$$

Note we have given the redefined Thiele modulus a slightly different symbol.

where Vp is the geometric volume

Ap is the external surface area.

If a flat plate (slab) has a thickness of Rp and a sphere has a radius of Rp, what is the value of $\frac{Vp}{Ap}$ for each of these shapes.

For the slab, since Vp = Ap x Rp (Area x Depth)

then $\frac{Vp}{Ap} = Rp$

For the sphere since $Vp = \frac{4}{3}\pi Rp^3$

$Ap = 4\pi Rp^2$

then $\frac{Vp}{Ap} = \frac{Rp}{3}$

Note that $\Phi = \phi$ for a slab and $\Phi = \frac{\phi}{3}$ for a sphere.

The whole point of redefining the Thiele modulus in this way is that plots of η against Φ for different geometrics become more or less coincide.

There is also a simple relationship between η and Φ. At high ϕ values $\eta = \frac{1}{\Phi}$ for any shaped particle.

Π Can you write down a brief explanation why $\eta = \frac{1}{\Phi}$ at high ϕ values for any shaped particle? Look at the definition of the Thiele modulus given in Equation 7.38, it may help you answer this question.

The way you may like to think about this is to think about a series of biofilms or spheres of increasing thickness or radius. Where does the reaction take place? We have drawn a series in which we have shaded the region in which substrate is available for reaction.

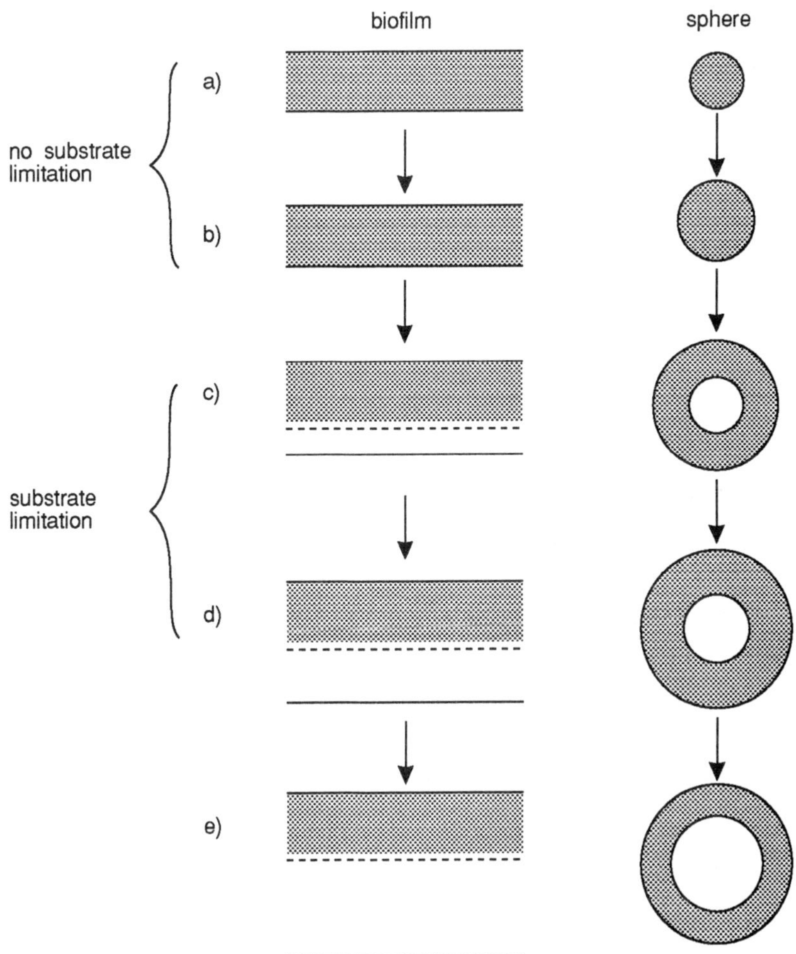

Can you see that at low thickness or radius, all of the biocatalyst is being used and η approaches 1 (see situations a) and b)). As the thickness increases, the rates of reaction at a site removed from the particle/bulk liquid interface is either stopped (because of substrate exhaustion (in the case of zero order kinetics) or greatly reduced (in the case of first order kinetics)). Thus even though we may increase the volume of catalyst from c) $\rightarrow$ e) in the above figure, we do not increase the amount of catalyst which is used. Thus from c) to e) the value of η declines.

For example by doubling the thickness (Rp) of the biofilm from b) to d) the effectiveness factor will fall from 1 to 0.5 (η is inversely proportional to Rp).

Now let us return to our question why is $\eta = \frac{1}{\Phi}$ at large values of Φ.

Φ we defined as $\frac{\text{geometric volume}}{\text{external surface}} \cdot \sqrt{(k_n/D_e)}$

Π Large Φ values can be achieved in one of 3 ways, list them.

Effectively large ϕ values can be obtained by:

- having large particles (ie $\frac{Vp}{Ap}$ is high);
- having high consumption rates (k_n is high);
- slow diffusion rates (D_e is low).

Whatever the reason(s) for a high Φ values, what effectively is achieved is that substrate is only available in a small region of the particle close to the surface.

By defining the Thiele modulus in the way we did in Equation 7.38 the curvature of the surface is unimportant.

Thus the key point to remember is that at very low Φ values η is close to unity and that at high ϕ values $\eta = \frac{1}{\Phi}$.

You should, however, be aware of two major points. Firstly there is a transitional area from $\eta = 1$ to $\eta = \frac{1}{\Phi}$ where neither of these relationships particularly hold and that the extent of this transitional area depends both on the geometry of the biocatalyst and on the kinetics of the reaction taking place within the biocatalytic matrix. We have shown this in a stylistic way in Figure 7.10.

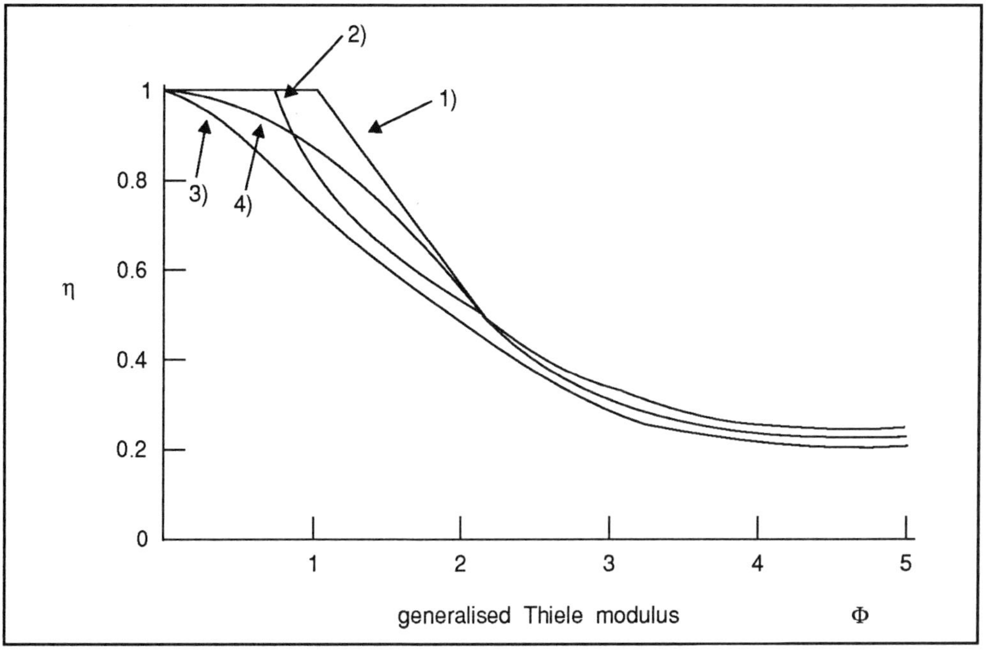

Figure 7.10 Stylised effectiveness factor as a function of Thiele modulus. 1) slab with zeroth order reaction; 2) sphere with zeroth order reaction; 3) sphere with first order reaction; 4) slab with first order reaction.

Note that at values of Φ between about 0.3 and 2, there is some divergence of the relationship between η and Φ, depending upon the geometry of the biocatalyst and the order of the reaction.

You should conclude that for Φ values below about 0.3, $\eta = 1$ and for values of ϕ above 2, then $\eta = \frac{1}{\Phi}$.

Note, other definitions of the Thiele modulus have been produced in the literature in an attempt to produce better generalised relationships. For example one form of generalised Thiele modulus is given as:

$$\Phi = \frac{Vp}{Ap}\sqrt{\left(\frac{n+1}{2}\,\frac{k_n\,S_b^{(n-1)}}{D_e}\right)}$$

where n = the order of the reaction kinetics.

Despite such elaborations, the relationships established between η and ϕ show differences in the region of $\Phi = 0.3 \rightarrow 2$. We will not consider these elaborations further in this text.

Let us now apply what we have learnt about the Thiele modulus and its relationship to the effectiveness factor by attempting SAQ 7.12.

SAQ 7.12

1) A biocatalyst is produced in the form of spherical beads. These beads have been known to use oxygen by zeroth order kinetics with a reaction rate constant of 3.6 kg^{-1} h^{-1}. If the radius of the beads is 6 mm and the effective diffusion coefficient for oxygen in the beads is 1×10^{-9} m s^{-1}, will the availability of oxygen become a limiting factor in this system?

2) The system for producing the beads described in 1) has been modified to produce small strips. These strips each have a volume of 6×10^{-9} m^3 and a surface area of 3×10^{-6} m^3. Will the change in shape of the biocatalyst increase or decrease the effectiveness factor of the biocatalyst?

3) The strips produced in 2) are made to different dimensions such that their volume is now 1×10^{-9} m^3 and their surface area is 6×10^{-6} m^2. Will these changes increase or decrease the effectiveness factor of the biocatalyst?

By attempting SAQ 7.12, you will see that the Thiele modulus enables us to predict the effectiveness factor of a system using a few, simple parameters. This is, of course, important in biotechnological processes. Often, the production of the biocatalyst (cells or especially, isolated and purified enzymes) is expensive. It is, therefore, imperative that we make maximum use of the catalyst (ie that is, operate with an effectiveness factor close to unity).

We have seen that the effectiveness factor can be related to the Thiele modulus and that this can be calculated from the $\frac{\text{volume of the biocatalyst}}{\text{surface area of the biocatalyst}}$, the reaction rate constant (k) and the effective diffusion coefficient (D_e). In most cases there is little we can do to increase D_e to increase η. The most obvious solution to increasing the use of the catalyst is to increase S_b. In systems displaying zeroth order kinetics, increasing S_b may increase η_o by reducing (preventing) substrate limitation. In systems displaying first order kinetics, increasing S_b does not increase η_1. However $\frac{ds}{dr}\Big|_{Rp}$ is increased and thus the substrate conversion rate is increased. There are two other options. One is to decrease k (that is reduce the concentration of biocatalyst in the matrix or reduce the efficiency). Thus we could for example reduce the concentration of enzyme we use in the immobilisation matrix such that k is lowered, η is kept close to unity and thus all of the enzyme is used.

The second option is to change the ratio of the volume of the biocatalyst to the surface area of the biocatalyst. The smaller the biocatalyst beads or the thinner the biofilm, the lower the volume to surface ratio. The smaller the volume:surface ratio, the lower the Thiele modulus value. The lower the Thiele modulus value the higher the effectiveness factor.

Let us put this into practice by helping to design an immobilised enzyme reactor by attempting SAQ 7.13.

SAQ 7.13

The enzyme we need to use is produced in purified form but it expensive. The design process is to entrap the enzyme in spherical gel beads. In the first trial experiments beads were produced which had a radius of 6 mm. Entrapped within the beads was 1g enzyme kg^{-1} of bead. This could catalyse the conversion of 7.2 kg of substrate kg^{-1} of beads h^{-1}. The substrate has an effective diffusion coefficient of 0.5×10^{-9} m^2 s^{-1}. (Assume zero order kinetics for the enzyme catalysed reaction and that the zeroth order rate constant is directly proportional to the concentration of the enzyme in the beads).

1) Determine the effectiveness factor for this system and determine how much of the enzyme would be used in this system.

2) Suggest how the concentration of the enzyme in the beads may be altered so that all of the enzyme is used for catalysing the reaction.

3) Using the same enzyme concentration in the beads as described at the beginning of the SAQ, what would the size of the beads need to be to ensure that all of the enzyme present in the beads was involved in catalysis?

7.5 Effects of external mass transport

In the previous section we have made an assumption that the substrate concentration at the interface between liquid and biocatalyst is the same as the concentration of the substrate in the bulk liquid. In practice, this is not always so. Often a stagnant film, separates the biocatalyst from the bulk liquid. This stagnant layer may act as a barrier liquid into the biocatalyst. It is, therefore, important that we consider this situation in a little more depth.

We have drawn a stylised substrate concentration profile for an immobilised (agglomerated) biocatalyst which is separated from the bulk liquid by a stagnant (unmixed) layer (see Figure 7.11). It can be seen from this figure that the substrate concentration is lower at the interface than in the bulk. We do not intend to explore film theory in depth here except to say that film theory predicts that there is a linear increase in the substrate concentration from the interface to the bulk liquid. Film theory also describes the transport at the interface as:

$$D_e \left.\frac{dS}{dr}\right|_{Rp} = k_L (S_b - S_i) \qquad \text{(E - 7.39)}$$

where k_L is the external mass transfer coefficient.

A detailed discussion of film theory is provided in the BIOTOL text 'Bioprocess Technology: Modelling and Transport Phenomena'.

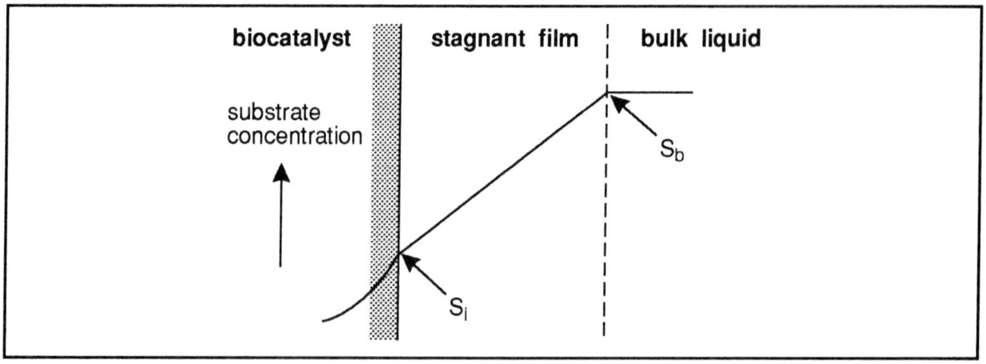

Figure 7.11 Substrate concentration profile at a biocatalyst interface in which the external mass passes through a stagnant film by diffusion.

Effectively what the stagnant film does is to place a barrier around the biocatalyst, thereby lowering the substrate concentration within the biocatalyst. This in turn may reduce the effectiveness factor of the catalyst. Thus it is worthwhile to further evaluate this situation.

For further evaluation three regions around the biocatalyst can be considered.

- from the matrix of the biocatalyst to the bulk solution;
- from the matrix of the biocatalyst to the interface;
- from the interface to the bulk solution.

For all these regions effectiveness factors can be defined. The first is called the overall effectiveness factor (η_{ov}) which describes the total performance of the biocatalyst with respect to the bulk substrate concentration. The second is called the internal effectiveness factor (η_i) which describes the performance of purely the biocatalyst with respect to the interface substrate concentration. The third describes the performance of the biocatalyst as if it is assumed that the whole biocatalyst is subject to the substrate interface concentration compared to the assumption that the whole biocatalyst is subject to the bulk substrate concentration and is called the external effectiveness factor (η_e). The equations for the discussed effectiveness factors now are as follows:

The overall effectiveness factor:

$$\eta_{ov} = \frac{As \,.\, D_e \,.\, \left.\frac{dS}{dr}\right|_{Rp}}{Rp \,.\, v(S_b)} \qquad \text{(E - 7.40)}$$

The internal effectiveness factor:

$$\eta_i = \frac{As \,.\, D_e \,.\, \left.\frac{dS}{dr}\right|_{Rp}}{Rp \,.\, v(S_i)} \qquad \text{(E - 7.41)}$$

The external effectiveness factor:

$$\eta_e = \frac{v\ (S_i)}{v\ (S_b)} \qquad \text{(E - 7.42)}$$

The effectiveness factors are related through:

$$\eta_{ov} = \eta_i \cdot \eta_e \qquad \text{(E - 7.43)}$$

Π Why is the actual reaction rate, as defined by the flux at the interface, different from the reaction rate as if the whole biocatalyst was subject to the interface substrate concentration?

It can be concluded from Equation 7.43 that the overall effectiveness factor is determined by both the internal and the external effectiveness factor. If the internal effectiveness factor is close to one, for example due to a low reaction rate, and the external effectiveness factor is about 0.5, due to a low external mass transfer coefficient, the overall effectiveness factor will be 0.5. In the opposite case due to, for example, a high external mass transfer coefficient and a high reaction rate, the overall effectiveness factor will also be 0.5.

In Section 7.4 we assumed that $\eta_e = 1$ since we assumed that $Si = S_b$ and thus $\eta_{ov} = \eta_i$. In situations in which the value of η_e is not equal to 1 (when Si is considerably smaller than S_b) then we would have to calculate the values of η_i and η_e in order to establish the overall effectiveness of the system. In principle η_i can be calculated from the relationships established in Section 7.4 except, of course, S_b would need to be replaced by S_i. We will not, however, extend the discussion to the mathematical derivation of substrate profiles and effectiveness factors for situations in which mass transfer are the stagnant film significantly alters the effectiveness factor.

However, if you would like to pursue the aspect further we have recommended further reading at the end of this text.

SAQ 7.14

Consider the graph in Figure 7.12 of the effectiveness factors for slabs with zeroth order reaction rate as a function of a generalised Thiele modulus ($\Phi = \frac{Vp}{Ap}\sqrt{(k_0/D_e)}$)).

By answering the following questions, you will learn how to identify situations in which external mass transfer greatly reduces the effectiveness of a biocatalyst.

1) Which of the two systems appears to have an external effectiveness factor of 1?

2) What is the apparent external effectiveness factor of the other system?

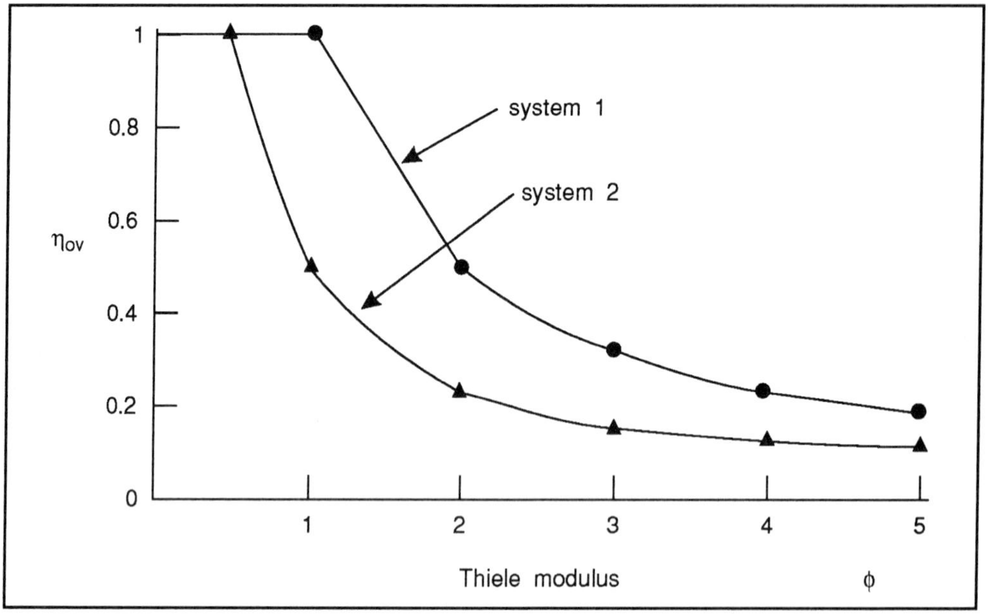

Figure 7.12 See SAQ 7.14 for details.

7.6 Michaelis Menten kinetics and immobilised biocatalysts

7.6.1 Introduction to Michaelis Menten kinetics in heterogenous systems

We have shown in section 7.4 that for zeroth and first order heterogenous biocatalysis we can determine analytical solutions for the overall effectiveness factor. In Chapter 5 we mentioned, however, that biocatalysis usually does not follow simple zeroth or first order kinetics. In fact, we described enzymes as usually following Michaelis Menten kinetics in which the velocity of the reaction was given by:

$$v_{(s)} = \frac{V_{max}[S]}{K_M + [S]}$$

where $v_{(s)}$ = velocity of reaction at a substrate concentration (s)

V_{max} = maximum velocity of the reaction

[S] = substrate concentration

K_M = Michaelis constant.

For biocatalysts consisting of whole cells, we described that the growth of these cells (and their substrate consumption) followed similar kinetics, the so called Monod kinetics.

Thus:

$$\mu = \frac{\mu_{max}[S]}{K_S + [S]}$$

where μ = specific growth rate

μ_{max} = maximum specific growth rate

K_s = substrate saturation constant.

We also describe the kinetics of these processes in the following way. When [S] is low (relative to K_s) then the kinetics are more-or-less first order. At very high values of [S] then the kinetics are zeroth order. At intermediate values a mixture of zeroth and first order kinetics occurs. Thus these types of kinetics are non-linear. This, of course, makes this situation mathematically extremely difficult to handle. We can, however, make some generalisations.

7.6.2 Some generalisation about the application of Michaelis Menten kinetics to heterogenous catalysis.

Although enzyme and cell mediated catalysis follows complex kinetics, we can make some simple approximations. If the substrate concentration throughout the biocatalyst is maintained at a high level compared to K_M (or K_s) then the system behaves with zeroth order reaction kinetics. If on the other hand, the substrate concentration is maintained at a low level compared to K_M (or K_s), then the system behaves with first order reaction kinetics. The key question is, do we meet these conditions in practice?

Let us consider the situation in which the substrate concentration is maintained at a high level compared with K_M throughout the biocatalyst. We can represent this situation by the following diagram.

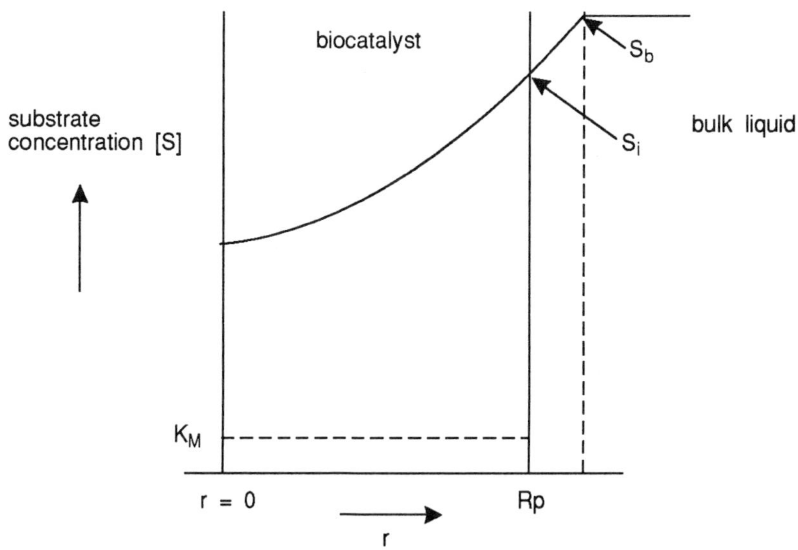

Π Make a list of conditions in which the substrate concentration profile drawn above is most likely to occur.

We would anticipate that you would have suggested that this might occur with:

- high S_b values;
- low consumption rates (k);
- small Rp values (thin films or small particles);
- high relative diffusion rates (D_e);
- low K_M (K_s) values.

Thus if we can provide the biocatalyst with high concentrations of the substrate in the bulk liquid, or we use thin films or the biocatalyst has a high affinity (low K_M or K_s) for the substrate, then these would follow the system operating with zeroth order reaction kinetics.

This situation is most probable with substrates which are relatively cheap and readily water soluble (eg glucose). We can often use, for example, a bulk liquid containing perhaps 10 15 kg m^{-3} glucose and the cells in the biocatalyst layer have a K_s for glucose of about 10 g m^{-3}. This means that for glucose we can have a considerable drop in concentration over the biocatalyst layer but still be operating at a level well above the K_s value (that is, with zeroth order kinetics).

On the other hand with poorly water soluble substrates such as oxygen or with substrates which are only supplied at low concentration (for example expensive precursors) it may prove impossible to maintain [S] well above K_M (K_s) values. This is especially true for substrates, such as oxygen, which may be consumed at high rates but have low solubility. Thus with oxygen we might anticipate there to be a transition from zeroth order kinetics near to Rp, to first order kinetics as we approach r = 0. Diagrammatically, we can represent this as:

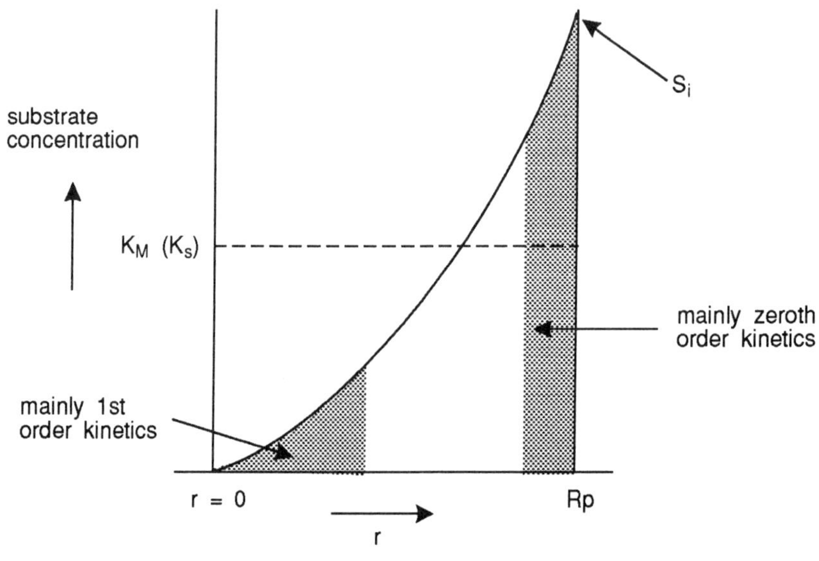

Another possibility is that the substrate in the bulk liquid is at a concentration well below K_M (K_s). For example, this might be a compound in some waste water that we would like to remove using an immobilised biocatalyst. We can represent this situation by:

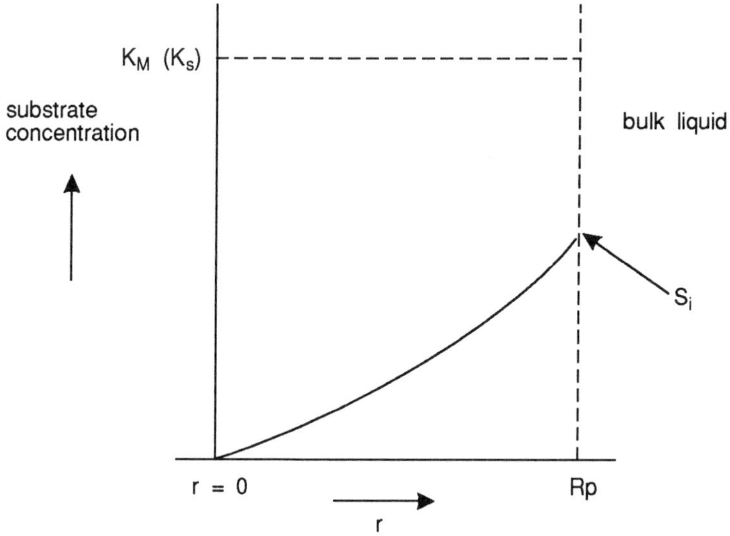

In this case, the kinetics approach first order kinetics throughout the biocatalyst.

Thus we have identified three types of circumstances we can conclude that:

- if S(r) is large compared to K_M (or K_s) through the catalyst then the system approximates to zeroth order kinetics;
- if S(r) is at all times much smaller than K_M (or K_s), then the system approximates to first order kinetics.

Both of these situations we have given analytical solutions for in earlier sections.

It is only in those circumstances where S(r) approaches K_M (or K_s) somewhere in the biocatalyst that we need new solutions. Mathematically this is quite difficult to handle which we will not describe in the body of the text. We have, however, provided this in an appendix. (Appendix 1). In this appendix, we show that the overall effectiveness of spheres in which the reaction kinetics are following Michaelis Menten kinetics can be determined from:

$$\eta_{ov} = \frac{3D_e\, Z|\,Rp}{Rp\; v_{(Sb)}}$$

where $Z\,|\,Rp = \frac{dS}{dr}$ at the interface and $v_{(sb)}$ = velocity of the reaction at the bulk liquid substrate concentration.

SAQ 7.15

Assume we have a slab biocatalyst, for each of the following circumstances, decide whether the biocatalyst will show approximately zeroth order kinetics, or approximately first order kinetics, or neither of these two.

Give reasons for your choice.

1) The biocatalyst consisting of cells consumes the amino acid arginine at a fairly slow rate. Arginine is supplied in bulk solution at 1 kg m^{-3}. The K_s of the cells for arginine is 1 g m^{-3}.

2) A biocatalyst consisting of immobilised cells oxidises a pesticide. The K_s of the cells for the pesticide is 0.1g m^{-3}. The concentration of pesticide in the bulk liquid is 1 part per million. (Assume the bulk liquid has a density of 1000 kg m^{-3}).

3) A culture of a bacterium is shown to grow at ½ μmax when its primary substrate concentration is 4 g m^{-3}. The cells of this culture are immobilised as a thin film and immersed in a solution of the primary substrate. This solution, which is vigorously mixed, contains 1 kg m^{-3} of the substrate.

4) The K_s of a culture for oxygen is 2% of the saturation concentration of oxygen in the aqueous medium (under air at atmospheric pressure). The system maintains the oxygen in the bulk solution at about 40% of the saturation concentration of oxygen.

Summary and objectives

This has been a long chapter, but it has covered some very important aspects of modelling of immobilised biocatalysts. The discussion in this chapter has predominantly focused onto the two most common geometries for immobilised systems - flat layers (slab or biofilm) and spheres (beads). We have also mainly considered the consumption of substrate by zeroth and first order kinetics. Although most biocatalysts consume substrates with Michaelis Menten kinetics we learnt that in many instances, because of operating conditions, the biocatalyst behaviour can be approximated by zeroth order or first order kinetics. It is only under circumstances where substrate concentrations approach K_M or K_s values that these approximations are invalid.

Now that you have completed this chapter, you should be able to:

- calculate substrate concentration profiles in slab and spherical biocatalysts for a wide variety of conditions from supplied data;
- determine if substrate limitation occurs in slab and spherical biocatalysts from supplied data;
- calculate the effectiveness factor (η) from supplied data;
- determine parameter values to ensure maximum effectiveness factors;
- use Thiele modulus: effectiveness factor relationships to predict effectiveness factors from supplied data;
- use Thiele modulus: - effectiveness factor relationships to calculate parameters to achieve design objectives;
- describe the effects of external mass transport on effectiveness factors;
- explain why approximations to zeroth and first order kinetics can be applied to immobilised biocatalysts;
- select circumstances in which zeroth, first order and Michaelis Menten kinetics can or should be applied to immobilised biocatalysts.

Examples of the applications of modelling of immobilised biocatalysts

Examples of the applications of modelling of immobilised biocatalysts

8.1 Introduction

In the previous three chapters, we have explored immobilised biocatalysts. Especially important were the models described in Chapter 7 which enables us to make predictions about substrate concentration profiles and the effectiveness of immobilised biocatalysts in various geometric configurations. Although immobilised biocatalysts have been used for a considerable time (for example in trickle filters in waste water treatment), recent years have seen a considerable increase in interest in the use of immobilised systems. The development of such systems usually passes through a modelling stage and a research and development stage before being employed on a production (commercial) scale.

We might anticipate, therefore, that within the biotechnology industries, immobilised systems are at various stages of development. Some, like the removal of pollutants from water, may still be at the modelling/research stages while others, such as the use of immobilised glucose isomerase to produce high fructose corn syrups, have already reached full scale operation. The orientation of this text is to consider the underpinning principles behind the various modes of operation in which biocatalysts are used. This, of course, includes important aspects of modelling. Thus by heading this chapter 'Examples of the application of modelling of immobilised biocatalysts' we mean the application of the modelling of immobilised systems rather than actual application of immobilised biocatalysts at a commercial level. Naturally, however, the application of the models is to ultimately bring processes to commercial realisation. The application of immobilised biocatalysts at a commercial level is largely the province of the BIOTOL texts which focused onto the application of biotechnology in various business sectors (for example health care, food, agriculture, waste treatment and chemical synthesis - see list of BIOTOL texts at the front of this book). Also important examples of immobilised biocatalysts are to be found in the BIOTOL text 'Technological Applications of Biocatalysts'.

Here we deliberately focus onto the modelling: research interface. We have selected three examples of published projects to exemplify the application of the modelling process. These include the removal of waste organic material from a gas stream, a membrane fixed enzyme reactor and a packed bed reactor designed to use immobilised yeast to produce ethanol.

Thus you should visualise the importance of this and Chapter 7 in the development of knowledge in immobilised biocatalysts technology by the BIOTOL series in the following way.

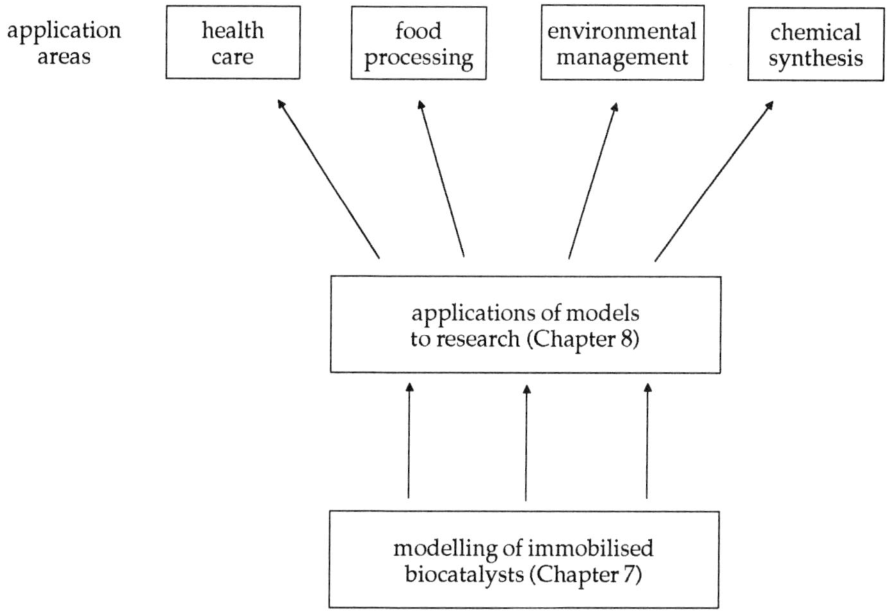

Unlike earlier chapters, this chapter is not designed to develop large amounts of new knowledge. It is, however, designed to give you insight of the range of immobilised systems that are being developed and to show how models can be important in developing new processes. For examples of the processes in operation within industry, we recommend the BIOTOL texts dealing with the specific business sector.

8.2 Example 1: Waste gas cleaning

The following example is taken from a paper by Ottengraft, SPP and Oever, van Dam MC, published in the journal Biotechnology and Bioengineering (Vol 25, 1983 Pp 3089-3102) under the title 'Kinetics of Organic Compound Removal from Waste Gases with a Biological Filter'.

8.2.1 Introduction to the project

The aim was to remove organic compounds from waste gases with a biological filter. In an experimental set up a waste gas stream mainly consisting of toluene was dosed to a packed bed inoculated with peat compost, (see Figure 8.1).

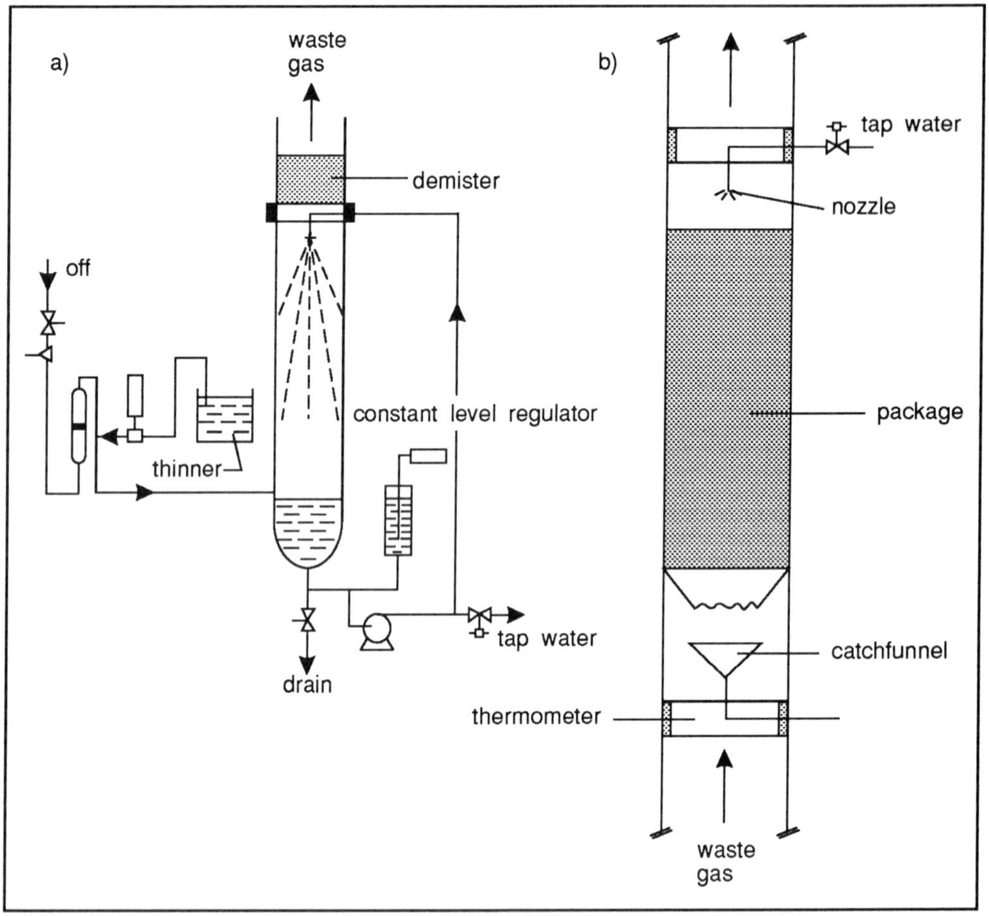

Figure 8.1 Experimental set up. a) General layout. b) Detail of the packed peat bed.

The performance of the column was modelled by modelling the kinetics of the toluene removal by the biofilm developing on the compost surface.

For modelling of the intrinsic kinetics it was assumed that:

- interface resistance in the gas phase can be neglected and the concentration at the interface (ie that is; $S_b = S_i$) may be assumed to be in equilibrium with the concentration of the gas (Figure 8.2);
- in the biolayer only the organic nutrient (ie toluene) is limiting and it is transported by diffusion described by an effective diffusion coefficient D_e;
- the micro kinetics of the substrate elimination reaction in the biolayer are of zeroth order with respect to the toluene concentration;
- the flow of the gas phase through the filter bed is of the plug flow type.

The zeroth order approach was experimentally verified by performing batch experiments in shaking flasks containing samples of the biolayer and toluene as the carbon source.

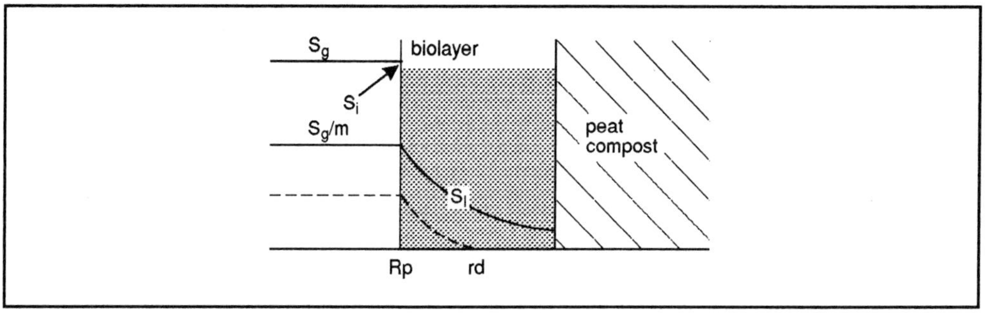

Figure 8.2 The model for the biological waste gas cleaning. S_g = concentration in the gas phase; m = distribution coefficient between gas and liquid phase; S_l = concentration in the biolayer liquid (see text for discussion).

8.2.2 Macrokinetics

In Chapter 7 we established relationships for the concentration of substrate in the biolayer. We remind you of some of these relationships by applying these specifically to the system under consideration.

The differential equation describing the concentration of toluene in the biolayer is:

$$\frac{d^2S}{dr^2} = \frac{k_0}{D_e} \qquad \text{(E - 8.1)}$$

Where k_0 is the zeroth order reaction rate constant. The boundary conditions are as follows:

$$\frac{dS}{dr} = 0, \quad x = rd$$

$$S_l = S_g/m, \quad x = Rp \qquad \text{(E - 8.2)}$$

in which m is the distribution coefficient between the gas and liquid phase with respectively concentrations S_g and S_l.

The solution of Equation 8.1 becomes:

$$S_l = 0.5\frac{k_0}{D_e}(r^2 - Rp^2) - \frac{k_0}{D_e}(rd.r - rd.Rp) + S_g/m \qquad \text{(E - 8.3)}$$

$$rd = Rp - \sqrt{(\frac{2D_e \,.\, S_g/m}{k_0})} \qquad \text{(E - 8.4)}$$

Thus using the modelling approach described in Chapter 7 we can predict the substrate concentration in the biolayer (S_l) and its penetration depth into the biolayer (Rp - rd).

From our previous experience we should be able to identify two different situations. These are:

- there is no diffusion limitation in the biolayer. This means that the biolayer is fully active and the conversion rate is only controlled by the reaction rate (reaction rate limitation);

- diffusion limitation occurs in the biolayer. This means that the biolayer is not fully active, the depth of penetration in the biolayer is smaller than the layer thickness. In this case the conversion rate will be controlled by the rate of diffusion (diffusion limitation).

For the first case rd equals zero, for the second case rd is given by Equation 8.4.

Π Is the ability to predict S_i and rd (and therefore the effectiveness factor η) sufficient in itself? (Think about what we are trying to achieve).

We are trying to remove organic compounds (eg toluene) from waste gases. Thus what we are really interested in is the concentration of the organic compounds in the gas phase. We can imagine that the extent to which these compounds will be removed is not only dependent upon the effectiveness of the biofilm, but also on the gas phase (for example flow rates). In the next section we will take this aspect into consideration.

8.2.3 Modelling the gas concentration in the column

Figure 8.3 schematically represents a part of the column is shown.

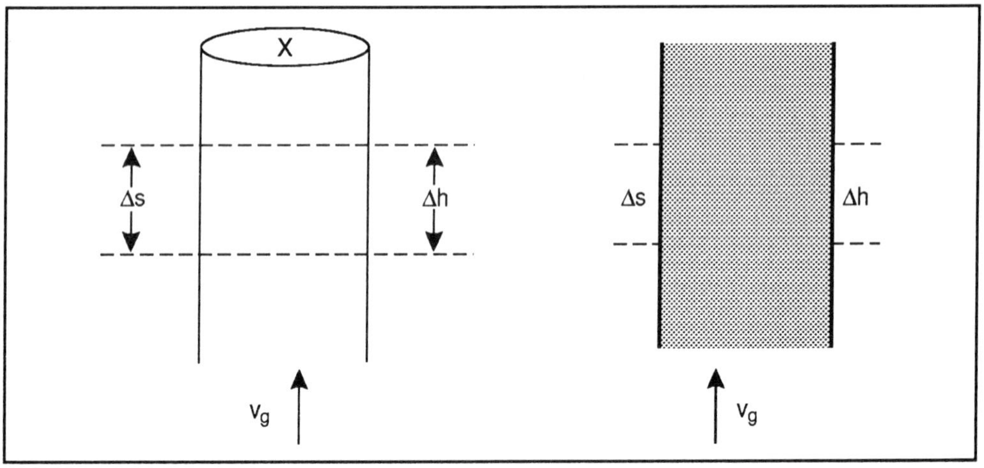

Figure 8.3 Model of the gas phase in the column. where v_g = superficial gas velocity; Δs = change in substrate concentration over change in position on column of Δh.

The difference of the gas flux over height dh is due to the conversion in the biolayer in the volume X.dh in which X is the area of the column:

$$X \,.\, v_g \,.\, dS_{g = X.dh\,.\,R} \qquad \text{(E - 8.5)}$$

v_g is the superficial gas velocity and R the conversion rate. The conversion rate R is equal to the specific surface per column volume (A_s) multiplied by the flux per unit area into the biolayer (N):

$$X.v_g \,.\, dS_g = -X.dh \,.\, N \,.\, A_s \qquad \text{(E - 8.6)}$$

If an infinitely small part of the column is chosen Equation 8.6 yields the following differential equation:

$$\frac{dS}{dh} = \frac{N \cdot A_s}{v_g} \qquad \text{(E - 8.7)}$$

This differential equation can be solved with the following boundary condition:

$S = S_i$, $h = 0$; the concentration at the beginning of the column equals the incoming gas concentration

This results in:

$$\frac{S}{S_i} = 1 - \frac{N \cdot A_s}{S_i \cdot v_g} \cdot h \qquad \text{(E - 8.9)}$$

Equation 8.9 gives the toluene concentration as a function of the column height. For the concentration of the out coming gas stream (S_o) at the top of the column ($h = H$) it follows:

$$\frac{S_o}{S_i} = 1 - \frac{N \cdot A_s}{S_i \cdot v_g} \cdot H \qquad \text{(E - 8.10)}$$

Now let us calculate the flux (N) into the biofilm. Let us assume that the biofilm is working under reaction limitations, rather than substrate limitations.

Π What is the value of effectiveness factor for a reaction limited systems?

Reaction limitation, means that substrate is available at all layers of the biofilm, thus all of the catalyst is being used and therefore, $\eta = 1$.

Π What will be the total rate of consumption of substrate by the biofilm (assume zeroth order kinetics) when the effectiveness factor = 1?

You should conclude that:

Total rate of consumption = volume of biofilm . rate constant of substrate conversion.

Total rate of consumption = $V \times k_0$.

In the column the rate of consumption of the substrate matches its rate of uptake (flux into) the biofilm.

Thus for the whole column, the rate of uptake by the biofilm = $V \times k_0$.

By assuming the thickness of the biofilm is constant we can calculate the rate of uptake per unit area.

Thus $N = \frac{volume}{area} \times k_0 = thickness \times k_0 = Rp\ k_0.$

Thus the flux into the biofilm under reaction limited conditions equals:

$$N = k_0 . Rp \qquad (E - 8.11)$$

Assuming that $A_s . k_0 . Rp$ is a constant (k) it follows that:

$$\frac{S_o}{S_i} = 1 - \frac{H . k}{S_i . v_g} \qquad (E - 8.12)$$

For the degree of conversion $f = (S_i - S_o)/S_o$:

$$f = \frac{H . K}{S_i . U_g} \qquad (E - 8.13)$$

From Equation 8.13 it can be concluded that the conversion is a linear function of the height of the column. In Figure 8.4 the experimentally and theoretically (Equation 8.13) obtained conversion rate of toluene is shown as a function of the height of the column.

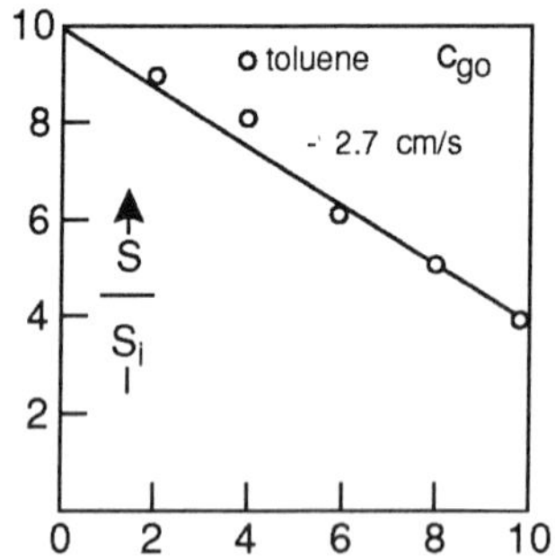

Figure 8.4 Dimensionless toluene concentration (S/S_i) as a function of the dimensionless height of the column (h/H).

Π From the data presented in Figure 8.4, would you conclude that the model adequately represents the removal of toluene from the gas phase?

It can be concluded that Equation 8.13 is valid, the reaction order is according to zeroth order kinetics and the biofilm is fully penetrated so no toluene depletion occurs.

Π From what you know about the kinetics of biocatalysed processes, do you think that Equation 8.13 will be valid at all toluene concentrations?

You should have concluded that at very low concentrations the consumption of toluene is likely its approach first order kinetics (remember the Michaelis Menten kinetics). Thus when the concentration is close to the K_M (or K_s) value, then a mixture of zeroth and first order kinetics will be found.

The elimination capacity of the filter bed will not be constant below a critical value of the gas phase toluene concentration. This phenomenon is shown in Figure 8.5 in which the elimination capacity for toluene is plotted as a function of the influent gas concentration.

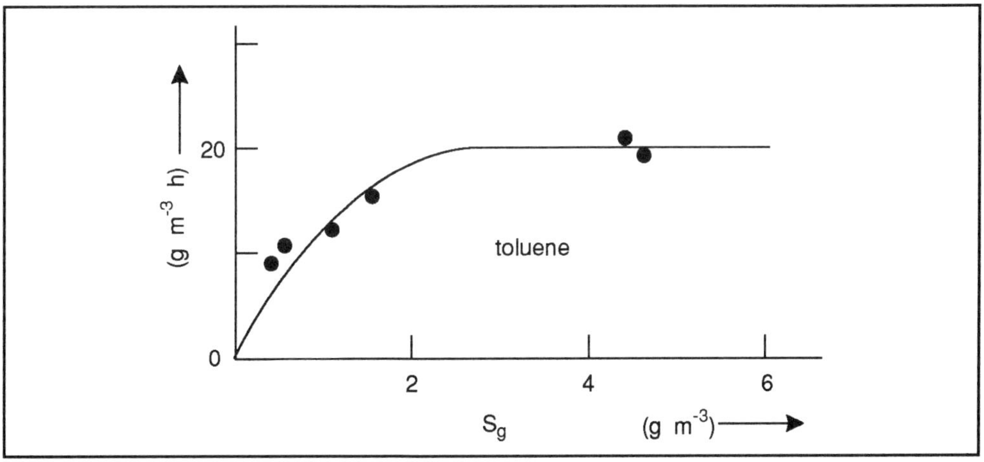

Figure 8.5 Elimination capacity as a function of the inlet gas concentration.

Under these circumstances the conversion rate is diffusion limited (rd > 0). The flux into the biofilm under diffusion limited conditions approximate to:

$$N = k_0 \cdot (Rp - rd) \tag{E - 8.14}$$

(This assumes zeroth order kinetics at all substrate concentrations - experimentally checked).

Diffusion limitation occurs if $Rp^2 \dfrac{k_0\, m}{D_e\, S_g} >= 2$.

SAQ 8.1

1) Calculate the critical penetration depth (that is when rd = 0) of toluene in the column when the incoming gas concentration = 1 g m^{-3}. Use the data presented in Figure 8.5. Additional data are $D_e = 0.85 \times 10^{-9}\ m^2\ s^{-1}$ and m = 0.27. The reaction rate per unit volume of biolayer = 20 g $m^{-3}\ h^{-1}$.

2) Explain why the value you have calculated may not properly reflect the real situation.

3) Estimate the actual thickness of the biofilm used for the experiments to obtain the data in Figure 8.5.

Let us explore the consequences of substrate limitation on the composition of the gas leaving the column.

Substitution of Equation 8.4 in differential Equation 8.7 describing the concentration dependence of height yields:

$$\frac{dS}{dh} = - \sqrt{\left(\frac{2De \cdot S}{m \cdot k_0}\right)} \cdot \frac{k_0 \cdot A_s}{v_g} \qquad \text{(E - 8.15)}$$

The solution with the boundary condition $S = S_i$ at $h = 0$ yields for the concentration at the outlet:

$$\frac{S_o}{S_i} = \left\{1 - \sqrt{\left(\frac{D_e \cdot k_0}{2m \cdot S_i}\right)} \cdot \frac{A_s \cdot H}{v_g}\right\}^2 \qquad \text{(E - 8.16)}$$

It can be seen from Equation 8.16 that the outlet concentration is no longer a linear function of the height of the column. The dependency of height under diffusion limited conditions was experimentally determined as is shown in Figure 8.6.

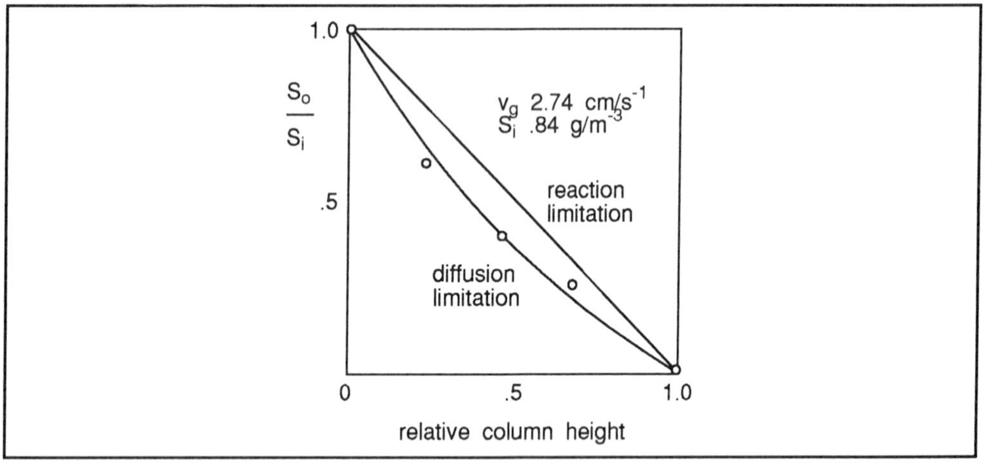

Figure 8.6 Toluene concentration as a function of the column height in the case of diffusion limitation in the biolayer (see text for discussion).

Two lines are shown. The straight line is a plot of $\frac{S_o}{S_i}$ against column height for a reaction limited system. The curved line is a plot of $\frac{S_o}{S_i}$ against column height for a diffusion limited system. The points marked o are experimentally determined values at the given gas velocity and inlet concentration. Note the close agreement between these points and the diffusion limited model. In the two models (reaction limited and diffusion limited) zeroth order reaction kinetics were used.

It was concluded that the measured points were in good agreement with the expected theoretical curve.

Thus the model provides relationships between the concentration in the inlet gas (S_i), the concentration in the outlet gas (S_o), the height of the column (H), the superficial velocity of the gas (v_s), the partition coefficient of the gas (m) and the effective diffusion coefficient. This relationship would, of course, be a immense importance in any scale up process. We might anticipate that in practice, S_i may well be fixed and that S_o would be set by some regulatory standard. Likewise the volume of the waste gas may well be predetermined. In principle, therefore, we could choose a whole variety of different shaped columns to achieve a satisfactory $\frac{S_o}{S_i}$ ratio. A column with a large cross-sectional area would produce a situation in which v_g was low and thus H could also be low. Conversely, a small cross-sectional area, would mean a high value for v_g and thus we would need a long column. The actual efficacy of such columns in removing substrate could be predicted from Equation 8.16. This example should have proven to you the value of being able to model a system properly. Let us now turn our attention to another example.

8.3 Example 2: Membrane fixed enzyme reactors

Thus example is taken from a paper by Prenosil, JE and Hediger, T, entitled 'Performance of membrane fixed biocatalyst Reactors 1: Membrane reactor systems and modelling' (Biotechnology and bioengineering, 1988 31, pp 913-921 and Desalination, 53, 1985).

In the two past decades research into immobilised enzyme and microbial technology was enjoyed a rapid development. This period has also been marked by a number of successful industrial applications of membrane technology. The excellent separation properties of the new synthetic membranes can be advantageously combined with chemical reaction in a membrane reactor. The yield of some reactions can be increased by continuous product withdrawal or by increase of reactant concentration within the membrane reactor. The development of a model for the performance of a membrane reactor is discussed below and some experimental results are given proving that the model describes the experimental results quite satisfactorily.

8.3.1 The membrane fixed enzyme reactor

The interest is focused on the diffusion membrane reactor, a hollow fibre membrane system was chosen in particular for its easy handling and favourable surface-volume ratio. The main purpose of the membrane in the diffusion type of membrane reactor is to provide a matrix or carrier for cell or enzyme immobilisation and to allow easy exchange of substrate and product with the bulk liquid in the hollow fibre lumen by diffusion through the membrane pores.

8.3.2 The reactor model

An asymmetric hollow fibre membrane system is considered. The biocatalyst is kept in the porous structure of the capillary shell side and the substrate solution is fed continuously into the capillary lumen, while the shell side is drained and closed. The substrate diffuses towards the biocatalyst porous structure, reacts and the product is removed by diffusion in the opposite direction. The scale-up of the reactor is based essentially on the model predictions. Figure 8.7 shows schematically the transport processes in one hollow fibre with the biocatalyst immobilised in the porous structure (region 3) and the substrate flowing into is interior called the lumen (region 1). Substrate comes into contact with the enzyme by diffusion across the membrane skin (region 2). The products of an enzymatic reaction diffuse back into the bulk solution in the lumen.

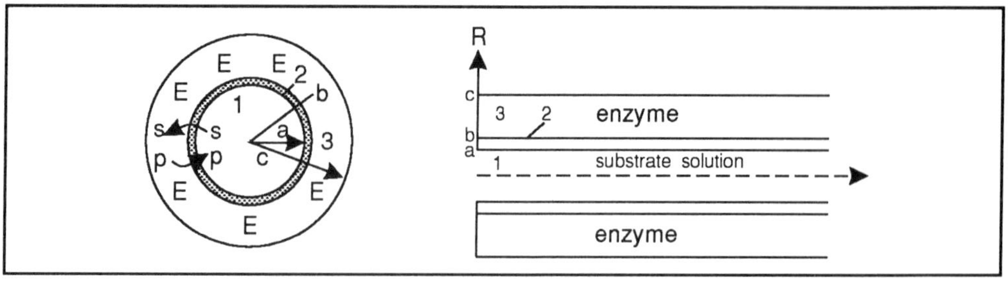

Figure 8.7 Geometry and transport processes in one membrane capillary (see text for details).

The principle of the model is basically a hydrodynamic reactor model, describing the flow of fluid, coupled to a model accounting for the diffusion and reaction. A steady state mass balance in all three regions results in a system of differential equations written in cylindrical coordinates:

$$D_{e1}\left(\frac{\partial^2 S_1}{\partial r^2} + \frac{1}{r}\frac{\partial S_1}{\partial r}\right) = vz\frac{\partial S_1}{\partial r} \qquad \text{region 1} \qquad \text{(E - 8.17)}$$

$$D_{e2}\left(\frac{\partial^2 S_2}{\partial r^2} + \frac{1}{r}\frac{\partial S_2}{\partial r}\right) = 0 \qquad \text{region 2} \qquad \text{(E - 8.18)}$$

$$D_{e3}\left(\frac{\partial^2 S_3}{\partial r^2} + \frac{1}{r}\frac{\partial S_3}{\partial r}\right) = v \qquad \text{region 3} \qquad \text{(E - 8.19)}$$

With D_e diffusion coefficient, - S substrate concentration, r and z radial and axial coordinates, v_z axial liquid velocity and v the reaction rate. The terms on the left hand sides represents the radial diffusion of substrate. The right hand side of Equation 8.17 is the convection term, where for laminar flow:

$$v_z = v_o\,(1 - r^2/a^2) \qquad \text{(E - 8.20)}$$

This equation represents a Poiseulle profile where v_o is the maximal axial fluid velocity. On the right hand side of Equation 8.19 is a kinetic expression for the considered reaction. In this case Michaelis Menten kinetics with product inhibition was used:

$$v = \frac{V_{max} \cdot S_3}{S_3 + K_M(1 + S_0 - S_3)/K_i)} \qquad \text{(E - 8.21)}$$

In which V_{max} is the maximum reaction rate, K_M and K_i are the Michaelis Menten and the inhibition constants.

We have dealt with Michaelis Menten kinetics earlier in this text in which we described the velocity of the reaction as being related to:

$$v = \frac{V_{max}\,[S]}{K_M + [S]}$$

However, we did mention that enzymes may be inhibited, especially by the product of their reactions. Inhibition may be basically one or two types (competitive and non-competitive). We do not intend to derive mathematical relationships between velocity, substrate concentration, K_M and inhibitors in this text. These are derived and explained in the BIOTOL text 'The Molecular Fabric of Cells'. The relationship used by Prenosil and Mediger (1988) in their model (Equation 8.21) is a typical velocity, substrate concentration relationship in which inhibition is taking place.

The boundary conditions are:

$S_1 (R, z) = S_0, z <= 0$ The inlet concentration.

$\partial S_1/\partial r = 0, r = 0$ No transport at the centre of the lumen, symmetry in the radial profile.

$D_{e1} . \partial S_1/\partial r = D_{e2} . \partial S_2/\partial r, r = a$ Diffusion transport out of region 1 equals transport into region 2.

$D_{e2} . \partial S_2/\partial r = D_{e3} . \partial S_3/\partial r, r = b$ Diffusion transport out of region 2 equals transport into region 3.

$\partial S_3/\partial r = 0, r = 0$ No transport beyond the membrane, symmetry in the radial profile.

It is obvious that this model can not be solved by analytical techniques, a numerical approach using finite difference or orthogonal collocation methods has to be used. Do not be put off if you are not familiar with these mathematical devices. The aim of this example is not the solution of the model but the principles behind the model.

8.3.3 Experimental and theoretical results

Experiments were performed with membranes for hydrolysis of lactose by galactosidase. The reaction is lactose → galactosidase + glucose. The kinetic coefficients of the reaction rate Equation 8.21 were determined ny batch experiments. A good correlation was achieved as can be seen in Figure 8.8.

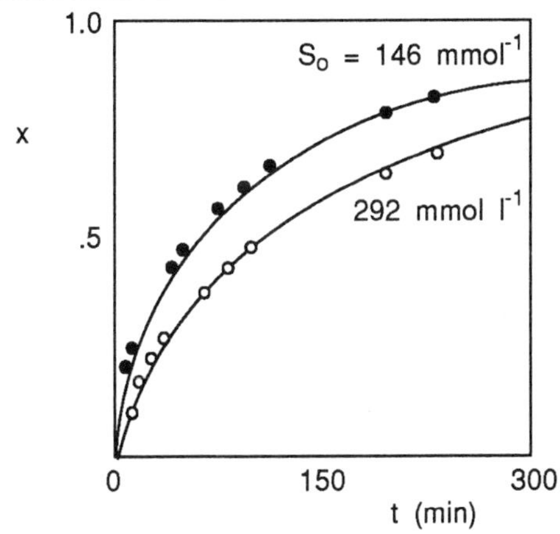

Figure 8.8 Batch lactose hydrolysis experiments as compared with the model, Michaelis Menten kinetics with product inhibition. The figure shows the conversion relative to starting concentration S_0 against time. Two starting concentrations (146 mmol l^{-1} and 292 mmol l^{-1}).

Some continuous hydrolysis experiments were performed with the membrane. The kinetics parameters together with the diffusion coefficients taken from literature, were used to run the model. In Figure 8.9 the results of a few experiments together with the model predictions are shown.

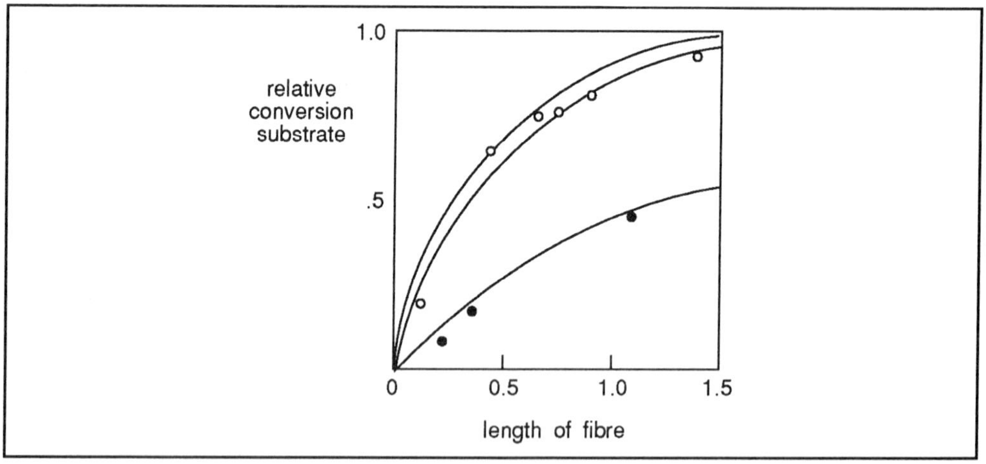

Figure 8.9 Model predictions compared with experimented results. The three lines represent experiments with three different inlet concentrations. Conversion is expressed as a fraction of the inlet concentration.

It was concluded that the agreement between the model and the experiments was good and that the model can be applied successfully.

8.4 Example 3: Packed bed reactor performance with immobilised yeast for ethanol fermentation.

This example is taken from a paper by Godia, F, Casas, C and Sola C, (1987, Biotechnology and Bioengineering 30, pp 836-843) entitles 'Mathematical modelisation of a packed bed reactor performance with immobilised yeast for ethanol fermentation'.

The performance of a continuous vertical packed bed reactor with yeast immobilised in carrageenan gel beads was reported. The yeast produces ethanol from glucose. In outline the study focuses on the mathematical modelling of the steady state fermenter behaviour by means of a tanks in series model which includes the intrinsic kinetics, the external mass transfer and the internal diffusion - reaction conditions in the beads.

8.4.1 The reactor

By means of entrapment in carrageenan a cell suspension of yeast was immobilised and spherical particles were obtained. continuous fermentation was carried out in a vertical packed bed reactor filled with gel beads, see Figure 8.10.

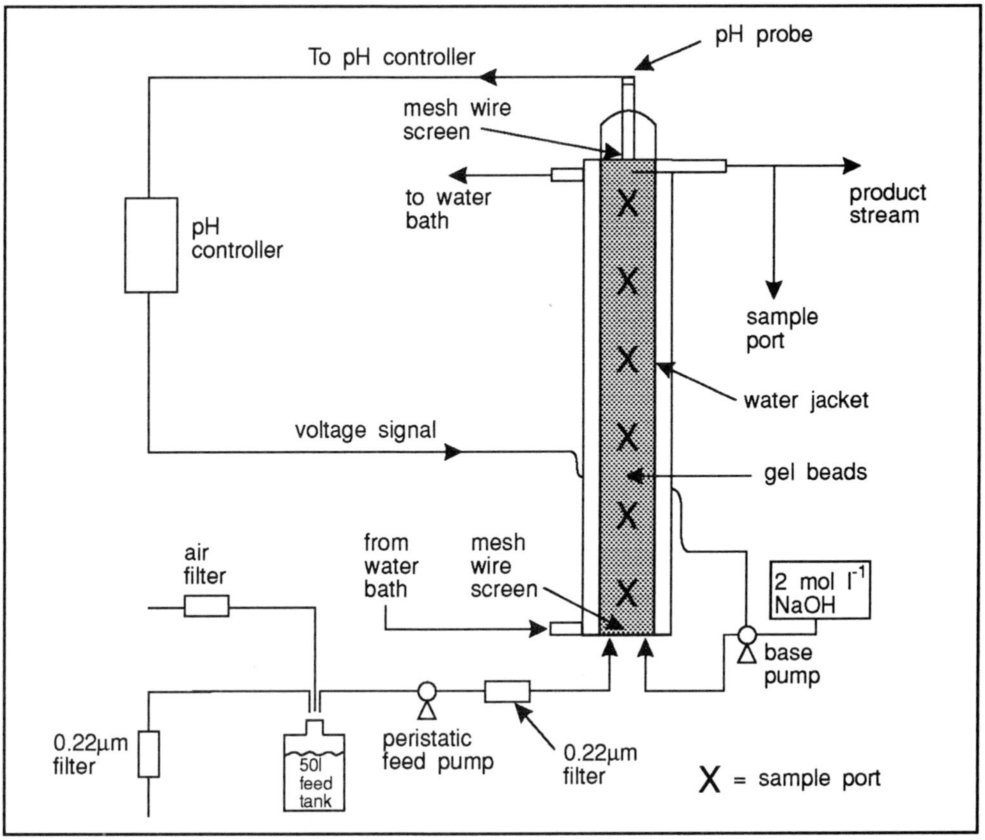

Figure 8.10 Packed bed reactor filled with gel beads, X-column sample port.

Π Examine Figure 8.10 carefully to ensure that you understand what is happening.

Begin in the bottom left hand corner. Note that media was filter sterilised by 0.22 μm filters before being pumped into the base of the bed reactor. The product stream was removed at the top. There was also a water jacket to keep the temperature of the reactor constant. What other parameter was kept constant?

The one we anticipated you would spot is pH. pH is measured at the top of the column and , via a pH controller, switches the 'base pump' off and on in order to maintain the pH. You should conclude from this set up that yeast in culture have a tendency to cause the medium to become acidic. Note also the multiple sampling ports.

8.4.2 Modelling of the reactor performance

Intrinsic kinetics coupled to diffusion and external mass transport of the beads enabled the development of a model to describe the performance of the column.

By studying the batch fermentation of free cells the following intrinsic fermentation kinetic equation was obtained:

$$v = V_{max} \frac{S}{K_s + S} \frac{P}{K_p + P} \qquad \text{(E - 8.23)}$$

in which v and V_{max} are the actual and maximal reaction rate, S the substrate (glucose) concentration, P the (product) ethanol concentration and K_s and K_p the kinetic constants. The equation reflects the non-competitive inhibition by ethanol, the fitting of the model to experimental data was very satisfactory.

Due to the characteristics of the immobilised cell packed bed reactor, the mass transfer should include the effects of external mass transfer, internal diffusion and reaction within the gel beads simultaneously. For transport and reaction inside the gel beads at steady state conditions this leads to the following equations for glucose and ethanol:

$$D_{es} \left(\frac{d^2 S}{d r^2} + \frac{2}{r}\frac{dS}{dr}\right) = \frac{X}{Y_{sp}} v(S, P) \qquad \text{(E - 8.24)}$$

$$D_{ep} \left(\frac{d^2 P}{d r^2} + \frac{2}{r}\frac{dP}{dr}\right) = - v(S, P)X \qquad \text{(E - 8.25)}$$

in which r is the radial coordinate, D_{ep} and D_{es} are the effective diffusion coefficients for ethanol and glucose, Y_{sp} is the yield factor (ie amount of cells produced per amount of substrate used) and X is the cell concentration in the gel.

The boundary conditions are:

$$\frac{dS}{dr} = 0, \quad r = 0$$

$$\frac{dP}{dr} = 0, \quad r = 0$$

$$D_{es} \frac{dS}{dr} = k_{Ls} (S_b - S_i), \quad r = Rp$$

$$D_{ep} \frac{dP}{dr} = k_{Lp} (P_p - P_i), \quad r = Rp \qquad \text{(E - 8.26)}$$

In which k_{Ls} and k_{Lp} are the mass transfer coefficients for glucose and ethanol, S_b and P_b are the bulk substrate and product concentration, S_i and P_i are the interface substrate and product concentration and Rp is the radius of the gel beads. Equations 8.26 reflect that the flux of sugar from the fluid phase to the external bed surface is equal to the diffusion flux into it, and the diffusion flux of ethanol out of the beads is equal to its transfer rate from the surface to the liquid phase, furthermore the concentration profiles in the beads of substrate and product are symmetric.

The actual reaction rate R can be calculated from the derivative at the surface of the particle:

$$R_s = D_{es} \left.\frac{dS}{dr}\right|_{Rp} \cdot \frac{3}{Rp} Y_{sp} \qquad \text{(E - 8.27)}$$

$$R_p = D_{ep} \left.\frac{dP}{dr}\right|_{Rp} \cdot \frac{3}{Rp} \qquad \text{(E - 8.28)}$$

These equations reflect that in the steady state, with no accumulation of glucose or ethanol, the rate of glucose conversion is equal to the rate of diffusion into the beads and the rate of ethanol production is equal to the rate of diffusion out of the beads. The fermentation rate if the concentration in all the beads were those of the interface Ri is given by:

$$R_i = X \,.\, V_{max} \frac{S_i}{K_s + S_i} \; \frac{P_i}{K_p + P_i} \qquad \text{(E - 8.29)}$$

Which allows to calculate the (internal) effectiveness factor:

$$\eta = R_a / R_i = R_p / R_i \qquad \text{(E - 8.30)}$$

You should note that alternative sugars to glucose can be used (eg lactose). These other sugars may well have different D_e and Y_{sp} values.

The model was developed assuming that substrate flow in the bed was by plug flow. However, when a lactose feed was perturbed in a step wise manner, the reactor did not perform according to the models predictions. In other words, the model did not adequately represent the situation.

The deviation was in fact explained by the fact that the conversion of lactose to ethanol was also accompanied by CO_2 gas in the fermenter causing mixing. This caused the authors to use a different model for the mixing process (a tanks in series model). This model led to a good correlation between predicted and actual results.

The point we are attempting to make here is that it is important to check if a model accurately represent a situation. This can only be done by performing appropriate experiments.

The outcome of Godia *et al*'s study is that they produced a model which fitted well to experimental data. Such a model would be of great importance to any scale up. It was also important in understanding flow patterns within the packed bed.

8.5 Immobilised systems and industry

Immobilised systems have been used extensively for quite some considerable time. The trickle filter beds used in water treatment used a mixed culture biofilm as a biocatalyst to remove undescribable material from the water. Likewise a river with its incumberent micro-organisms in the sediment is also effectively an immobilised biocatalyst. Such mixed systems, although extremely useful are only poorly defined in both biological and engineering terms. The new era of biotechnology has seen new impetus put into using immobilised biocatalysis in a more defined manner. These new processes often involve a single type of organism or, even a single enzyme and frequently involve using valuable substrates to make expensive products. The need to reduce microbial contamination together with the high standards of quality that are applied to many of these products (eg food, medicines) means that these are invariable strongest, and most costly, operational parameters. This means that such processes must be operated efficiently. This in turn, means that the system needs to be properly modelled at the research stage if sound predictions are to be made about production scale performance. Thus the new era of using immobilised systems is characterised by particular attention being paid to modelling such processes. The experience of modelling provided by Chapter 7, together with the glimpse of some applications of such modelling to quite different processes in the chapter, will enable you to contribute to this important area of development.

Already many immobilised biocatalytic systems are in operation including the production of semi-synthetic penicillin using immobilised penicillin acylase, a high fructose corn syrup using glucose isomerase as well as a variety of enzymes used for example, for the synthesis of specific and valuable amino acid derivatives. We should anticipate many more such processes being developed over the next few years.

Summary and objectives

Our aim in this chapter was to use a few specific examples to illustrate the importance of being able to monitor immobilised (heterogenous) biocatalysis. The intention was not to provide any particular 'new knowledge' but to ensure that you understood the usefulness of modelling. The three examples used included quite different types of reactor configurations including a surface film, packed bed reactor; a cylindrical, hollow fibre, membrane reactor and a bead filled, packed bed reactor. The success of the modelling process in each case illustrates that the types of strategy developed in Chapter 7 can be used in a wide variety of circumstances. The three cases examined are only a fraction of the published examples in which modelling of immobilised biocatalysis has been used.

Now that you have completed this chapter you should be able to:

- to explain the importance of modelling and its value in scale-up;
- understand the objectives and strategies used in published papers modelling heterogenous biocatalysis.

Strategies of sterilisation

Strategies of sterilisation

9.1 Introduction to sterilisation

Sterilisation is the process by which micro-organisms are either killed or removed from material or equipment. In the strictest sense, a sterile material has no live organisms present. There is no such thing as a partially sterilised system. The presence of a single live organism renders the system non-sterile. In this context, we mean by 'live organism', organisms that are capable of growth and reproduction. In practice, in some systems however, we can tolerate the presence of some live organisms. Such systems should be more properly regarded as pasteurised, if it has been treated to destroy a proportion of the micro-organisms (especially disease-producing) that are present. In this chapter we are, however, primarily concerned with sterilisation.

Micro-organisms are more-or-less ubiquitous. The equipment and feedstock we use in biotechnological processes are more-or-less inevitably contaminated by micro-organisms. These may be derived from the workers handling them, from soil and from the air. Amongst these may be non-toxic saprophytic organisms and potential disease-causing types.

Sterilisation is therefore necessary to:

- ensure that a process or experiment is carried out with the desired micro-organisms;
- ensure that the products which are made are of predicted quality;
- the environment is protected from undesirable contamination;
- prevent deterioration (microbial spoilage) of products.

In this chapter we shall learn about:

- the types of sterilisation procedures commonly used in bioprocess technology;
- how effective these procedures can be modelled and tested;
- how these procedures are used to sterilise different items.

We begin the chapter by examining sterilisation by heat and show how this can be used to sterilise media. We will then examine chemical sterilisation especially by using ethylene oxide and β-propiolactone. We will then examine sterilisation using ionising radiation. These procedures have a common strategy namely to kill (inactivate) the micro-organisms that are present. An alternative strategy (removal of all of the micro-organisms present) is dealt with under the heading filter sterilisation. We will also examine the sterilisation of air and other gases. In the final part of the chapter we will briefly consider asepsis (the state of being free of micro-organisms) and bioreactor design.

9.2 Sterilisation by heat

When a pure culture of micro-organisms is heated above the maximum growth temperature the micro-organisms will start to die. The same holds for endospores when the maximum resistance temperature is reached.

When we follow this process by taking samples at different time intervals we will see that the number of survivors decreases in time. By plotting the logarithm of the number of survivors against the time we get a straight line, the survival curve (Figure 9.1). The problem with these survival curves and especially with the determination of the number of surviving micro-organisms is that growth parameters influence the outcome of the determinations. Sublethal damaged cells can survive under favourable culture conditions.

survival depends on media composition

We will illustrate this with an example. If an organism can either be grown in a simple medium containing just mineral salts and glucose or in a complex medium containing a mixture of amino acids and vitamins, then transfer of organisms to these media after heat treatment may result in quite different observed rates of survival. For example, let us assume that the most heat liable enzymes in the organism are those involved in the biosynthesis of the amino acid leucine. After heat treatment, these enzymes may be completely inactivated. This means that if we transfer the cells to a glucose/mineral salts medium they will not be able to grow as they would produce no leucine and could not, therefore make proteins. On the other hand, transfer to a medium containing leucine means that these cells could continue to make proteins because their inability to make leucine would be circumvented by the exogenous source. Amongst the proteins that would be made would be the enzymes of leucine biosynthesis. The organism would, therefore, recover. Thus the survival curve of the type shown in Figure 9.1 is dependent upon the medium in which the heat-treated cells are placed.

Π In the legend to Figure 9.1 we pointed out that *Bacillus stearothermophilus* spores are used as the test organism in many heat-based sterilisation procedures. Write down a reason why this is so. (A clue is to be found in the name of the organism).

Bacillus stearothermophilus spores are particularly heat tolerant. In testing a heat-based sterilisation procedure, the inactivation of these spores is a good indication that other, more heat sensitive, organisms would be also inactivated.

The straight line of the survival curve from t = 0 onwards points to a first order kinetic or 'single hit' reaction. Which is described by the equation:

$$N_t/N_o = e^{-\delta t} \qquad \text{(E - 9.1)}$$

N_t = number of surviving micro-organisms at time t.
N_o = initial number of micro-organisms
δ = specific death rate
t = time

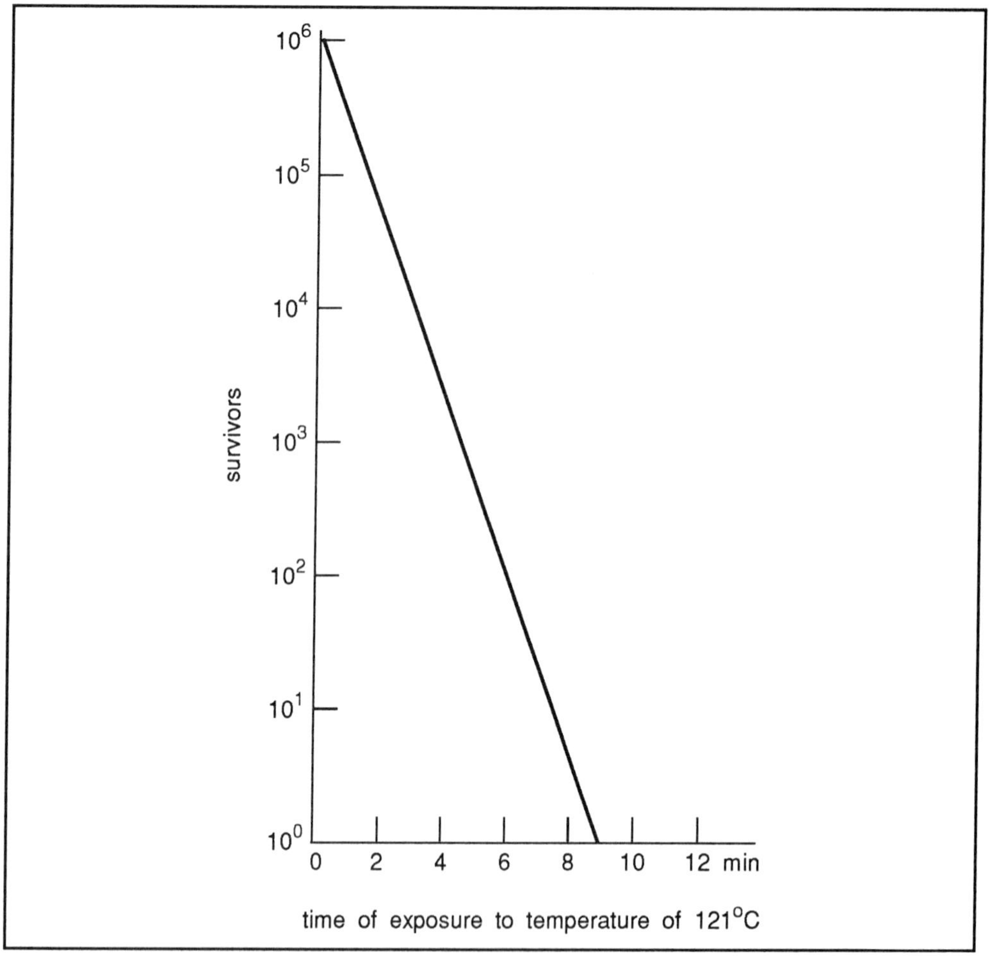

Figure 9.1 Example of a survival curve of a micro-organism. The example shown is the survival curve or *Bacillus stearothermophilus* spores in saturated steam at 121°C. Note that *B.stearothermophilus* spores are used as the test organism in many heat-based sterilisation procedures.

specific death rate

The specific death rate depends on the type of micro-organism and the environmental parameters during sterilisation. Not all microbial populations exhibit a first order kinetics reaction. The result is that we do not get a straight line for the survival curve. Sometimes a shoulder appears first. In that case the survival kinetics can be expressed as:

$$N_t/N_o = 1 - (1 - e^{-\delta t})^n \qquad \text{(E - 9.2)}$$

n = extrapolation number equal to the intercept on the N_t/N_o axis, which gives 'the number of hits' required for lethality. The reason for this is that there can be a multi hit requirement for lethality, or that each viable unit of micro-organisms consists of several separate cells or spores and inactivation of the unit is only observed after the death of every cell or spore in the unit. Besides shoulders, tailing is very often observed in survival curves.

The most important environmental parameter in heat sterilisation is the temperature. The specific death rate is also temperature dependent: the higher the temperature, the higher the value of δ. This relationship can be described by a normal Arrhenius equation:

$$\delta = a \cdot e^{-E/RT} \qquad \text{(E - 9.3)}$$

δ = specific death rate
E = activation energy
R = universal gas constant
T = absolute temperature of the medium
a = reaction constant

SAQ 9.1

In a sterilisation experiment, carried out in an autoclave at 121.1°C, the number of viable bacteria was determined at different time intervals. The results were as follows:

t = 0 seconds	number of viable bacteria = $1.0 \cdot 10^6$
t = 15 seconds	number of viable bacteria = $7.5 \cdot 10^5$
t = 30 seconds	number of viable bacteria = $3.9 \cdot 10^5$
t = 45 seconds	number of viable bacteria = $1.5 \cdot 10^5$
t = 60 seconds	number of viable bacteria = $5.5 \cdot 10^4$
t = 75 seconds	number of viable bacteria = $2.0 \cdot 10^4$
t = 90 seconds	number of viable bacteria = $6.5 \cdot 10^3$
t = 105 seconds	number of viable bacteria = $2.8 \cdot 10^3$

Draw a survival curve for this experiment. Calculate δ and what can be said about the morphology of the bacteria?

9.2.1 D-value; Z-value and F-value

It has been shown that the time necessary to destroy all bacterial spores present increases with an increasing initial number of spores. It has also been observed that the time of heating needed to reduce the initial numbers of spores or organisms by a factor 10 is independent of the absolute number of spores. The time which is needed to reduce the number of spores or organisms to 10% of the initial number was later called the decimal reduction time or D-value. It is generally expressed in minutes.

decimal reduction time (D)

The D-value can be calculated from the N_o and N_t values in the following way:

For N_t/N_o in the formula (9.1) 1/10 is substituted and D is substituted for t. We then get:

$$\ln 1/10 = -\delta D \text{ or } 2.303 \log 0.1 = -\delta D \text{ or } D = 2.303/\delta \qquad \text{(E - 9.4)}$$

SAQ 9.2

Calculate D for the bacterium from the SAQ 9.1 and compare the value with results obtained with the survival curve.

The z-value is the temperature difference by which the D-value is changed by a factor of 10.

The relationship between temperature and D-values can be determined by plotting the log D-values against the temperature. This results in a thermal death curve (Figure 9.2) from which the z-value can be directly derived by interpolation.

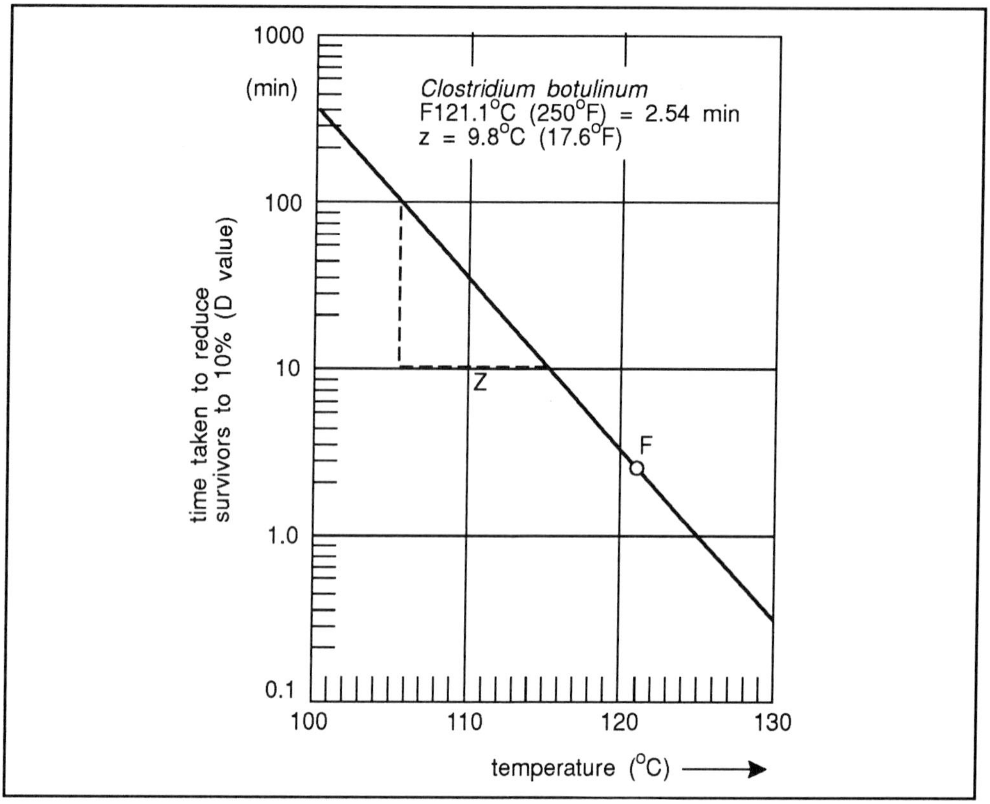

Figure 9.2 Thermal death rate curve for spores of *Clostridium botulinum* with z- and F-value. Data from Pflug (1973), Industrial Sterilisation Edited by G B Phillips, Duke University Press, Durham (see text for details).

The thermal death time curve is based on the relationship described by the equation:

$$(T_1 - T_2)/z = \log D_2 - \log D_1 \qquad \text{(E - 9.5)}$$

T = absolute temperature
D = D-value
z = the temperature difference by which the D-value is changed by a factor 10

In the literature one can also find the term Q_{10}. This is a quotient indicating how much more rapidly death of a culture occurs at a temperature T_1, when $T_2 - T_1 = 10°C$.

Since $(T_1 - T_2)/z = \log D_2 - \log D_1$

Thus $\text{antilog}\left(\frac{T_1 - T_2}{z}\right) = \frac{D_2}{D_1}$

But $\frac{D_2}{D_1} = Q_{10}$ when $T_1 - T_2 = 10°C$

Hence $z = 10/\log Q_{10}$. In the canning industry in the UK and USA z-values are often expressed in °F.

As Q_{10} is always expressed in °C, z (°F) becomes $18/\log Q_{10}$.

SAQ 9.3

Another sterilisation experiment with the same bacterium as that used in SAQs 9.1 and 9.2 was conducted at 115°C gave D-value of 2.2. Calculate the z-value (Note the temperature difference, and use the D-value calculated in SAQ 9.2).

The F-value denotes the effectivity of a sterilisation process and can be defied as the time in minutes, required at a designated temperature in the case of heat sterilisation, or any other lethal action to kill all the spores of a specified *Bacillus* species (the test spores) taking in account the initial spore number ('bioburden') and a chosen survival probability (for instance 10^{-6}).

The F-value can be expressed as a number of D-values:

$$F = n \cdot D \tag{E - 9.6}$$

For instance when we want to sterilise a product or medium with an initial spore count of 10^6/unit and after the treatment we want a spore count of 10^{-6}/unit the equation results in: F = 12 D. The only thing we have to do now is to substitute the D-value of the particular spore species (eg *Bacillus stearothermophilus* has a $D_{121.1}$ of 1.5 min. So F = 12 . 1.5 min = 18 minutes).

SAQ 9.4

Calculate the F-value for the sterilisation experiment of SAQ 9.1 when the probability of one surviving spore should be 1:100.000. D = 0.69 min (from SAQ 9.2).

Because a z-value of 10°C is frequently observed in spores and 121.1 °C is commonly adopted as a reference temperature in sterilisation, these conditions are used by the FDA to define F_0. An $F_0 = 1$ means a procedure of 1 minute at 121.1°C or the equivalent time at other temperatures and a z-value of 10°C.

For temperature other than 121.1 °C the F-value can be recalculated according to the formula:

$$F_{121.1^\circ C}^{10^\circ C} = F_T^z \cdot 10^{\left(\frac{T-121.1}{z}\right)} \quad \text{or}$$

$$F_T^z = \frac{F_{121.1^\circ C}^{z^\circ C}}{10^{\left(\frac{T-121.1}{z}\right)}} \tag{E - 9.7}$$

SAQ 9.5

Calculate the F-value for the experiment in SAQ 9.1 conducted at a temperature of 105°C. We remind you that z = 12.1 (from SAQ 9.3). $F_{121.1}^z = 7.59$ min (from SAQ 9.4).

When F_T is known, it becomes possible to check the sterilisation process in an autoclave or bioreactor with the equation:

$$F_T^z = \int L \cdot dt \qquad \text{(E - 9.8)}$$

What this means is that during a sterilisation process every lethal temperature executed during a certain time results in a lethal effect. This lethal rate can be characterised with the equation:

$$L = 1/F = 10(T - 121.1/z) \qquad \text{(E - 9.9)}$$

T = core temperature measured at the coldest place.

9.2.1 Sterilisation of media

lethality S

The theoretical analysis of this process was elaborated by Deindorfer and Humphry (1959). The total lethality S, which they called sterilisation criterium, may be determined from the equation:

$$S = \ln N_0/N_t \qquad \text{(E - 9.10)}$$

S should be a positive number and can be used as a measure sterilisation quality. Theoretically N_t can never be zero. When N has a value of 0.01 there is a change of 1:100 that the sterilisation is successful. That is we have $\frac{1}{100}$ chance of a live organism being present. Likewise with a chance of 1:1000, N_t = 0.001. Depending on the wanted effectiveness of the procedure N_t should be chosen. S is also dependent on the volume to be sterilised, because N_0 denotes the total number of bacteria or spores present in the total volume of medium to be sterilised. Combination of the Equation 9.10, 9.3 and 9.1 results in:

$$S = \delta \cdot t = a \cdot e^{-E/RT} \cdot t \qquad \text{(E - 9.11)}$$

Equation 9.11 holds only for a constant temperature. In a batch sterilisation, however, the temperature during heating and cooling is not constant. The difference in heat transfer depends on the type of heating and the size of the bioreactor. So the total lethality should be calculated as an integration of the Equation 9.11 in the time:

$$S = \int_0^t e^{-E/RT} \cdot dt \qquad \text{(E - 9.12)}$$

When we split up the sterilisation time into a heating, a holding and a cooling period the resulting equation should be:

S = [sterilisation during heat] + [sterilisation during holding] + [sterilisation during cooling]

$$S = \int_0^{t_1} e^{-E/RT} \cdot dt + a \cdot e^{-E/RT_2} \cdot t_2 + a \int_0^{t_3} e^{-E/R} \qquad \text{(E - 9.13)}$$

Because the lengthy calculations involved in such an analytical solution simpler methods have been devised.

The graphical method

This is the oldest method. The procedure is as follows. The temperature is measured at the coldest point in the product to be sterilised. During the sterilisation process a temperature/time curve is established. After that, every corresponding lethal rate whose values can be found in published tables, is plotted against time. An example of such published tables is provided in Table 9.1.

The surface below the resulting curve is the total lethality for the entire sterilisation process and can be determined planimetric or simply by counting the squares on the graph paper used. The number of cm^2 has to be multiplied by the surface of the 'unit square', which is the surface of the square corresponding with the ordinate 121.1°C and the length of 1 minute on the time axis. This 'unit square' corresponds by definition with an F-value of 1 minute at 121.1°C (Figure 9.3)

T°C	z-value = 6°C (*Bacillus stearo-thermophilus*)	z-value = 10°C (*Clostridium botulinum*)	T°C	z-value = 6°C (*Bacillus Stearo-thermophilus*)	z-value = 10°C (*Clostridium botulinum*)
95	0.000	0.003	113	0.068	0.200
96	0.000	0.004	114	0.100	0.251
97	0.000	0.005	115	0.147	0.316
98	0.000	0.006	116	0.215	0.398
99	0.000	0.008	117	0.316	0.501
100	0.000	0.010	118	0.464	0.631
101	0.000	0.012	119	0.681	0.794
102	0.001	0.016	120	1.000	1.000
103	0.001	0.020	121	1.468	1.259
104	0.002	0.025	122	2.154	1.585
105	0.002	0.032	123	3.162	1.995
106	0.005	0.040	124	4.642	2.512
107	0.007	0.051	125	6.813	3.162
108	0.010	0.063	126	10.000	3.981
109	0.015	0.079	127	14.680	5.012
110	0.022	0.100	128	21.540	6.310
111	0.032	0.126	129	31.620	7.943
112	0.046	0.159	130	46.420	10.000

Table 9.1 Lethal rates (L) (minutes at 120°C reference temperature/min at T°C) $F^{z}_{120}/F^{z}_{T} = 10^{\frac{T-120}{z}}$ for z = 6°C and z = 10°C. Data derived from Pflug (1973) in Industrial Sterilisation, Edited by G B Philips, Duke University Press, Durham.

Π From the data presented in Table 9.1 which organism (*B.stearothermophilus* or *Clostridium botulinum*) is more likely to be killed at temperatures below 120°C?

Which organism is killed most quickly at 130°C?

The lethal rates for *Clostridium botulinum* are greater than those of *B.stearothermophilus* at low temperatures indicating that it is more likely to be killed at these temperatures. On the other hand the lethal rates for *B.stearothermophilus* are over four times greater than those of *Clostridium botulinum* at 130°C indicating that at this elevated temperature, *B.stearothermophilus* is killed more rapidly than the *Clostridium.*

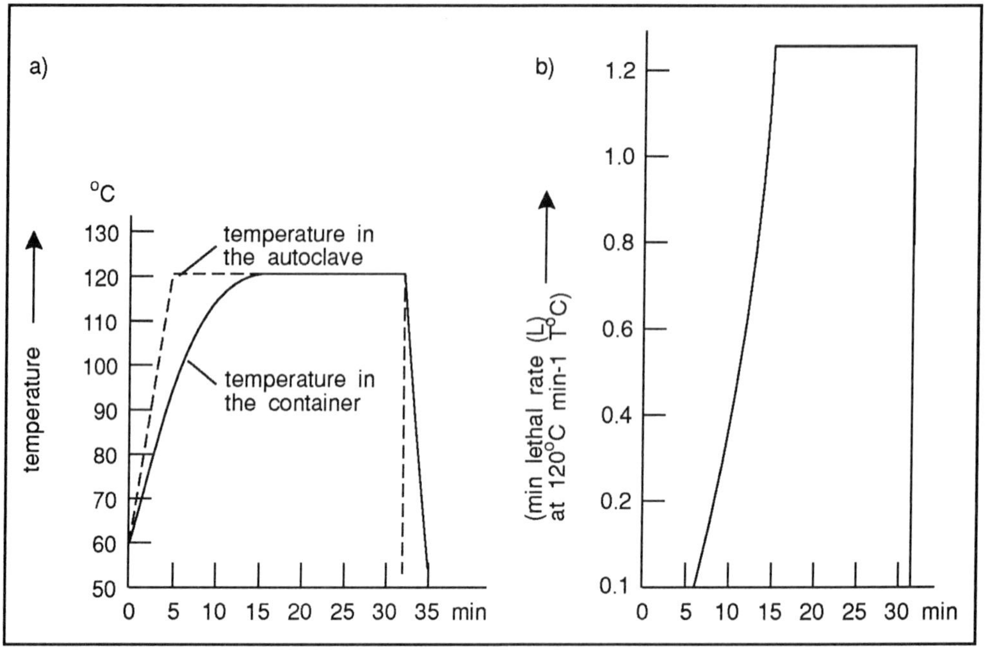

Figure 9.3 a) Temperature/time curve of a sterilisation process. b) Lethality curve constructed by means of Table 9.1 from Figure 9.3a. Surface of the unit square is 0.1 . 5 min = 0.5 min. L = lethal rate = minutes at 120°C reference/min at T°C.

The addition method

In this case the surface under the curve described above is divided in small parallel streaks, with a time interval of one minute. The surface of the curve can be constructed by:

$$0 = t(L_1 + L_2 + L_{n-1}) \qquad \text{(E - 9.14)}$$

The L value denotes the lethal rate: $L = 10(T - 121.1/z)$, as shown in Equation 9.9. For every temperature the lethal effect can be calculated or looked up in a specific table (Table 9.1).

SAQ 9.6

Determine (via the addition method) the total lethality for a sterilisation process with a z-value of 6°C and in which the temperature/time curve has the following shape:

time in minutes	temperature in °C	time in minutes	temperature in °C
5	95	11	116
6	101	12	117
7	107	13	118
8	110	14	119
9	112	15	120
10	114	16-30	121
		31	110

The calculation of the F-value for a sterilisation process can easily be computerised when the lethal rates at specific temperatures are known. The heating and cooling phase can then also be taken into account and after reaching the desired F-value the computer can terminate the heating.

9.2.2 Sterilisation of large volumes

As we have seen in the previous paragraph, a sterilisation process consists of three periods: the heating, holding and cooling period. It is clear that the temperature/time course during the heating and cooling period is dependent on the volume of the medium to be sterilised (Figure 9.4).

Besides the volume, the geometry of the autoclave or bioreactor heat transfer also influence the temperature/time course. While the heat transfer during the sterilisation of small volumes (5 - 50 ml) is almost immediate so that the temperature in the medium container is practically the same as the temperature read by the autoclave thermometer, this is not the case when larger volumes are sterilised. When these large volumes are sterilised there is a lag in heat transfer. This lag is also influenced by the place in the medium container (Figure 9.5) and the material of that container. To prevent the complications which arise during sterilisation of large volumes and also in the case that the medium contains solid particles, the medium should be sterilised with constant agitation. In practice media in large bioreactors are usually sterilised with stirring.

9.2.3 Medium quality and sterilisation

By heat treatment during sterilisation not only are the micro-organisms and spores killed, but heat liable nutrients and growth factors can also be destroyed. A well known example is the Maillard reaction between reducing sugars and the ammonium ions or the amino groups of amino acids or proteins. On the other hand there are some reports of positive growth effects in propionic bacteria. Another example is the destruction of vitamins by heat. When we look at the Equation 9.3 we see that the activation energy (E) has a great impact on the specific death rate. The same holds for the rate of denaturation of vitamins. The Q_{10} for bacterial spores, however, is around 10 whilst for many chemical reactions a Q_{10} of 3.2 is not uncommon. This means that vitamins are less damaged by shorter periods at higher temperatures than by longer periods of heating at lower temperatures.

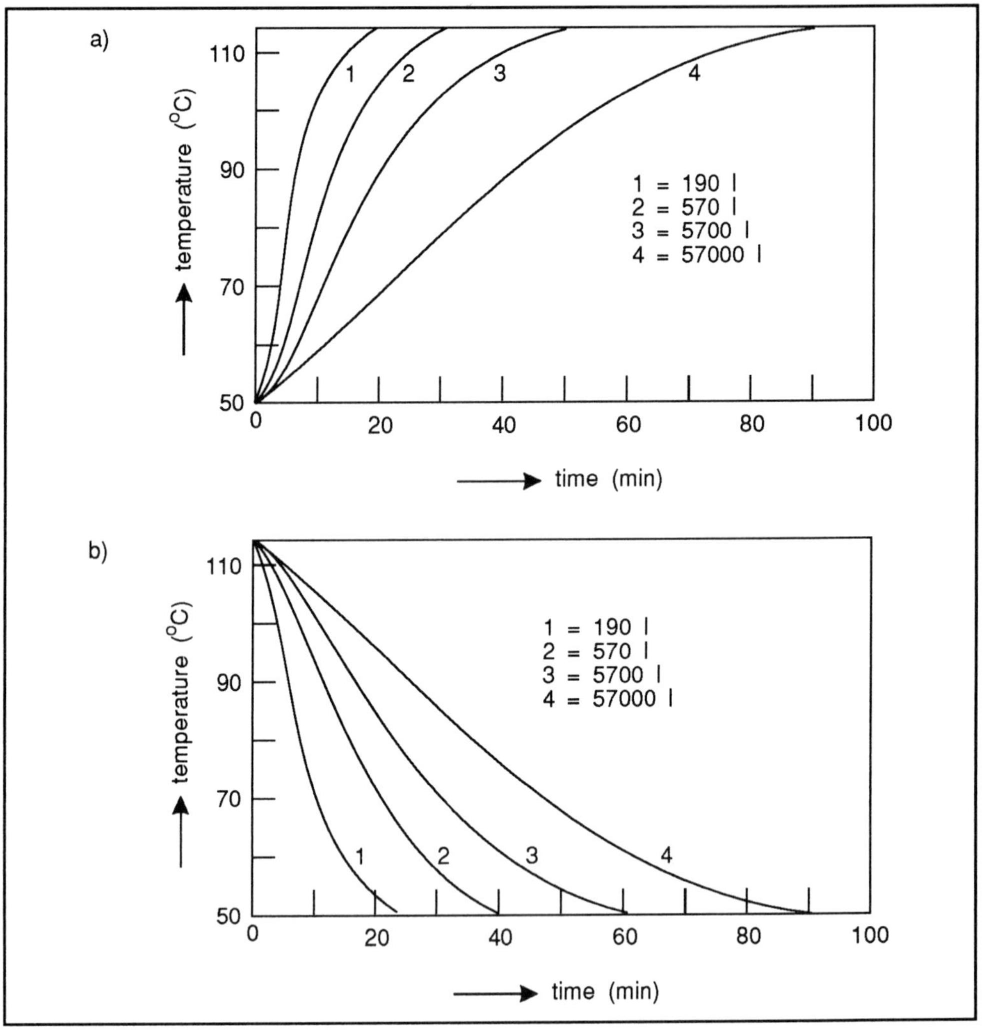

Figure 9.4 a) Temperature/time curves during heating. b) Cooling of different volumes of media during sterilisation. Redrawn from Sikyta (1983), Methods in Industrial Microbiology, Chichester.

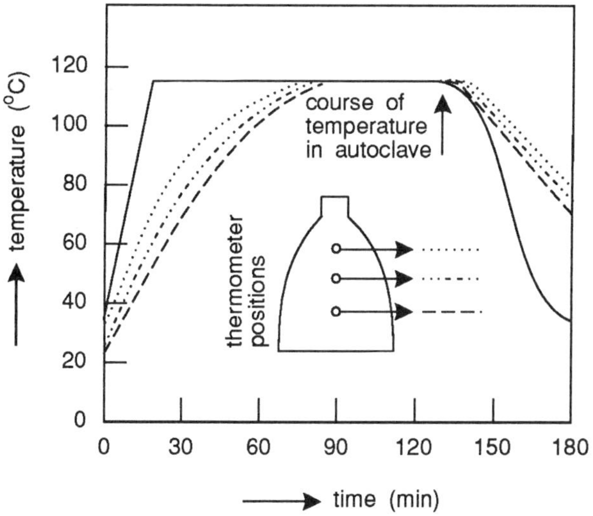

Figure 9.5 Differences in the course of temperature measured in different points during the sterilisation of 20 1 medium in an autoclave.

Π Can you suggest a practical way in which we could make use of this difference in Q_{10} values?

The answer is, of course, to use a high temperature:short time protocol for sterilisation. In this way we would do minimal damage to the desirable nutrients in the medium.

This type of sterilisation is often referred to as HTST (High Temperature Short Time) processing and is very common is the food industry. As we can see in Table 9.2 the activation energy for the killing of the two *Bacillus* species is 4-6 times higher than that for the destruction of the vitamins.

Vitamin or organism	Activation energy (kJ . mol^{-1})
Bacillus stearothermophilus	283.45
Bacillus subtilis	318.19
Folic acid	70.35
Cyanocobalamin	96.70
Thiamine	92.10

Table 9.2 Activation energies for the destruction of vitamins and bacterial spores.

9.2.4 Continuous steam sterilisation

HTST processing is best achieved by a continuous sterilisation process. Batch sterilisation in bioreactors by direct steam injection may approach this type of sterilisation. There are two techniques for continuous sterilisation: the heat exchanger process and the steam injection process. In principle we can distinguish in both devices three basic units. These units are:

- a heating unit in which the medium is heated to the required temperature as quickly as possible;
- a holding vessel in which the medium is kept at a constant temperature for as long as is needed for complete sterilisation;
- a cooling unit in which the medium is cooled down to the desired temperature.

In case of the heat exchanger process (Figure 9.6a), the heating is performed by the use of some kind of heat exchanger, which can have a plated, tubular or spiral form and which is heated by steam or oil. The holding unit is either a tube or a stirred vessel. The cooling unit can be of various different types and involves usually a heat exchanger, in which cold medium is preheated before sterilisation.

Steam injection (Figure 9.6b) is a process by which the medium is very rapidly heated by mixing with high pressure steam in a venturi-type mixer. The sterilised medium from the holding coil comes via a throttling valve in an expansion chamber which is kept at atmospheric pressure, or under vacuum. The steam is immediately flashed off and the temperature of the medium very rapidly drops. The remaining heat can also be used for preheating the cold unsterile medium.

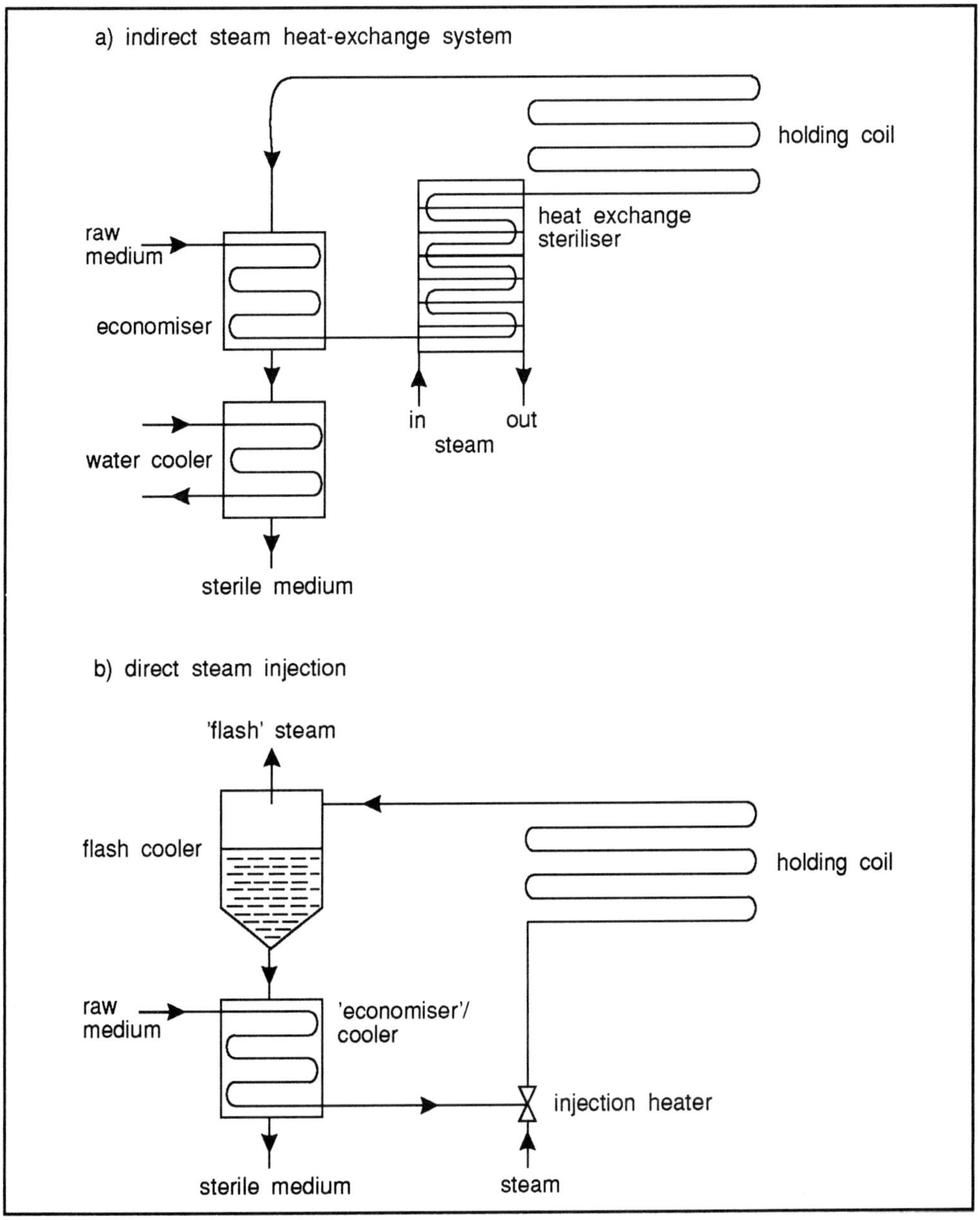

Figure 9.6 Schematic diagrams of continuous flow sterilisation. Redrawn from Bu'lock (1987), Basic Biotechnology Academic Press, London (see text for a description).

9.3 Chemical sterilisation

Chemical sterilisation, in contrast to heat sterilisation has only a very limited use. It is mostly used in surface sterilisation and sterilisation of medical products in cases where no other alternatives are available.

Π Make a list of the criteria you think should be used to assess the suitability of a chemical as a sterilising agent. Then compare it with our list below.

Any substance used for chemical sterilisation should fulfil the following conditions:

- appreciable biocidal effect, even at low concentrations;
- no adverse effect on the medium or product;
- residual concentrations of the agent must be readily removed or must have no effect on the growth of micro-organisms and on product synthesis;
- effective at normal temperatures and pressures;
- easy application, low price, commercial availability and rapid effect;
- the agent should not be inflammable, explosive or toxic.

Sterilisation by chemical agents is influenced by a number of factors such as time, temperature, concentration, pH, humidity and surface tension. The appropriate agent must be chosen according to these factors.

The product mostly used for chemical sterilisation are ethylene oxide, formaldehyde, β-propiolactone and ozone. From which ethylene oxide and β-propiolactone are the most important.

The physical properties of these two agents are summarised in Table 9.3.

	ethylene oxide	β-propiolactone
Formula	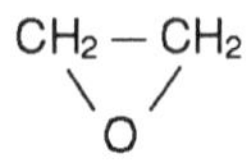	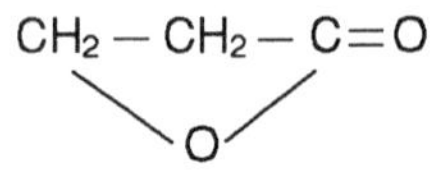
Molecular weight	44.05	72.06
Boiling point	10.7°C	155°C
Melting point	-112°C	-33.4°C
Refractive index	4_p 1.3614	20_p 1.4131
Density (d^{20})	0.891	1.149
Flammability point	-10°C and less	74°C
Solubility	completely soluble in most organic solvents and in water	soluble in most organic solvents; in water 37% (v/v) at 25°C

Table 9.3 Physical constants of β-propiolactone and ethylene oxide.

Π Look down the list of physical properties of ethylene oxide and β-propiolactone noting especially their boiling point, flammability points and solubility.

9.3.1 Ethylene oxide

At atmospheric pressure and ambient temperature ethylene oxide is a colourless gas with a faint etherical odour. Mixed with air it is explosive and flammable in concentrations from 3 - 100%.

Besides this explosiveness, ethylene oxide is also toxic, MAC's ranging from 5 ppm in Sweden and France to 50 ppm in the USA and Japan. The acute toxicity is at the same level as ammonia. Symptoms of intoxication are irritation of the conjunctiva (membrane around the eyes) and the respiratory tract, headaches, dizziness and vomiting. Concentrations of 100-200 ppm are, in a few cases lethal.

Because of the hazards involved in handling pure ethylene oxide sterilisation is always carried out with gas mixtures; the composition of which depends on the type of sterilisation procedure. The effect of ethylene oxide on micro-organisms is probably two-fold. First there are reactions with functional groups in the side chains of proteins and secondly there is alkylation of the nucleic acids, both of which can lead to destruction of the micro-organism. All kinds of micro-organisms, including viruses and endospores of bacteria, are inactivated by ethylene oxide.

Factors affecting the sterilisation process

One of the most important factors in ethylene oxide sterilisation is the free access of the gas to the micro-organisms. When the micro-organisms are packed in crystals or proteins or other material, the effect of the sterilisation procedure is greatly diminished. Other factors affecting the sterilisation procedure are: ethylene oxide concentration, humidity, process time and the type of wrapping in which the apparatus or product to be sterilised are packed.

In the case of free access of the ethylene oxide to the micro-organisms and sufficient humidity the D-value for *Bacillus subtilis var niger* under conditions of 1200 mg l^{-1} and a temperature of 54.5 °C is about 2 minutes.

Sterilisation procedure

Sterilisation procedures using ethylene oxide can be divided into three groups procedures under decreased pressure; under normal pressure and procedures under increased pressure. Which type of sterilisation procedure is chosen depends on what is to be sterilised and also the type of sterilisation gas to be used. Sterilisation gas is a mixture of ethylene oxide and some other gas, mostly CO_2. The concentration of ethylene oxide used varies from 10-15% for pressurised procedures to 90% in vacuum procedures.

In general a sterilisation process with ethylene oxide consists of the following steps:

- a preconditioning phase in which vacuum is applied to the steriliser to remove the air from the load. After that a humidification step, in which steam is applied, to bring the load to the desired humidity;

- a sterilisation phase in which the sterilisation gas is applied at the desired pressure;

- a drying phase in which the load is dried and the ethylene oxide is, as much as possible, removed from the load by alternatively applying a light vacuum and letting in sterile air.

9.3.2 β-propiolactone

This agent has about many similar properties and a similar working spectrum as ethylene oxide. Although in the early 1960's there were schemes made for continuous sterilisation plants based on the use of β-propiolactone as a sterilising agent, it is nowadays hardly used anywhere because of the unwanted side effects such as the cancerogenic and mutagenic properties of this compound.

9.4 Sterilisation using radiation

Radiation sterilisation is a good alternative to chemical sterilisation as it is also a low temperature sterilisation. Mostly used for large operation sterilisation such as disposable medical, laboratory and pharmaceutical products it makes use of energy rich (ionising radiation). Besides the ionising radiation ultraviolet (UV) radiation can also be used for the destruction of micro-organisms but because of the low penetration depth the application of this type of radiation is very limited. UV light is mostly used for the reduction of the bioburden in room air in aseptic filling rooms and airlocks. Sterilisation of smooth surfaces such as wrapping foils used for aseptically produced pharmaceuticals, cosmetical and food products and shallow layers of clear fluids can also be sterilised using UV light. Figure 9.7 gives a summary of the different types of radiation that can be used for killing micro-organisms.

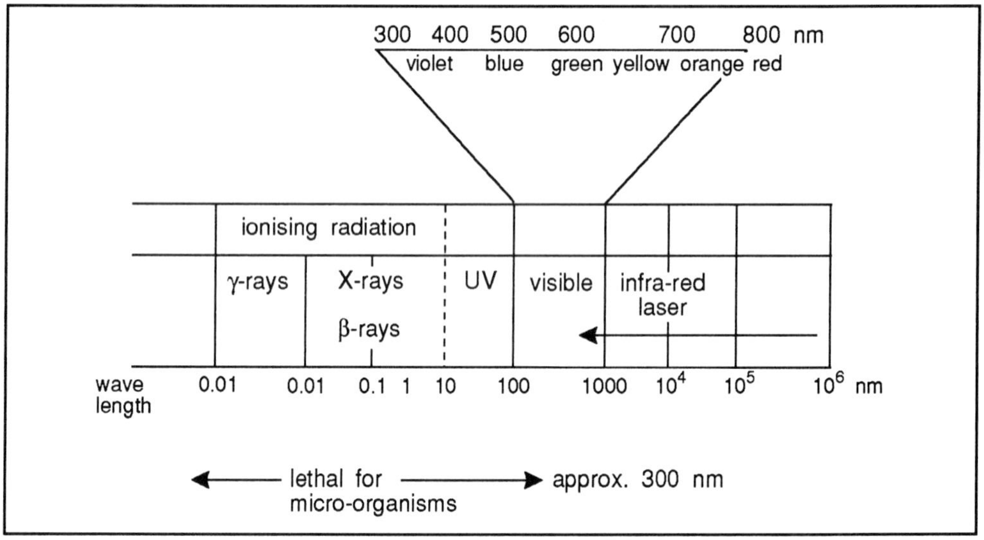

Figure 9.7 Wavelengths that can be used for sterilisation (nm = nanometer or 10^{-9}m).

We remind you that 1 nm = 10Å and 1 nm = 10^{-3} µm.

9.4.1 Ionising radiation

For sterilisation procedures the electron radiation (β radiation) and gamma radiation are used. For the production of electron radiation electron accelerators are used. The accelerated electrons can transfer some millions of electronvolts (Megaelectronvolt = MeV). The binding energy of a covalent chemical bond is mostly below 12 electronvolt, so during sterilisation an immense variety of chemical reactions in the micro-organisms and the product can occur.

Gamma radiation is produced by radioactive elements. For sterilisation purposes only ^{60}Co sources are used. The radiation from a ^{60}Co source consists of two photons (1.33 and 1.17 MeV respectively) and one electron (0.31 MeV). The half-life of such a source is 5.3 years.

The radiation types differ greatly in penetration depth. For example, electrons of 1 MeV are completely absorbed by a water layer of 0.5 cm whilst gamma radiation of 1 MeV is only reduced by half by a waterlayer of 30 cm.

The effect of both types of radiation on micro-organisms is probably two-fold. One effect is on the nucleic acids from which the phosphodiester backbone is broken. Another is by the formation of radicals. Both can lead to the death of the cell.

The molecular events occurring in exposure to radiation are described in the BIOTOL texts 'Genome Management in Prokaryotes' and 'Genome Management in Eukaryotes'. The former of these focuses on micro-organisms, the latter gives an account of the metabolic and physiological consequences of exposure to radiation by higher (including man) organisms.

Factors affecting the sterilisation process

In contrast with the chemical sterilisation there is quite a difference in resistance against radiation between organisms. The most radiation resistant bacteria we know is *Deinococcus radiodurans*, a Gram positive coccus, first isolated from irradiated meat, for which the lethal doses of radiation is 6000 krad. Note *Deinococcus radiodurans* is also known as *Micrococcus radiodurans*. Other factors determinating sterilisation effectiveness are:

- the oxygen effect. In the presence of oxygen, bacteria and yeasts are two to threefold more sensitive to radiation than in its absence;

- humidity has also a positive effect on the destruction of micro-organisms by radiation.

Both oxygen and humidity can contribute to the destruction of micro-organisms by the formation of H_2O_2. Radiation protection is given to micro-organisms by reducing substances especially those which contain sulphydryl groups such as cysteine, glutathione and thiourea.

DNA repair mechanism

Reactivation of micro-organisms after radiation is mostly carried our by a DNA repair mechanism called the SOS system. The high resistance of *Deinococcus radiodurans* is probably based on this system. (Again this is described in detail in the BIOTOL text 'Genome Management in Prokaryotes').

Temperature has a positive effect on the destruction of micro-organisms by radiation. This effect is the basis for the process of thermoradiation.

Radiation sterilisation processes can also be characterised by the D-value. Some D-values are summarised in Table 9.4.

Gram negative bacteria	D-values in Mrad	Gram positive bacteria	D-values in Mrad
Escherichia coli	0.009	*Staphylococcus aureus*	0.02
Salmonella typhi	0.02	*Streptococcus faecium*	0.28
Salmonella typhimurium	0.02	*Micrococcus (Deinococcus)* radiodurans	0.22
Pseudomonas spp.	0.003-0.006	*Micrococcus spp.*	0.4 - 0.5
Flavobacterium spp.	0.003-0.006	*Lactobacillus brevis*	0.12
		Lactobacillus plantarum	0.008
Yeast		Spore formers	
		Bacillus pimilus	0.3
		Bacillus subtilis	0.06
Saccharomyces cerevisiae	0.05	*Bacillus brevis*	0.05
Torulopsis candida	0.04	*Clostridium botulinum*	0.12 - 0.33
		Clostridium sporogenes	0.16 - 0.22
Fungi		*Clostridium welchii* (perfringens)	0.12 - 0.20
Penicillium notatum	0,02	*Clostridium tetani*	0.24
Aspergillus niger	0.047	Viruses	
		MKS	1.3

Table 9.4 D-values for some micro-organisms in Megarads (Mrad).

SAQ 9.7

Why is it technically difficult to sterilise animal food products by ionising radiation?

9.4.2 UV Sterilisation

production of UV dimers

Radiation of wavelengths from 200 to 280 nm has the strongest effect on bacterial survival. The maximum effect being observed at a wavelength of 253.7 nm. The reason why this is so is that the genetic material (DNA) absorbs light maximally at this wavelength. As a source of this type of radiation low pressure mercury vapour lamps are used. The radiation capacity of such lamps is in the order of 0.5-1.5 $\mu W\ cm^{-2}$ at a distance of approximately 10 cm. The radiation dose is the radiation capacity multiplied by the time. The UV radiation, however, is very rapidly absorbed by a lot of different materials, including fluids. Thus the microbicidal effects are very superficial. The effect of the UV radiation on the cell is mainly an effect on the DNA. Under the influence of UV radiation covalent bonds are formed between adjacent pyrimidine residues in the DNA, (Figure 9.8). These pyrimidine dimers induce a distortion in the DNA molecule which in turn inhibit DNA replication and leads to disturbance of the protein synthesis which in the end results in cell death. Some micro-organisms, however, possesses repair systems for UV damage.

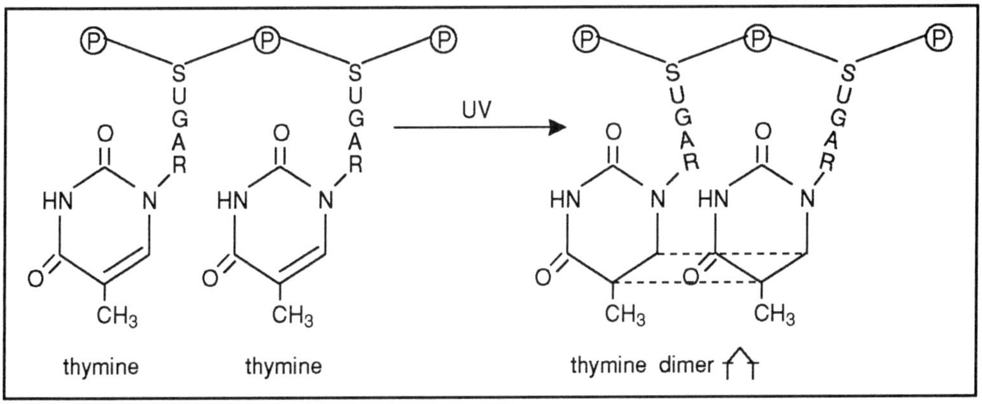

Figure 9.8 Thymine dimer formation by UV-light.

Factors affecting sterilisation procedures

Besides the already mentioned differences and the effect of repair systems there are also effects of humidity, temperature, anti-oxidants, repair and the 'age' of the micro-organisms. (In some cases cells in the stationary growth phase are less sensitive then cells in the logarithmic phase).

We perhaps should not be surprised that micro-organisms have repair mechanisms to protect themselves against UV damage. UV is the only mutagen that is abundant in the natural environment - it is, after all, a component of sun light. Biologically, therefore, it makes sense to evolve a mechanism to repair damage from exposure to UV light. The reader should note that UV light sources age and do not give out the same energy at a given wavelength after a relatively short time. This of course influences the efficacy of the sterilisation procedure.

Let us now turn our attention to a completely different strategy for sterilisation. This far we have described methods in which the intent was to kill all of the micro-organisms present. In the next section we will use a different strategy altogether - namely to remove all of the micro-organisms present in a system. To do this we have to use a system of filters.

9.5 Filter sterilisation

Bacterial filters were already known in the time of Pasteur. The first filters were made by Chamberland, a colleague of Pasteur. Because of the shape, these filters were called Chamberland candles. They were made of unglazed porcelain. The next type of filter developed was the so called Berkefeld filter which consisted of compressed diatomaceous earth. Asbestos filters, similar to the Berkefeld filters, were also developed in Germany mainly by Seitz and under this name they were well known. However, since the discovery that asbestos is carcinogenic, this type of filter is hardly used any more. In the nineteen twenties Schott in Jena invented the sintered glass filter. The grade 5 of this filter has a pore size of 1-2.5µm. The size of the pores is determined by the particles size of the powered glass used to make them. Besides glass, metal and ceramic materials such Al_2O_3 as AlO_3 and ZrO_2 are used to make this type of filter. At the same time as Schott introduced the sintered glass filter, Sartorius manufactured the first membrane filters based on the work of Zsigmondy and Bachman.

materials used for filter construction

9.5.1 Membrane processes

Separation with membranes can be accomplished because of differences in size, affinity, charge, chemical structure and vapour pressure. Besides sterilisation, desalination of seawater by reverse osmosis (hyperfiltration) and kidney dialyses are applications of membrane processes. In all these processes the separation is achieved by a membrane. The membrane can be considered as a barrier between two homogeneous phases, determining the rate of transfer from one phase to the other. Transport through the membrane takes place when a driving force is applied to the components of one phase. In most membrane processes the driving force is a pressure, concentration or activity difference across the membrane. These parameters (pressure, concentration, activity) are covered by one parameter, the chemical potential μ. This chemical potential can be defined as:

$$\mu_i = \mu^o_i + S_iT + V_iP + RT \ln a_i + Z_iF \psi + ... \quad \text{(E - 9.15)}$$

μ^o_i = chemical potential of the standard state
S_i = partial molal entropy
V_i = partial molal volume
R = ideal gas constant
Z_i = valence
F = Faraday's constant
T = temperature
P = pressure
a_i = chemical activity, T_iC_i
ψ = electrical field potential

Note that μ_i is related to concentration (C_i).

The driving forces are usually pressure or concentration or temperature. Two or more can act simultaneously. A classification of membrane processes according to their driving forces is given in Table 9.5.

pressure difference	concentration (acitivity) difference	temperature difference	electrical potential difference
microfiltration	pervaporation	themoosmosis	electrodialysis
ultrafiltration	gas separation	membrane distillation	electro-osmosis
hyperfiltration	dialysis		
piezodialysis	liquid membranes		

Table 9.5 Classification of membrane processes according to their driving forces. Mulder M, (1989), Basic principles of membrane technology, Enschede.

Π Which of the driving forces described in Table 9.5 is used in filter sterilisation.

D'arcys law

Pressure since microfiltration is the process used for filter sterilisations. The volume flux through these microfiltration membranes can be described by D'arcys law in which:

$$\phi_v = Lp\, \Delta P \quad \text{(E - 9.16)}$$

ϕ_v = volume flux

Lp = phenomenological hydraulic permeability = $\frac{k}{\eta \, l}$

k = hydraulic permeability
η = viscosity
l = length of the capillary
ΔP = pressure

The hydraulic permeability is a characterisation of the membrane, in which the surface porosity ε (the fractional pore area times the number of pores) is the main parameter.

Hagen Poiseuille relationship

Depending on the pore geometry the convective flow can be characterised by the Hagen Poiseuille relationship. In the case of a membrane consisting of straight capillaries this is given by:

$$\phi = \frac{\varepsilon \, r^2 \, \Delta P}{8 \, \eta \, \tau \, l} \qquad \text{(E - 9.17)}$$

ϕ = effluent flux
ε = porosity = $n \, . \, \pi \, . \, r^2 / A_m$
n = number of pores
r = pore radius
A_m = membrane surface

Kozeny-Carmen equation

τ = pore tortuosity, in case of cylindrical non-angular pores = 1. In case of capillaries formed by the package of spherical particles the Kozeny-Carmen equation can be used:

$$\phi = \frac{\varepsilon^3 \, \Delta P}{\eta \, (1 - \varepsilon)^2 \, . \, l} \qquad \text{(E - 9.18)}$$

9.5.2 Characterisation of membranes

pore sizes

Membrane characterisation is a way of correlating membrane properties such as pore size, pore size distribution and free volume to membrane separation properties. In sterilisation procedures the main object is to hold back micro-organisms. These micro-organisms differ in size from 30 nm for the smallest viruses to 10 000 nm for yeasts and fungi. The type of membranes used in sterilisation procedures are the so called porous membranes. The pore sizes of these varying from 10 to 0.1 µm and differ also in respect to the pore geometry (Figure 9.9).

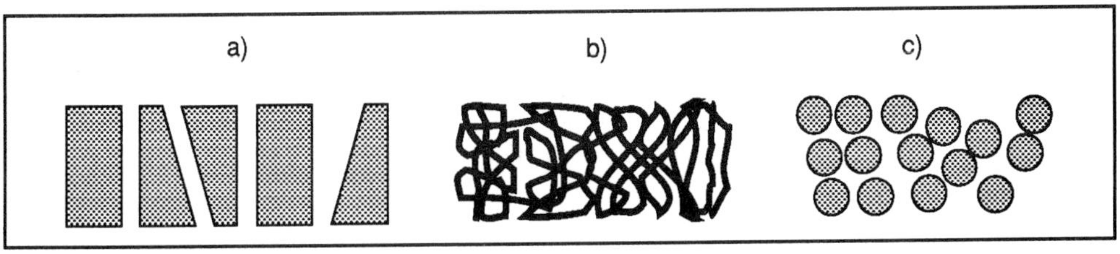

Figure 9.9 Some characteristic pore geometries that can be found in porous membranes.

Π In the examples shown in Figure 9.9, do they have a pore tortuosity (τ) = 1?

The answer is no. None of the pore geometries shown are cylindrical, non-angular pores.

A range of organic polymers and other materials are used for the manufacturing of membranes (Table 9.6). The main characterisation techniques for these microfiltration membranes are: electron microscopy, the bubble point method, mercury intrusion porometry and permeation measurements. We will examine each in turn.

Membrane type	Materials used
polymeric membranes (hydrophilic)	cellulose esters
	(aliphatic) polyamide (PA)
	polycarbonate (PC)
	polyimide/polyetherimide (PI/PEI)
	polysulfphone/polyethersulfone (PSf/PES)
polymeric membranes (hydrophobic)	polytetrafluoroethylene (PFTE, teflon)
	polypropylene (PP)
	polyvinylidenefluoride (PVDF)
ceramic membranes	alumina (Al_2O_3)
	zirconia (ZrO_2)

Table 9.6 Some materials for the manufacturing of porous membranes.

Electron microscopy

SEM

TEM

There are two types of electron microscopy: Scanning Electron Microscopy (SEM) and Transmission Electron Microscopy (TEM). The main advantage with SEM is that the reliefs of the membrane can be visualised. The resolution limit of the simple scanning electron microscope lies in the order of 0.01 μm, while the pore diameter for microfiltration membranes is from 0,1 to 10 μm as we have seen. The main problem with this technique is the preparation of the sample. Because of the high kinetic energies of the electrons used (1-15 kV) the samples are easily damaged or burned. To prevent this damage the samples are covered with a thin layer of heavy metal, mostly gold. The coating procedure is rather critical and can give rise to artefacts, which blurr the real picture.

The bubble point method

Cantor's relationship

Bechhold first evaluated pore size by measuring the pressure necessary to blow air through a water filled membrane. He employed Cantor's relationship.

$$r = 2\sigma/P \qquad \text{(E - 9.19)}$$

in which:

r = radius of the capillary
σ = surface tension water/air
P = pressure

In practice, the filter apparatus is placed upside down so that the air can impinge on the membrane from beneath. Bubbles of air are then observed as they penetrate the membrane into the overlying layer of water. Because the larger pores are open at lower pressures, this method tends to yield high values. This, however, can be used for estimating the pore distribution because if the number of pores, which are permeable to air increases substantially with a small increase in pressure, a narrow pore size distribution is indicated. On the other hand, a gradual increase in the number of air-permeable pores is an indication for a broad pore size distribution. This bubble point method is only valid when the imbibed medium completely wets the membrane and when the ratio of the pore diameter to that of the permeant species is very large. As this ratio diminishes the validity of Equation 9.19 decreases, particularly when there is an interaction between the permeant species and the membrane. Moreover, because of the high water/air surface tension of 73 dyne cm^{-1}, narrow pores require a relatively high pressure before air will permeate. High pressure, however, can lead to plastic flow in the membranes, which results in pore size changes in a time-dependent manner. Because of these difficulties other fluids have been used. Today mostly kerosene is used as the wetting fluid. When pressures are measured in psi and the pore diameter is expressed in µm the following formula holds:

$$\mu m = 5/P \qquad \text{(E - 9.20)}$$

Another convenient relationship is that the kerosene bubble point is exactly half the water bubble point.

Mercury intrusion method

A modification of the bubble point method is the mercury intrusion method. Porometers for this estimation are commercially available. Equation 9.18 is modified to take into account the contact angle θ: between mercury and the membrane.

$$r = \frac{2\,\sigma \cos\theta}{P} \qquad \text{(E - 9.21)}$$

A membrane is placed in a porometer chamber, and pressure is applied, to force the mercury into the pores. It is assumed that all void space is full at the highest pressure, usually about 75 atm. This assumption is of course only valid for cases in which all the voids are of the open cell variety. From the weight of the membranes at the lowest and the highest pressures, bulk densities are obtained from which the void volume (porosity) can be obtained by difference.

Permeability method

Hagen Poiseuille relationship

The pore radius r can also be obtained by measuring the volume of water permeating in a given time at constant pressure from the Hagen Poiseuille relationship (Equation 9.16). For laminar flow in which a Reynolds number Re < 2300 is assumed, Equation 9.16 can be solved for r:

$$r = \sqrt[4]{(8\,\phi\,\eta\,d)/(n\,\pi\,A\,P\,T)} \qquad \text{(E - 9.22)}$$

Further if the void volume $V = n\pi r^2$:

$$r = \sqrt[2]{(8\,\phi\,\eta\,d)/(V\,A\,P\,t)} \qquad \text{(E - 9.23)}$$

Note that Reynolds number is a measure of the ratio of convective mass transfer and diffusion mass transfer. Values of Re below 2300 are typical of laminar flow, higher values indicate turbulent flow (Reynolds numbers are discussed in the BIOTOL text 'Bioprocess Technology: Modelling and Transport Phenomena').

Therefore, pore radii can be determined from permeability measurements in two ways: using the pore density n or using the void volume. The value of n established by (electron) microscopy, however, is not necessary equal to the effective pore density. Because some surface opening can lead to dead ends. The same problem holds for the void volume. When closed voids are present or when the membrane framework is porous, the observed void volume can differ from the effective void volume.

To establish the pore radius the permeability method can, however, be used in combination with the bubble point method to measure the pore size distribution.

SAQ 9.8

1) Using the water bubble point the mean radius of the capallaries of a membrane was found to be 0.1 mm. Given that the surface tension of air/water surface is 73 dyne cm^{-1}, what pressure of air was needed to produce bubbles of air in water with this membrane?

2) If kerosene had been used instead of water in the determination what pressure of air would have been needed to produce bubbles?

3) In the determination carried out in 1), the rate of which bubbles were produced increased rapidly over a very small range of pressures. Are the capillaries of the membrane of similar or diverse radius?

4) Using an electron microscope, the volume of the capillaries in a membrane have been estimated to have a value of x. Using the mercury intrusion method, the volume of the capillanes has been estimated to be about $\frac{1}{2}x$. Explain why different results were obtained by the two methods.

9.6 Air sterilisation

There are several ways in which air can be sterilised, for example by heat and UV radiation. Most widely used, however, are filtration techniques, using porous or fibrous materials as filters. The two types of filters most widely used differ in the packing of the filter material:

- compressed in a cylinder creating a homogenous layer of considerable thickness;
- fine fibres are layered on a firm substratum (mostly textile) so the filtering unit consists of a relatively thin layer.

materials used for filter construction

Materials used for the type 1 filters are cellulose, viscose or (mostly) glasswool. In most cases a pre-filter is installed to remove dust and oil droplets coming from air compressors. The quality of the filter bed is very important since damage of the filter results in a sharp deterioration of performance. For the type 2 filters, mostly plastic fibres, with a diameter of less than 1 mm, are used. These diameters are far less than those of cotton or glasswool. Before using these filters, sterilisation is necessary and mostly performed by passing steam through the filters. Sometimes formaldehyde vapour is also used. It is then necessary before using the filters that the filters are dried or that the formaldehyde is effectively removed by passing a stream of hot air through the filter.

Successful filtration depends largely on the choice of suitable filter material and filter design. The efficiency of industrial filters is difficult to access directly since high flow rates are mostly used. Usually, the Boudillon method is used for this purpose. With this method a fraction of the sterilised air is, at the high flow rate, forced through a narrow slit against the surface of a solid culture medium placed at short distance from the slit.

The efficiency percentage of the filter E, is defined as:

$$E = 100\frac{N_1 - N_2}{N_1} \qquad \text{(E - 9.24)}$$

N_1 = number of incoming micro-organisms

N_2 = number of micro-organisms leaving the filter.

For the ratio of incoming to leaving micro-organisms the following equation holds:

$$\ln\frac{N_1}{N_2} = kL \qquad \text{(E - 9.25)}$$

L = depth of the filter
k = parameter independent of L but dependent on air velocity, filterbed packing etc.

From which can be read that each element of the filter retains only a limited number of micro-organisms. Thus of course the micro-organisms migrate inside the filter. Restriction of migration is achieved by keeping the filter dry.

The optimum filer thickness or the required number of filter layers can be determined in three ways:

- from experiments with small models of the same material and filter bed under the same conditions as in the actual device;
- from data supplied by the manufacturer;
- by calculation from the theoretical relationship.

A useful characteristic of the filter efficiency is the so called decimal reduction filter thickness L_{90}. With Equation 9.25 we get:

$$\log\frac{N_1}{N_2} = \frac{k}{2.3}L \qquad \text{(E - 9.26)}$$

According to the decimal reduction we get:

$$1 = \frac{k}{2.3}L_{90} \qquad \text{(E - 9.27)}$$

and:

$$\log \frac{N_1}{N_2} = \frac{L}{L_{90}} \quad \text{(E - 9.28)}$$

$$L = L_{90} \log \frac{N_1}{N_2} \quad \text{(E - 9.29)}$$

This equation defines the dependence of the filter efficiency on the filter layer thickness or the number of filter layers under constant conditions. The determination of this filter efficiency is complicated by the fact that the decimal reduction thickness is not a constant but depends on the air flow rate. This indicates that the mechanism of particle trapping in the filter depends on flow rate. A filter which is reliable under all kinds of conditions should be designed on the maximum L_{90} value, (that is, the value determined at the critical flow rate at which the filter efficiency is the lowest).

SAQ 9.9

A single layer of a filter membrane has a thickness of 1mm. In an experiment air carrying 10^6 viable spores m^{-3} was passed through the membrane. In the outflow of this layer, there were 10^2 viable spores m^{-3}. What is the decimal reduction thickness (L_{90}) of this filter material?

9.7 Bioreactor design and asepsis

Fermentation processes vary widely in their requirements for asepsis and also in the degree to which the environment must be protected from escape of the micro-organisms and fermentation products (containment).

For containment, especially for work with genetically manipulated micro-organisms, regulation and legislation have been devised both on a supernational and at a national level. Once these levels of asepsis and containment were established it became possible to translate them into construction and engineering demands. In general these demands are:

- no direct connections between sterile and non-sterile parts of the system - Micro-organisms are known to grow through closed vents;
- flange connections should be used as least as possible, because these connections can move because of vibrations and thermal expansion;
- dirty condensation areas are to be avoided, because they can harbour dust particles, which can lead to problems with sterilisation;
- it should be possible to sterilise different parts of the bioreactor system separately;
- it should be possible to steam sterilise the connections to and from the bioreactor so contamination through these entrances can be avoided;
- dead alleys in pipelines and valves are to be avoided because contaminants can accumulate there and cause problems with sterilisation;

- valves and stopcocks should be easily to clean and sterilise. Vessel and pipe work should be equipped with steam sealed drain valves at their lowest point;
- to prevent contamination of the bioreactor contents it is possible to keep an overpressure in the bioreactor, however, this can influence containment demands.

In Figure 9.10 hazards of asepsis and containment are visualised. As we have seen before the demands for physical containment differ slightly from country to country. We will illustrate the basis of physical containment using Holland as an example.

Π Figure 9.10 is quite complex - it might provide a helpful form of summary to make a table of the areas of hazard relating to asepsis and containment. For example:

Asepsis	containment
culture (inoculum)	culture (inoculum)

You should find at least 19 items from our figure. You could probably add some more of your own.

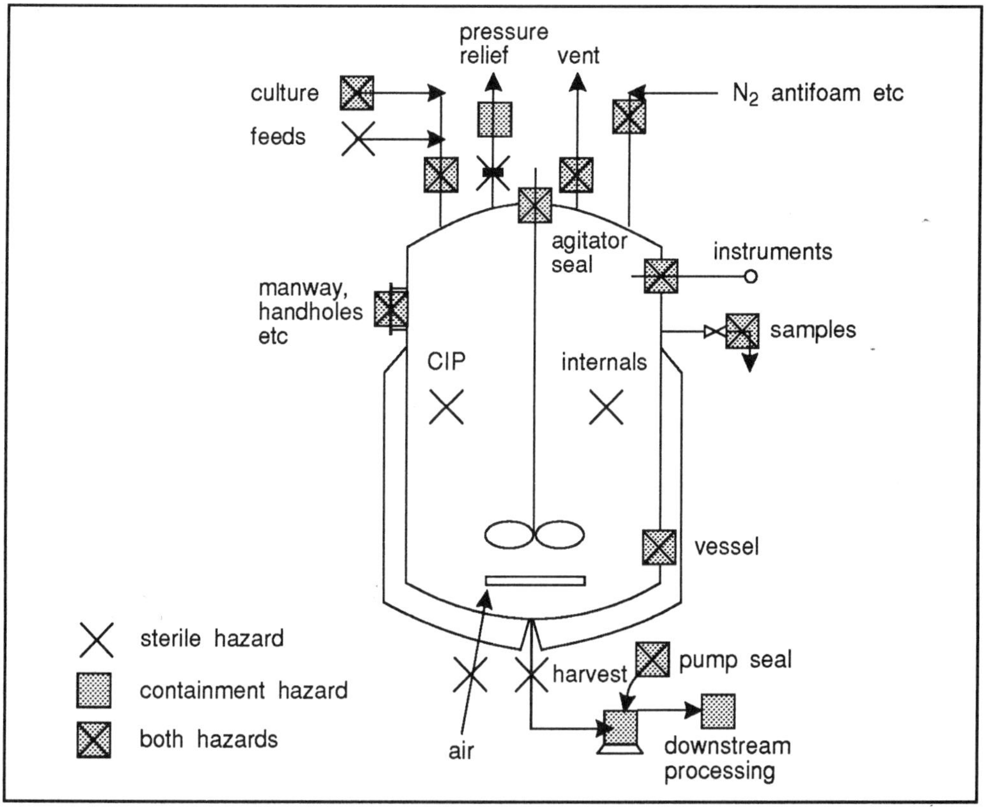

Figure 9.10 Hazard pathways in a fermentor system - adapted from Brooks and Russell (1986), Process Biotechnology. A unified design approach to fermentor sterility and containment.

The demands are in Holland set by the Nederlandse Vereniging Voor Microbiologie. For the cultivation of micro-organisms with a low individual and community risk in bioreactors these demands focus on the following aspects:

Tightness of the bioreactor and the appendages

The micro-organisms should not be able to escape from the bioreactor during cultivation, nor should 'foreign' micro-organisms be able to gain entrance to the bioreactor, (for example by migration through liquid films). In general there should be no more micro-organisms on the bioreactor and in the surroundings than is allowed for small scale laboratory work. The tightness of the bioreactor should be monitored frequently by means of leaktests and by means of air sampling. Also the surface of the bioreactor should be frequently monitored for the cultivated micro-organism. Rubber closing as O-rings etc should be checked and replaced if necessary. In closed cooling and heating systems for bioreactors disinfectants can be used to prevent the escape from micro-organisms via heat exchangers.

Exhaust gas

This gas should be free of micro-organisms, mostly one hydrophobic bacterial filter is sufficient. It is recommended that before the filter a reflux cooler is installed. In long term fermentations the cultivated micro-organisms can grow through humid bacterial filters. In such cases it can be necessary to heat the filter.

Foam control

The bioreactor should have a high level detector for foam, with a feed back to the gas inlet. In case of unexpected strong foaming, it is practically certain that contamination has occurred or that the fermentation for other reasons has failed. If that is so there is a risk of the air vents getting blocked, so that the pressure in the installation can rise to a level where fermentation fluid can be forced out of the bioreactor through valves, tubing etc. In such cases the gas inlet has to be shut off, preferably automatically by a signal from the high level foam detector. It is also possible to kill the micro-organisms by addition of a germicide.

Construction material

For small scale work glass of good quality can be used. For larger bioreactors stainless steel with viewing glasses of thick industrial glass are necessary.

Stirrers

Shafts should be closed by a 'double mechanical seal' or use magnetically driven stirrers.

Inoculation, sampling, harvesting and contamination risk of the surroundings

Spilling of culture fluid must be avoided and the formation of aerosols prevented.

Temperature regulation via a water circuit

This circuit has to be separated from the tap water circuit.

Downstream processing

All manipulations causing aerosols should be exercised in biosafety cabinet.

Emergency situation

An action plan should be prepared.

This may appear quite a long list. However, many of the items become second nature in process operation.

Π Look down this list and tick those items which influence the actual physical design of a bioprocess.

You probably came to the conclusion that all of the items listed will in some way or another, influence the design of the equipment used in a bioprocess. You would be quite right. Asepsis and containment pervades the whole operation of the bioprocess.

Summary and objectives

In this chapter we have learnt about a variety of sterilisation procedures used in bioprocess technology. We have examined relationships between sterilisation efficiency and a variety of parameters and explained how these procedures are used to sterilise different types of items (eg equipment, media, air).

Now that you have completed this chapter you should be able to:

- calculate sterilisation parameters (including D-values, z-values and F-values) from supplied data;
- explain how various parameters influence sterilisation efficency;
- describe how media sterilisation may be carried out on large volumes and explain why media quality influence sterilisation strategy;
- describe the methods used to evaluate filter membranes and calculate filter parameters (eg L_{90}) from supplied data;
- explain how asepsis and containment may influence bioreactor design.

Responses to SAQs

Responses to Chapter 2 SAQs

2.1 9.07 mg l^{-1}

H_g for oxygen is 1.4 mmol l^{-1}.Pa at 20°C (Table 2.1). Air contains 21% of oxygen, therefore, P_g = 0.21 Pa. The molar weight of oxygen is 32 g mol^{-1}. Using these data and Equation 2.1, C* = 9.41 mg l^{-1} in water. When 30 g l^{-1} of glucose is added, C_{gl}* can be calculated with Equation 2.5. Thus:

C_{gl}* = 9.41 (1 - 0.0012 x 30) = 9.07 mg l^{-1}, corresponding to a decrease of ca. 36% in comparison to pure water.

2.2 Barrier codings refer to Figure 2.2.

1) Substrate diffusion through the liquid phase inside the fungal pellet will constitute the most important barrier, ie barrier VII.

2) This is a low viscosity medium, a combination of the liquid phase resistance and the gas-liquid interface will be rate controlling, ie barriers II and III.

3) Diffusion of substrate through an almost stagnant liquid phase inside the alginate matrix will be rate controlling, ie barrier VII.

4) In the early stage, the medium is of low viscosity, therefore barriers II and III will be rate limiting. The viscosity gradually increases during the process and barrier IV will become increasingly more important.

5) There will be a shortage in gas-liquid contacting area, ie barrier II, and possible III as well due to slow mass transfer through a thick film resulting from low stirrer speeds.

2.3 The CO_2 concentration profile is the inverse of the gas concentration profile in Figure 2.2, ie the highest CO_2 concentration is observed in the cell aggregate and the lowest CO_2 concentration is measured in the gas bubble. The same mass transfer barriers are present.

Thus your figure should have looked something like this:

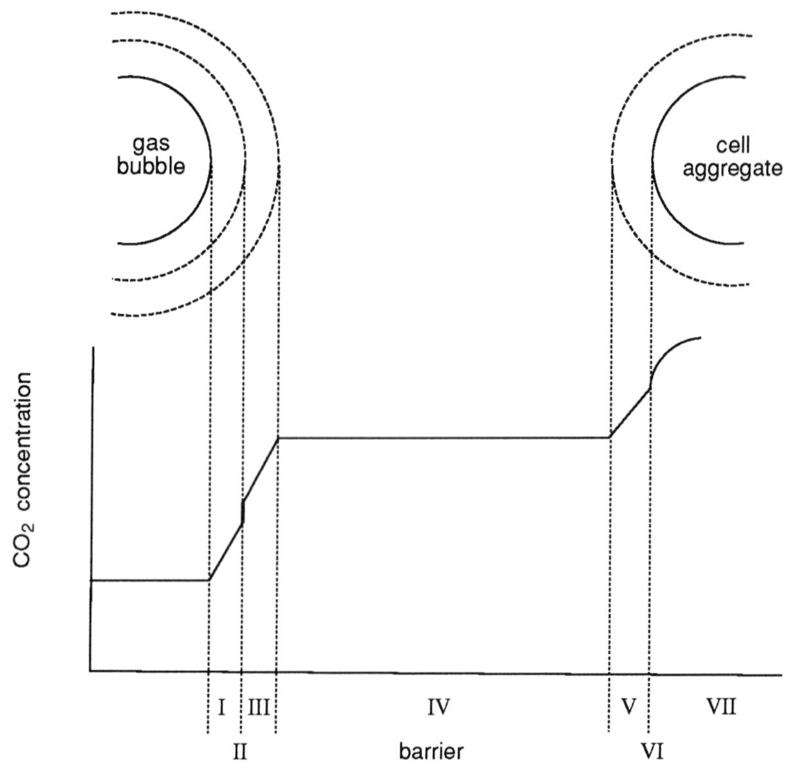

2.4 The maximum concentration that can be grown in the Erlenmeger flask is 2 g l^{-1}. In the continuous stirred tank reactor (CSTR) is 10 g l^{-1}.

OTR will be highest when $C_l = 0$, ie $(C^* - C_l) = C^*$.

At 30°C, $C^* = 0.21\ (67.75 - 56.61 + 33.21 - 8.34) = 7.56$ mg l^{-1} (use Equation 2.3). The OTR = 7.56 x 0.02 = 0.15 mg l^{-1}. s^{-1} corresponding to OTR = 0.54g l^{-1}.h^{-1}. This is the maximum amount of oxygen available to the cells. Thus, 0.54/0.27 = 2 g l^{-1} of *E. coli* dry weight can be adequately supplied with oxygen during growth in this Erlenmeyer. In the CSTR, k_La is five times as high, implying that ca. 10 g/l of cells can be obtained.

2.5 $2 \times 10^2\ m^{-1}$

The volume of the medium appears to increase because of the air bubbles.
The volume of the medium = 5 l
The volume of the gas (V_g) = 1 l
The total volume of the reactor medium = V_r = 6 l
The gas hold-up value ε_g is $V_g/V_r = 1/6$
$d_b = 5\ mm = 5 \times 10^{-3}\ m$
Thus $a = 6\varepsilon_g/d_b$

$$a = 6 \times \frac{1}{6} \div (5 \times 10^{-3}\ m) = 2 \times 10^2\ m^{-1}$$

2.6 1) $k_La = 0.30\ min^{-1}$ or $0.005\ s^{-1}$

Use Equation 2.28 and replace C by pO_2 (%) and C* by 100%.
Then, $\ln(1 - pO_2/100) = -k_La * t$

t(min)	0	1	2	3	4	5	6	7
pO_2 (%)	0	26	45	59	70	78	83	88
$(1 - pO_2/100)$	1	0.74	0.55	0.41	0.30	0.22	0.17	0.12
$-\ln(1 - pO_2/100)$	0	0.30	0.60	0.89	1.20	1.51	1.77	2.12

You will then need to plot a graph of $\ln(1 - pO_2/100)$ against t (see Figure 2.7).

From the graph of $\ln(1 - pO_2/100)$ versus t, k_La can be calculated to be about 0.30 min^{-1} corresponding to ca. 0.005 s^{-1}, which is a rather low value.

2) As electrode response times are usually in the order of 1-10 s, no substantial problems were to be expected in this experiment in which the time scale is in the order of minutes.

2.7

1) False. Long filaments effectively increase the viscosity of the medium and this reduces mixing which, in turn, limits oxygen transfer (ie k_La is lowered). (See Figure 2.3d).

2) True. See Equation 2.25, Table 2.4 and Figure 2.3b.

3) False. Samples have to be taken, there is no on-line use of this method.

4) True. (See 2.7.2)

5) True.

6) True. Over a long period the biomass concentration (X) in the vessel would increase (ie the cells would grow). Since the slope = OUR * X we would anticipate the slope would increase as X increases.

Responses to Chapter 3 SAQs

3.1

1) 10 000 W

The relationship to use is Equation 3.5

$$P_s = N_p \cdot \rho \cdot N^3 D^5$$

Thus $P_s = 10 \times 1000 \times 1^3 \times 1^5 = 10\,000$ W

2) 80 000 W

$$P_s = 10 \times 1000 \times 2^3 \times 1^5 = 80\,000 \text{ W}$$

3) 5000 and 40 000 W

Since $P_g \approx 0.5\ P_s$ (see Equation 3.6)

3.2

1) According to Reuss *et al* (1980) the $\frac{P_g}{P_s}$ ratio is proportioned to $\left(\frac{T}{D}\right)^{0.8}$ (Equation 3.7).

Thus doubling the value of D would decrease the $\frac{P_g}{P_s}$ ratio by a factor of $\left(\frac{1}{2}\right)^{0.8}$

2) Reuss *et al* (1980) indicates that the ratio of $\frac{P_g}{P_s}$ is proportional to the Flow number (Fl). In the following way.

$\frac{P_g}{P_s}\ \alpha\ Fl^{-0.38}$ (see Equation 3.7)

But flow number = ϕ_g/ND^3

thus $\frac{P_g}{P_s}\ \alpha \left(\frac{\phi_g}{ND^3}\right)^{-0.38}$

Hughmark (1980) on the other hand indicates that the

$\frac{P_g}{P_s}\ \alpha \left(\frac{\phi_g}{NV}\right)^{-0.25}$ (see Equation 3.8)

Thus both authors agree that the $\frac{P_g}{P_s}$ ratio will fall a ϕ_g rises, but there is some discrepancy between them over the factor (-0.38 and -0.25) describing this relationship.

3.3

Conditions 1, 3 and 4 would give complete circulation.

The T/D ratio for the vessel = 1.25/0.5 = 2.5 which is the ratio for the vessel described in Figure 3.11. This we can use this figure to solve this problem.

We next calculate the impeller tip velocity from πND for each impeller speed. Since D = 0.5m then $\pi D \approx 1.6$.

Thus:

Impeller speed (N s^{-1})	Impeller tip velocity (ms^{-1})
6	9.6
1	1.6
3	4.8
4	6.4
2	3.2

Reading from the graph shown in Figure 3.11, we can have ϕ_g greater than any of those represented on the graph when the impeller tip velocity = 9.6 ms^{-1}. So we would get complete circulation for condition 1.

By similar reasoning, condition 2 would produce incomplete gas circulation, condition 3 would give complete mixing, condition 4 would give complete circulation, condition 5 would give incomplete circulation.

3.4

1) Approximately 25s

We can use Mersmann's relationship (Equation 3.22) to determine this since T/D = 3 (ie 2 < T/D < 3.3). In this relationship

$$\frac{t_m^3\ P}{\rho T^5} = 300$$

$$\text{Thus } \rho T^5 = \frac{(20)^3 \times 500}{300} \text{ as } t_m = 20\text{s and } P_s = 500 \text{ W}$$

$$\rho T^5 = \frac{8000 \times 500}{300} = 13333. \text{ Now if P} = 250 \text{ W}$$

$$\text{then } \frac{t_m^3 \times 250}{13333} = 300. \text{ Thus } t_m^3 = \frac{300 \times 13333}{250} = 16000$$

t_m = approx. 25 s^{-1}

2) 16 s^{-1}

Since we are working with turbulent conditions (that is Re-number > 5000) then Nt_m = constant. Under the first set of conditions $N_1 = 20\ s^{-1}$ and t_m = 20s.

Under the second set of conditions t_m approx. 25 s^{-1} and N_2 = ?

Thus $N_1 \cdot t_{m1} = N_2 \cdot t_{m2}$ Therefore: $20 \times 20 = N_2 \times 25$ $N_2 = 16\ s^{-1}$

3.5

1) A foam is a gas:liquid mixture in which in 80% of volume of the mixture is a gas. 80% is not a right figure. Some foams have a higher gas content, others a lower one. The key point is that the gas content is high and it is present as more-or-less stagnant bubbles.

2) Typical values are 3-5 mm.

3) Should read as: Salts tend to decrease bubble coalescence, thus equilibrium bubble sizes are decreased.

4) Should read as: Gas hold-up is characterised by freely moving bubbles.

3.6

1) $\varepsilon = 0.167$

live $\varepsilon = \frac{V_g}{V_l + V_g}$ (see Equation 3.23) $= \frac{2}{10+2} = 0.167$

2) a) P/V will have a greater effect than v_s. At high salt concentrations ε is related to P/V and v_s as described in Equation 3.27. Thus ε is proportional to $\left(\frac{P}{V}\right)^{0.57}$ and to $v_s0.24$.

b) P/V will have a lesser effect than v_s. In water the relationship of ε to P/V and v_s is as described in Equation 3.24. Thus ε is proportional to $\left(\frac{P}{V}\right)^{\frac{1}{3}}$ and to $(v_s)^{\frac{2}{3}}$

3.7 1) 12.5 s^{-1}

Since $k_La = k_L20a \cdot \theta$ (T-20) with θ = 1.020-1.024 (Equation 3.32) then k_La increases about 2.5% $°C^{-1}$. Thus over 10°C rise k_La will increase by about 25%. 25% of $10s^{-1}$ = 2.5 s^{-1}. Thus the new rate will be 12.5 s^{-1}.

2) $k_La = K\left(\frac{P_g}{V}\right)^{\alpha} \cdot (v_s)^{\beta}$ (see Equation 3.28).

3) $\frac{1\,g\,l^{-1}\,s^{-1}}{0.1\,C^*}$. Since OUR = k_La (C* - C_l) and C_l = 0.9 C*. Then 1 gl^{-1} s^{-1} = k_La (C* - 0.9 C*)

$k_La = \frac{1\,g\,l^{-1}\,s^{-1}}{0.1\,C^*}$.

3.8 1) True.

2) Untrue. When the distance of the stirrer is less than the diameter of the stirrer from the bottom of the tank, the dissipated power decreases (see 3.6.3).

3) True. $k_La \sim T/H$.

4) True.

Responses to Chapter 4 SAQs

4.1 1) $v_{lc} = 0.9\,(0.1g)^{1/3}$

At gas flow rates of 0.1m s^{-1} we have heterogeneous flow condition. The relationship we need is given by Equation 4.2.

Thus $v_{lc} = 0.9\,(g.T.v_s)^{1/3}$

Substituting in the given values: $v_{lc} = 0.9\,(g.1.0.1)^{1/3} = 0.9\,(0.1\,g)^{1/3}$

2) $\phi_l = 0.3\,(g)^{1/3}$

The correlation that is required is given by Equation 4.3.

$\phi_l = 0.30 \cdot T^{\frac{5}{3}} \cdot \phi_g{}^{\frac{1}{3}} \cdot g^{\frac{1}{3}}$

Substituting in the given values $\phi_l = 0.30\ 1^{\frac{5}{3}}.\ 1^{\frac{1}{3}}.\ g^{\frac{1}{3}} = 0.3\ (g)^{1/3}$

3) 12.5

The number of mixing compartment n is given by

$n = \dfrac{H}{0.8 \cdot T}$ (Equation 4.7). Thus $n = \dfrac{10}{0.8} = 12.5$

4) $t_c = \dfrac{11}{(0.1g)^{\frac{1}{3}}}$

Since $t_c = \dfrac{H}{0.9(g * T * v_s)^{\frac{1}{3}}} = \dfrac{10}{0.9\ (g * 1 * 0.1)^{\frac{1}{3}}} = \dfrac{11}{(0 * 1g)^{\frac{1}{3}}}$

4.2 1), 2), 3) and 4) are all true. If you got any of these wrong, we suggest you re-read the section on bubble size at the sparger:

1) is stated on Equation 4.14.

2) can be shown to be true from the data in Figure 4.8.

3) can be shown to be true from the data in Figure 4.9.

4) is described in the text.

4.3 1) The two conditions we hoped you would identify are that either the system involves a liquid which allows coalescence, thus, irrespective of v_s, the bubbles coalesce to produce bubbles which at equilibrium have a $d_{vs} = 6$ mm or that non-coalescing liquid is involved but that the sparger produces bubbles with a $d_{vs} = 6$ mm. The first of these explanations is illustrated in Figure 4.10.

2) Non-coalescing

Clearly the bubble size depends upon the P/V ratio. In Figure 4.10 we showed that with a non-coalescing liquid, the bubble size (d_{vs}) decreases as the P/V ratio increases.

3) Homogeneous flow

For homogeneous flow conditions, $\varepsilon = \dfrac{v_s}{0.25}$ (see Equation 4.23).

Thus $\varepsilon = 4v_s$ or $v_s = \dfrac{1}{4}\varepsilon$.

4) a) 0.6

b) 4

Since for heterogeneous flow $\varepsilon = 0.6\ v_s^{0.7}$ (Equation 4.22) then $\varepsilon = 0.6\ (1)^{0.7} = 0.6$

For homogeneous flow $\varepsilon = \dfrac{v_s}{0.25}$ (Equation 4.23)

Thus $\varepsilon = \dfrac{1}{0.25} = 4$

5) Increase

Since this decreases liquid circulation, it increases hold-up.

4.4

1) 18.45% oxygen

We can use our rule of thumb that oxygen concentration is reduced by 0.55% for every m. Thus with 5 m column, 5 x 0.55 = 2.75% oxygen will be removed. Thus the oxygen concentration in the gas outflow = 21-2.75% = 18.25%.

2) No, the oxygen concentration would be virtually unaffected by this temperature change.

3) 16.25%

The conditions described are the same as those illustrated in Figure 4.14. With a column height of 5 m, the % oxygen removal m^{-1} = 0.95% (from Figure 4.14). Thus over 5 m, the oxygen removal would be equivalent to 5 times 0.95% = 4.75%. Thus the oxygen left in the outflow gas = 21-4.75% = 16.25%.

4.5

1) We would have selected a membrane reactor because the cells are grown in a protected environment while nutrients are supplied via media perculating on the other side of the membrane. You could, however, have chosen a cell lift reactor in which the cells would be grown attached to beads. The critical issue is that we need to protect the cells from shear forces.

2) We would have selected a membrane reactor because the lactic acid could have been removed from the medium via dialysis across the membrane.

3) This is a type of cell lift reactor but these devices can also be used in a fluidised-bed reactor.

4) We would anticipate that you would select the pumped tower loop reactor.

5) You should have remembered from 4.7.2 that the modular design of the mutli-purpose tower bioreactor allows for flexibility; a system can be assembled according to the demands of the bioprocess.

6) Here we believe there are two main options. One is the liquid impelled loop reactor; the other is a membrane reactor. The membrane reactor enables us to create two separate compartments (one aqueous, the other organic) separated by a membrane. The membrane allows passage of low molecular weight components between the

two compartments. The liquid impelled loop reactor achieved contact between the two phases by creating droplets of one solvent within the other, thus enabling diffusion between the two phases.

In real life the choice of reactor type depends upon many factors, the answers given here are rather simplistic and are rather generalised. In practice we would have to consider many more parameters such as the molecular weight of reactants/products, equipment costs, scale of operation and so on.

Responses to Chapter 5 SAQs

5.1

Feature	Immobilised system using an enzyme as the biocatalyst	Immobilised system using whole cells as the biocatalyst
Products are usually chemically very similar to substrates	✓	
The reaction rate is likely to slow down as time passes even though the substrate concentration may be maintained at a constant level	✓	
Energy is expended by the biocatalyst to maintain itself		✓
The products of the process are likely to contain a complex mixture of chemicals		✓

The first feature listed above is typical of system in which a single enzyme is used. A single enzyme will catalyse a single reaction step. Thus the product is likely to be similar to the substrate. With whole cells, the substrate is likely to be passed down a whole sequence of reactions (a metabolic pathway). The accumulation of chemical changes means that substrates and products are often quite different. You should note, however, that in some processes, although whole cells are used, only a limited number of enzymes within these cells are employed. In such a case, substrates and products are also chemically similar.

With the second feature, we concluded that this is typical of a single enzyme system. Enzymes are unstable and thus become progressively denatured. The more denaturation has occurred, the less active enzyme remains available for catalysis. Thus the reaction rate will diminish. With whole cells, they are capable of repair and replication. With such systems the rate may be maintained or even increased over a period of time.

Feature three, follows from the discussion above and is typical of whole cell systems.

Feature four is typical of whole cell systems as whole cells contain many different enzymes and the products of metabolism may be many and varied. Thus, often, whole cell systems produce a chemically complex mixture of products.

5.2 Like many aspects of life, the answers to these questions are not straightforward and thus we have provided a brief discussion with our responses to these questions.

1) Usually true. As the biocatalyst is retained in the reactor, the product leaves the reactor free (or largely free) of the biocatalyst. This generally simplifies the purification of the product. There are, however, circumstances where there is no need to remove the biocatalyst from the product and thus in these circumstances, we cannot regard the removal of the catalyst from the product an advantage.

2) True in almost all cases. With enzymes which require cofactors or metal ions, these additional requirements are usually water soluble and are removed with the products of the reaction. Thus they need to be continually renewed. For example consider the following reaction.

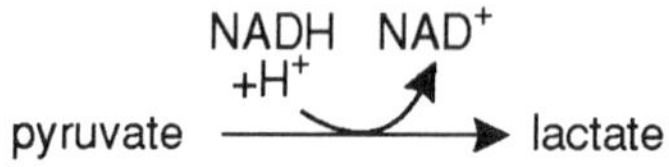

This reaction is catalysed by the enzyme lactate dehydrogenase. To proceed, it requires NADH. Thus the problem using this enzyme in an immobilised system is how to supply NADH. In principle we could collect the water soluble NAD^+ that is produced, reduce it to NADH and return it to the reactor. This is not an easy process. One solution is to use intact cells which retain the co-factors within the cell membrane and thus makes them available for enzyme activity.

3) We can not answer simply true or false to this question. Very often this immobilisation procedures will have a negative effect on enzyme activity, but on the other hand the enzyme can also be 'fixed' into a configuration with an increased activity or increased stability.

4) This is true, immobilisation presents washing out of the biocatalyst.

5) False. We cannot however give a simple answer to this. If the biocatalyst is expensive to prepare, immobilisation may be cheaper to operate because the biocatalysts can be used many times. However, the immobilisation procedure may be quite expensive and may even denature the biocatalyst. In such cases producing an immobilised system may be much more expensive than using the free enzyme. We must also consider the impact immobilisation has on the costs of product purification. We must conclude that in some cases immobilisation is cheaper than using free biocatalysts. In other cases the reverse is true.

Responses to Chapter 6 SAQs

6.1 1)

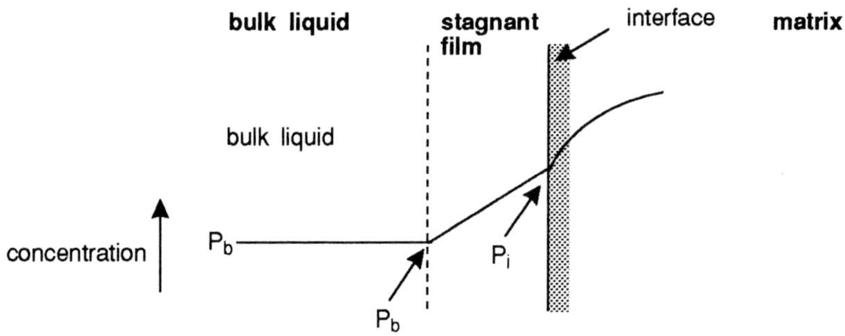

This is really the reverse of the profile shown for substrate. In this case the product (P) is made in the matrix and diffuses out of the matrix into the stagnant film and thence into the bulk liquid.

2) This is a little more difficult. The key is to look at the interface.

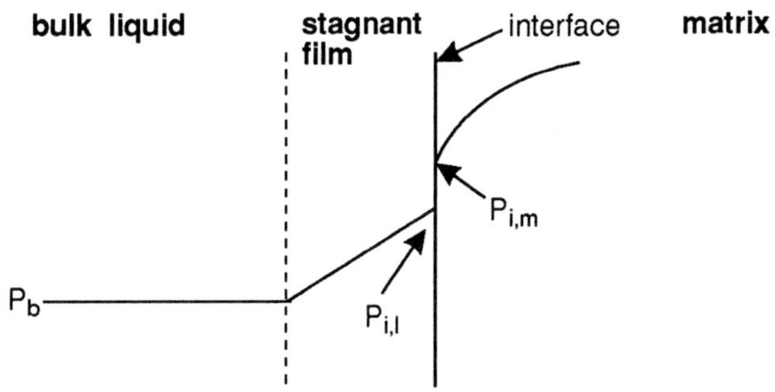

Note that on the matrix side of the interface the concentration of the product ($P_{i,m}$) is greater than on the liquid side ($P_{i,l}$). This reflects the partitioning of the product between matrix and liquid. The greater the partitioning, the greater the difference between these two. In most systems, however, the matrix and the liquid phase are very similar and there is only limited partitioning. Thus the condition described in 1) above usually occurs. Note that the effect of partitioning on mass transfer across interfaces is described in the BIOTOL text 'Bioprocess Technology: Modelling and Transport Phenomena'.

6.2 1) To prove the enzyme follows Michaelis Menten kinetics (ie $v(S) = \frac{V_{max} S_b}{K_M + S_b}$) it is easiest to plot $\frac{1}{v(S)}$ against $\frac{1}{S_b}$. A straight line would indicate Michaelis Menten kinetics apply.

Thus from the data given in the question we calculate

$\frac{1}{S_b}$ l mol^{-1}	$\frac{1}{v(S)}$ (min mol^{-1})
5.00×10^3	4.71×10^7
3.30×10^3	3.7×10^7
2.50×10^3	3.4×10^7
1.67×10^3	3.0×10^7
1.00×10^3	2.6×10^7

Now we plot the graph using these values

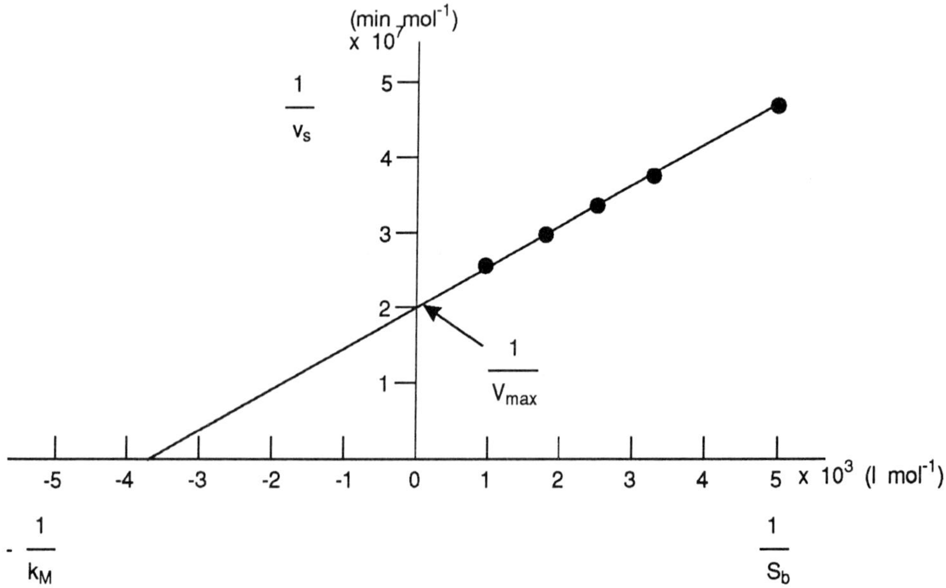

The straight line indicates that Michaelis Menten kinetics are being followed.

2) From the intercept $\frac{1}{V_{max}}$ = 2×10^7 min mol^{-1} thus $V_{max} = 5 \times 10^{-8}$ mol l^{-1}. From the intercept on $\frac{1}{S_b}$, $-\frac{1}{K_M}$ = -3.7×10^3 l mol^{-1} (approx). Therefore $K_M = 2.7 \times 10^{-4}$ mol l^{-1}.

Responses to Chapter 7 SAQs

7.1 1) 3×10^{-2} mol l^{-1}

Since $S(r) = S_b - 0.5 \frac{k_0}{D_e} (Rp^2 - r^2)$

Then in the first case

$S(r)_1 = 1 \times 10^{-2} = 4 \times 10^{-2} - 0.5 \frac{k_0}{D_e} (Rp^2 - r^2)$ ...... A

In the second case

$S(r)_2 = 6 \times 10^{-2} - 0.5 \frac{k_0}{D_e} (Rp^2 - r^2)$ ...... B

Since k_0, De, Rp and r are the same in both cases, we can use Equation A to eliminate $0.5 \frac{k_0}{D_e} (Rp^2 - r^2)$ from Equation B.

Thus $3 \times 10^{-2} = 0.5 \frac{k_0}{D_e} (Rp^2 - r^2)$

Thus $S_{r2} = 6 \times 10^{-2} - 3 \times 10^{-2}$

$= 3 \times 10^{-2}$ mol l^{-1}

2) It would rise to 4.5×10^{-2} mol l^{-1}

Since $0.5 \frac{k_0}{D_e} (Rp^2 - r^2) = 3 \times 10^{-2}$ (see 1) above)

In the new condition if we have half the amount of biocatalyst k_0 will be reduced to half its former value. Thus for this new k_0 $0.5 \frac{k_0(new)}{D_e} (Rp^2 - r^2) = 1.5 \times 10^{-2}$

Thus the concentration at the mid point

$S(r) = S_b - 1.5 \times 10$

$= 6 \times 10^{-2} - 1.5 \times 10^{-2} = 4.5 \times 10^{-2}$ mol l^{-1}

7.2 1) Profile b). Since the concentration at position r in the biofilm is given by:

$S(r) = S_b - 0.5 \frac{k_0}{D_e} (Rp^2 - r^2)$ (Equation 7.9) the higher the effective diffusion coefficient (D_e) the shallower the gradient will be.

2) Profile a). Since $S(r) = S_b - 0.5 \frac{k_0}{D_e} (Rp^2 - r^2)$, then the greater the value of k_0, the greater the gradient will be because of the faster consumption of substrate consumption within the biofilm.

3) Profile a). This is really the inverse of question 1): the lower the effective diffusion coefficient the greater the gradient in substrate concentration.

7.3 Your drawing should have looked like this

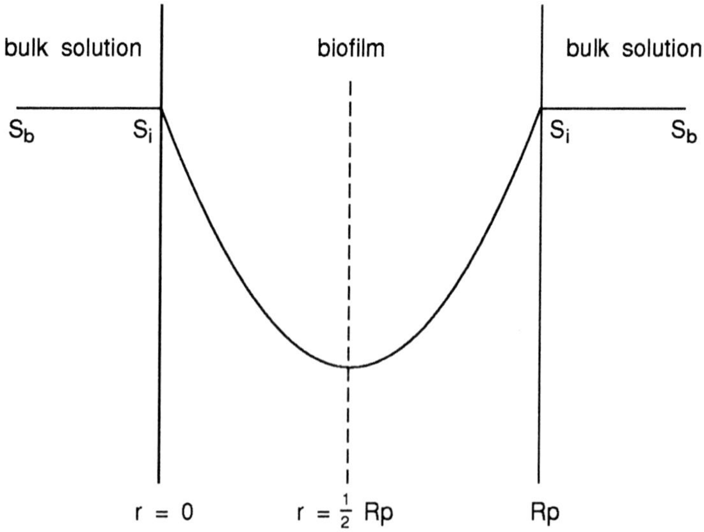

The substrate would diffuse from both faces. At the midpoint ($r = \frac{1}{2}Rp$) the diffusion from one side would balance diffusion from the other. At this point $\frac{dS}{dr} = 0$.

Hence in such a situation the boundary conditions are that at r = 0 and r = Rp; $S_i = S_b$ and at $r = \frac{1}{2}Rp$; $\frac{dS}{dr} = 0$.

7.4

1) The substrate does not penetrate the whole membrane. We know this from the following calculation.

   The total volume of the biofilm is Area x Thickness = 100×10^{-3} m^3 = $0.1 m^3$.

   But the zeroth order reaction constant = 1 mol m^{-3} h^{-1}. Thus 0.1 m^3 of membrane should consume 0.1 x 1 mol h^{-1} = 0.1 mol h^{-1}.

   However, it only consumes 0.08 mol h^{-1}. Thus it is working below capacity and this must reflect substrate depletion.

2) 0.8 mm.

   Since the substrate is consumed at 0.08 mol h^{-1} and the zeroth order reaction constant = 1 mol m^{-3} h^{-1} this must mean that 0.08 m^3 of the biofilm is consuming substrate.

   The area of the biofilm is $100 m^2$, so the effective working depth of the membrane is $\frac{0.08}{100}$ m = o.0008 m = 0.8 mm.

7.5

1) Volume = 1 m^3; area = 500 m^2.

A total amount of biocatalyst of 100 kg was used at a concentration of 100 kg m^3. Thus the volume of the biofilm must be 1 m^3. But, since the thickness of the biofilm was 2 mm (= 2 x 10^{-3}m), the area must be $\frac{1}{2 \times 10^{-3}}$ m^2 = 500 m^2.

2) Oxygen only penetrates to 1.2 mm from the bulk liquid. Since the reaction rate constant = 1 x 10^{-3} mol s^{-1} kg^{-1} . 100 kg of biocatalyst should consume 100 x 1 x 10^{-3} mol s^{-1} = 0.1 mol s^{-1}.

   However, it only consumes 0.06 mol s^{-1}. Thus only $\frac{6}{10}$ of the biofilm has oxygen available for consumption. Thus oxygen only penetrates $\frac{6}{10}$ of 2 mm = 1.2 mm.

3) Effectiveness factor = 0.6.

   Since $\eta = \frac{Rp - rd}{Rp}$ and Rp = 2mm and rd = 0.8 mm.

   We get $\eta = \frac{2 - 0.8}{2} = 0.6$

4) 0.327 mol kg^{-1}

   Since $\eta = \sqrt{(2S_b D_e / Rp^2 k_0)}$

   Then $0.6 = \sqrt{(2S_b \times 2.2 \times 10^{-9}/(2 \times 10^{-3})^2 \times 1 \times 10^{-3})}$

   Note we have converted Rp into meters to have consistant of units.

   Thus $0.36 = \frac{2S_b \times 2.2 \times 10^{-9}}{4 \times 10^{-6} \times 1 \times 10^{-3}}$

   $0.36 = 1.1\ S_b$

   Thus S_b = 0.327 mol kg^{-1}

   Note the rather strange units of concentration. This is because the reaction rate constant was expressed in mol s^{-1} kg^{-1}.

7.6 0.9 mol m^{-3}

At the point value where S(r) = 0 at r =0, $2\, D_e\, S_b / Rp^2 k_0 = 1$ (Equation 7.19).

Thus $2 \times 2.2 \times 10^{-9}\ S_b / (2 \times 10^{-3})^2 \times 10^{-3} = 1$.

Note we have converted all linear measurements to m.

Thus S_b = 0.9 mol m^{-3}

The key question is therefore can we dissolve oxygen in the bulk liquid to give a concentration of 0.9 mol m^{-3} if this cannot be achieved, then oxygen will become limiting unless we change one of the other parameters (eg Rp or k_0).

7.7 1) Your responses should have looked something like this:

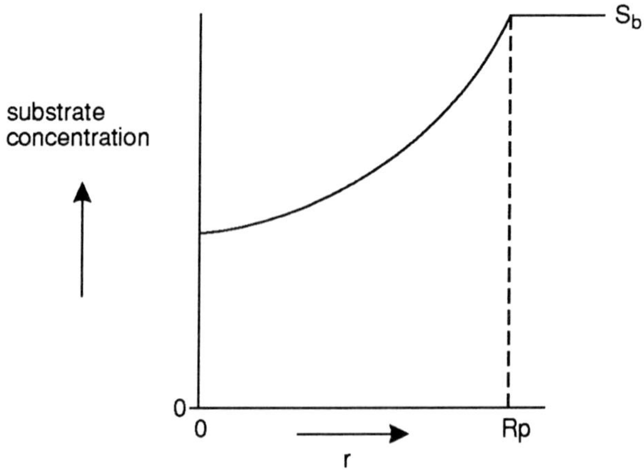

Since $Rp^2k_0/S_b D_e < 6$, then substrate limitation does not take place (see Equation 7.25) thus S(r) is finite when r = 0.

2)

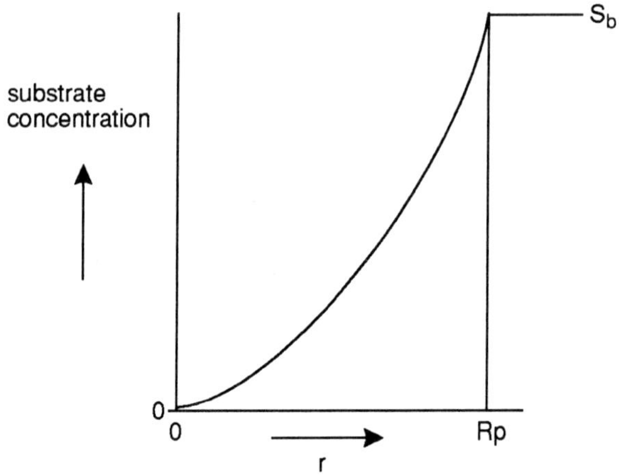

Here substrate consumption leads to S(r) = 0 when r = 0.

3)

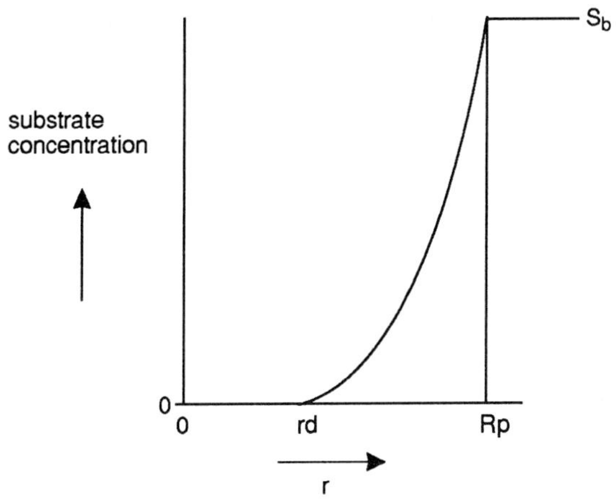

Because $Rp^2 k_0/S_b D_e > 6$, this means we get substrate limitation.

7.8 At least 4.44 mol m^{-3}.

Since the condition in which substrate just becomes limiting is given by:

$$\frac{Rp^2\, k_0}{S_b\, D_e} = 6 \text{ for a sphere (Equation 7.25)}$$

Then:

$$\frac{(4 \times 10^{-3})^2 \times 2 \times 10^{-3}}{S_b \times 1.2 \times 10^{-9}} = 6. \text{ Therefore: } S_b = 4.44 \text{ mol m}^{-3}$$

7.9 The beads should have a radius no larger than 0.17 mm.

The first step in the question is to bring all of the parameters into standard units.

Thus:

pollution concentration (S_b) = 1 mol m^{-3}

$$k_0 = 0.01 \text{ m mol cm}^{-3} \text{ min}^{-1} = \frac{0.01 \times 10^{-3} \times 10^6}{60} \text{ mol m}^{-3} \text{ s}^{-1} = \frac{1}{6} \text{ mol m}^{-3} \text{ s}^{-1}$$

$D_e = 0.8 \times 10^{-9}$ m^2 s^{-1}

If we are not to use too much catalyst, then essentially we need S(r) to = 0 when r = 0. We have learnt that the condition is reached when $\frac{Rp^2\, k_0}{S_b\, D_e} = 6$ (Equation 7.25).

$$\text{Thus } \frac{Rp^2 \times 1}{6 \times 1 \times 0.8 \times 10^{-9}} = 6$$

$Rp^2 = 28.8 \times 10^{-9}$ Therefore: $Rp = 1.69 \times 10^{-4}$ m = 0.17 mm

7.10

1) Since $\eta = \dfrac{\tanh(Rp\sqrt{(k_1/D_e)}}{Rp\sqrt{k_1/D_e}}$ (see Equation 7.31)

and $D_e = 1 \times 10^{-9}$ $k_1 = 6\ min^{-1} = 0.1\ s^{-1}$ and $Rp = 2 \times 10^{-3}$ then

$$\eta = \frac{\tanh(2 \times 10^{-3}\sqrt{(0.1/10^{-9})}}{2 \times 10^{-3}\sqrt{0.1/10^{-9}}} = \tanh\frac{(2 \times 10^{-3} \times 10^4)}{2 \times 10^{-3} \times 10^4} = \frac{\tanh 2 \times 10}{2 \times 10} = 1$$

2) In this case $k_1 = 0.06\ min^{-1} = 0.001\ s^{-1}$. Applying Equation 7.31

$$\eta = \tanh\frac{(2 \times 10^{-3}\sqrt{(10^{-3}/10^{-9})})}{2 \times 10^{-3} \times \sqrt{(10^{-3}/10^{-9})}} = \tanh\frac{(2 \times 10^{-3} \times 10)}{2 \times 10 \times 10} = \frac{\tanh(2)}{2} = 0.78$$

7.11

1)

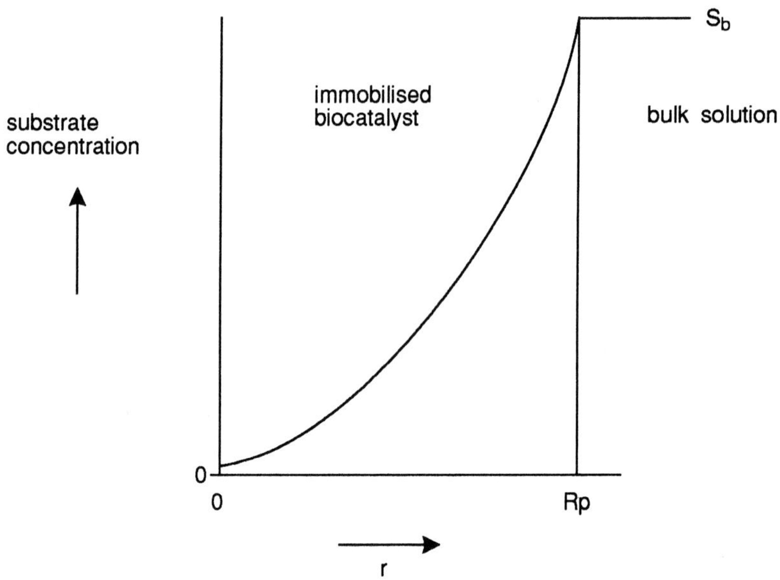

Since $Rp\sqrt{(k_1/D_e)}$ has a high value, the substrate concentration will fall to almost zero at r = 0. It will not, however, quite reach zero since the reaction is first order and substrate consumption is dependent upon substrate concentration.

2)

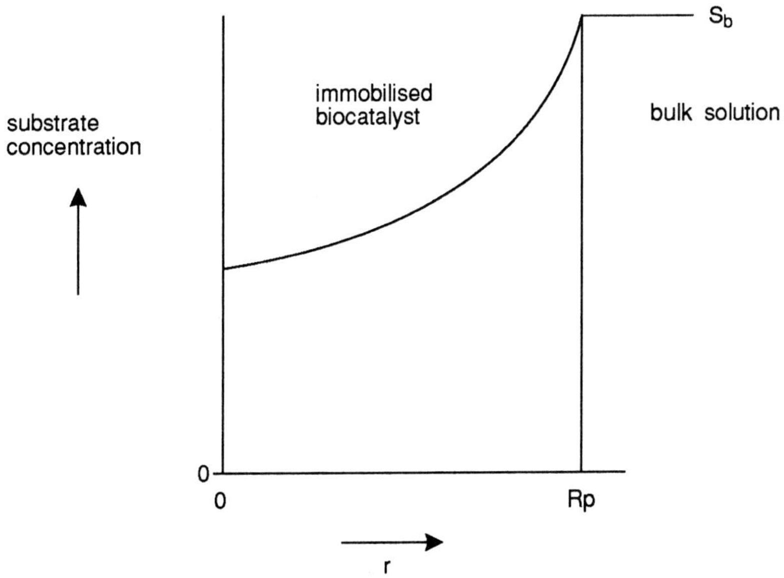

Since $Rp\sqrt{(k_1/D_e)}$ is low, the substrate concentration will remain fairly high even at the centre of the sphere.

7.12 1) Yes, oxygen is a limiting factor.

We can use the Thiele modulus approach to determine the effectiveness factor. If the effectiveness factor is 1, then oxygen availability is not a limiting factor. If it is less than 1 then oxygen is limiting factor.

$$\text{Thus } \Phi = \frac{Vp}{Ap}\sqrt{k_0/D_e}$$

$$= \frac{\frac{4}{3}\pi r^3}{4\pi r^2}\sqrt{k_0/D_e}$$

for $k_0 = 3.6\ kg^{-1}\ h^{-1}$ we obtain

$$k_0 = 3.6\ kg^{-1}\ h^{-1} = 0.001\ kg^{-1}\ s^{-1} \rightarrow \frac{r}{3}\sqrt{k_0/D_e} = \frac{6 \times 10^{-3}}{3}\sqrt{0.001/1 \times 10^{-9}} = 2$$

When Φ is equal to or greater than 2 then $\eta \rightarrow \frac{1}{\phi}$, thus the effectiveness factor is about 0.5. In other words only half of the biocatalyst is exposed to oxygen.

2) The effectiveness factor will hardly be altered.

Our reasoning is as follows.

We can calculate Φ for each of the two systems. For the spherical system we have already calculated this in 1) to be = 2.

For the new strips

$$\Phi = \frac{Vp}{Ap} \sqrt{k_0/D_e}$$

$$= \frac{6 \times 10^{-9}}{3 \times 10^{-6}} \sqrt{(k_0/D_e)} = 2 \times 10^{-3} \sqrt{10^6} = 2$$

Since Φ is the same in both cases, η will be the same and be approximately = 0.5 (since $\eta = \frac{1}{\Phi}$).

3) It will increase the effectiveness factor to 1 and oxygen will be no longer a limiting factor.

We can calculate a Thiele modulus value for the new dimensions.

As $\Phi = \frac{Vp}{Ap} \sqrt{(k_0/D_e)}$ (Equation 7.36)

then $\Phi = \frac{1 \times 10^{-9}}{6 \times 10^{-6}} \sqrt{10^6}$

$$= \frac{1}{6} \times 10^{-3} \times 10^{3} = 0.167$$

Since Φ is well below 1 and the system is essentially of slab geometry, then $\eta = 1$ (see Figure 7.10). Thus oxygen is no longer limiting.

7.13

1) $\eta = 0.25$ (approx) - this means only about a quarter (0.25) of the enzyme would be used.

Effectiveness factor can be derived from Φ.

$$\Phi = \frac{Vp}{Ap} \sqrt{(k_0/D_e)} = \frac{\frac{4}{3}\pi r^3}{4 \pi r^2} \sqrt{(k_0/D_e)}$$

Since $k_0 = 7.2$ kg kg^{-1} h^{-1} = 0.002 kg kg^{-1} s^{-1} and $r = 6 \times 10^{-3}$ $D_e = 0.5 \times 10^{-9}$ m^2 s^{-1}.

Then $\Phi = \frac{6 \times 10^{-3}}{3} \sqrt{2 . 10^{-3}/0.5 \times 10^{-9}} = \frac{6 \times 10^{-3}}{3} \sqrt{4 \times 10^{-6}} = \frac{12}{3} = 4$

when $\phi = 4$, $\eta = 0.25$ (approx) (since $\eta = \frac{1}{\Phi}$ at high Φ values).

2) Approximately 0.016 g enzyme kg^{-1} of beads.

Your instinct might have been to conclude that since only a quarter of the enzyme was being used in 1), reducing the concentration of enzyme by a factor of 4 would remove the excess enzyme. This would be erranoes. Remember that Φ is related to $\sqrt{k_0}$ not proportional to it.

The way we would attempt this is as follows. We would like to produce a system in which η is close to 1. This means that Φ has to be small, say up to about 0.5 (see Figure 7.10).

Thus we can calculate a new value for k_0 that will satisfy this condition.

Since $\Phi = \dfrac{Vp}{Ap} \sqrt{(k_0/D_e)}$ then $0.5 = 6 \times 10^{-3} \sqrt{(k_0/D_e)}$

$$\left(\frac{0.5}{2 \times 10^{-3}}\right)^2 = \frac{k_0}{0.5 \times 10^{-9}}$$

$k_0 = 31250 \times 10^{-9} = 3.125 \times 10^{-5}$ kg kg^{-1} s^{-1}

But k_0 for the original beads was 2×10^{-3} kg kg^{-1} s^{-1}

Thus we have to reduce the amount of enzyme by a factor of $\dfrac{3.125 \times 10^{-5}}{2 \times 10^{-3}}$ to achieve a Φ value of 0.5, thus by a factor 0.0156.

In the original beads we used 1g enzyme kg^{-1} of beads. If we are to operate at an η value close to one, we need to use 1 x 0.0156 g enzyme kg^{-1} of beads.

3) The beads would need to have a radius of no greater than about 0.75 mm.

To use all of the enzyme (ie have an effectiveness factor η close to unity), we need a Φ value of 0.5 or less (see 2) above).

Since $\Phi = \dfrac{Vp}{Ap} \sqrt{(k_0/D_e)} = \dfrac{\frac{4}{3}\pi r^3}{4 \pi r^2} \sqrt{(k_0/D_e)}$

$0.5 = \dfrac{r}{3} \sqrt{0.002/0.5 \times 10^{-9}}$ since $k_0 = 7.2$ kg kg h^{-1} = 0.002 kg kg^{-1} s^{-1}

$0.5 = \dfrac{r}{3} (2 \times 10^3)$ Therefore: $\dfrac{1.5}{2 \times 10^3} = r$ and $r = 0.75 \times 10^{-3}$ m.

Thus we need to produce beads with a radius of 0.75 mm.

7.14 1) We learnt in Section 7.4 that for a slab, $\eta = 1$ when the Thiele modulus value was less than 1, but was equal to $\dfrac{1}{\Phi}$ for higher Φ values. System 1 fits this condition.

But, you should recall that these relationships were established for the condition $S_i = S_b$. This, of course, describes the situation when $\eta_e = 1$. Thus we can conclude that system 1 has an external effectiveness factor of 1.

2) System 2 has an apparent external effectiveness factor of 0.5.

Since the overall effectiveness factor (η_{ov}) is given by $\eta_{ov} = \eta_i \cdot \eta_e$ and for slab geometry we would anticipate η against Φ to follow the line of system 1 in Figure 7.12, if $\eta_e = 1$, then we appear to have a 50% reduction in the anticipated values of η_{ov}. Thus it would appear that η_e has a value of approximately 0.5.

This question will have shown you that as the external effectiveness factor falls, so does the value of η. We could imagine therefore a whole series of curves for different values of η_e. Another way to think of this is that the greater the barrier the stagnant film offers to substrate entry into the biocatalyst (eg low diffusion coefficient of the substrate, thick stagnant layer) the lower the overall effectiveness of the biocatalyst.

Graphically we can represent this as:

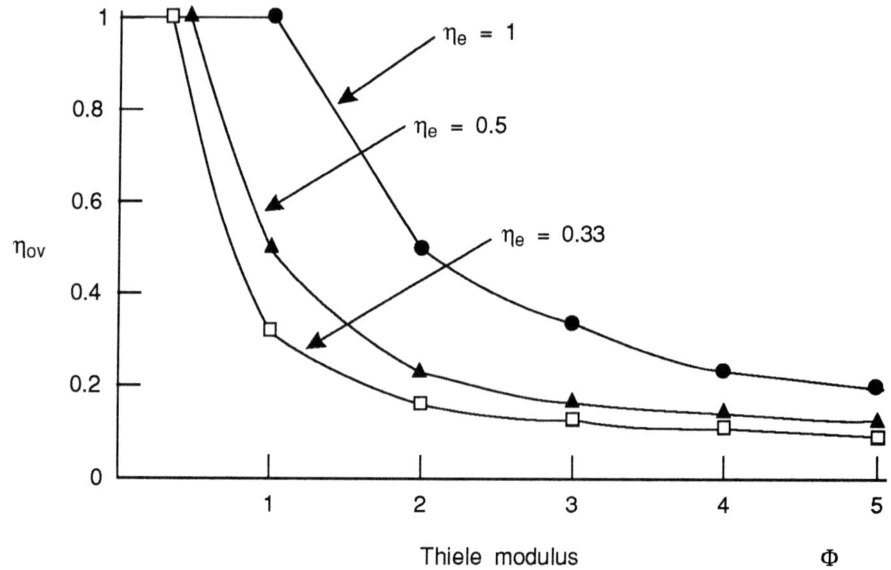

7.15 1) This will probably operate with zeroth order kinetics with respect to arginine.

The arginine concentration in the bulk liquid is 1000 time greater than K_s. Even though the concentration in the slab will decline away from the interface, the fact that this substrate is consumed at a fairly low rate and arginine is a fairly small molecule and will readily diffuse, means that the concentration is likely to be maintained well above K_s throughout the slab. Thus kinetically, arginine consumption will approximate to zeroth order kinetics.

2) This will probably operate with first order kinetics.

The concentration of pesticide (1 part per million) is equivalent to 1 mg m^{-3}. This is $\frac{1}{100}$ the K_s value. Thus we would be operating well below K_s and the reaction would approximate to first order kinetics. This reaction rate would, however, be extremely slow at this very low substrate concentration.

3) Possibly zeroth order, but you had best read our caution below.

This question tested your memory that $K_s = [S]$ when $\mu = \frac{1}{2}\mu_{max}$ (prove it to yourself by substituting $\mu = \frac{1}{2}\mu_{max}$ into the Monod relationship $\mu = \frac{\mu_{max}[S]}{K_s + [S]}$).

Thus $K_s = 4\ g\ m^{-3}$. The substrate concentration on the bulk liquid is 250x greater than this, thus it would appear that we are operating well above K_s and thus with zeroth order kinetics. We have, however, to be cautious. Because this substrate is a primary substrate it probably means that the bacterial culture (the biocatalyst) consumes this substrate at a high rate (ie k is high). We also have no knowledge about the molecule size of the substrate and thus have no knowledge it if diffuses slowly or rapidly. If it is large and thus D_e is small then this combined with a high k value will mean that we will have a very steep substrate concentration profile within the biocatalyst. Thus, although we might be operating at high [S] relative to K_s near the interface, then [S] could be much lower closer to r = 0. Thus this could approach or even fall below the value of K_s. Under these circumstances the kinetics of consumption would not approximate to zeroth order but would be much better described by Michaelis Menten kinetics.

4) Neither zeroth not first order, more likely Michaelis Menten.

The bulk solution has an oxygen concentration of 20 x that of the K_s. Thus at first glance this might appear to suggest zeroth order kinetics. But, let us consider what takes place within the biocatalyst layer. Let us assume that there is no external mass transfer restriction and that $S_i = S_b$. Thus oxygen at the interface has a concentration of 40% of a saturated solution. Oxygen is not very soluble and thus although this is 40% of saturation, it is still a low concentration. The consumption of oxygen by cells is usually very fast. Thus although oxygen can diffuse quite readily, the concentration in the biocatalyst declines rapidly away from the outerface. Thus we may represent the concentration profile as:

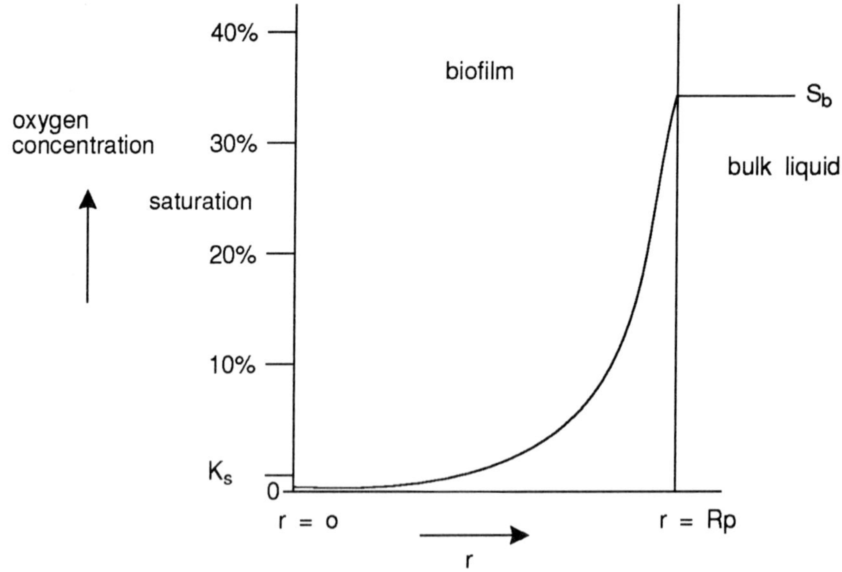

Thus we are likely to experience zeroth order kinetics near r = Rp and first order kinetics near r = 0. This is typically Michaelis Menten kinetics.

Responses to Chapter 8 SAQs

8.1 1) Penetration depth = 3.36×10^{-4}m

Method 1

For diffusion limitation $\frac{Rp^2\, k_0 m}{D_e\, S_g} > \; = 2$

$k_0 = 20 g\, m^{-3}\, h^{-1} = \frac{20}{360}\, g\, m^{-3}\, s^{-1}$

(We took these values from the point where consumption rate was independent of substrate concentration).

Thus $\frac{Rp^2 . 20 . 0.27}{360 . 0.85 . 10^{-9} . 1} = 2$ (ie the point at which rd just = 0).

$$Rp^2 = \frac{2 . 360 . 0.85 . 10^{-9}}{20 . 0.27} = 112 \times 10^{-9} \text{ m}$$

Rp = 3.36×10^{-4} m is the critical distance the gas volume will penetrate when supplied in a concentration (S_g) of 1 g m^{-3}.

2) The key point is that we have assumed zeroth order kinetics at all concentrations of substrate. This is unlikely to be accurate at low substrate concentrations. Thus instead

of a concentration profile predicted by zeroth order, we are likely to have a different curve.

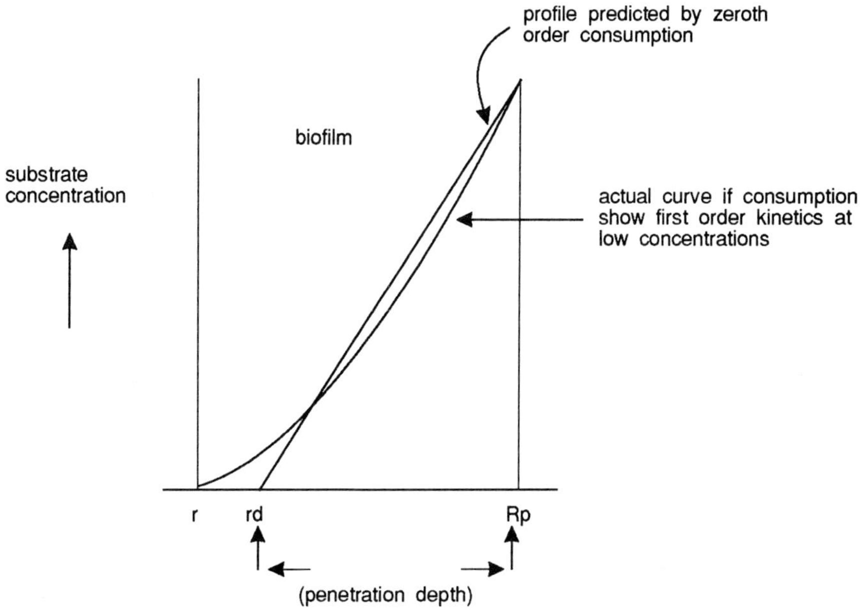

Thus the determination of the penetration depth is likely to be an underestimation. The shape of Figure 8.5 would suggest that toluene consumption becomes first order at low values of (S_g).

3) To do this we need to take two sets of values from the Figure 8.5.

If we assume that the physical dimensions of the column are constant, then we can assume that the flux per unit area (N) is proportional to the total flux per unit volume (V).

Thus N = fV, but N = k_0 (Rp - rd).

From Figure 8.5 we can take two values. Let us say one showing substrate limitation, the other not.

Thus at 4 g m^{-3} substrate, no substrate limitation

thus N = k_0 Rp

The rate of substrate consumption = 20 g m^{-3} h^{-1} (from Figure 8.5).

Thus N = f (20) g m^{-2} h^{-1} = k_0 Rp .... (A)

At 1 g m^{-3} substrate, with substrate limitation, the rate of substrate consumption = 12.5 g m^{-3} h^{-1} (from Figure 8.5).

Thus N = f(12.5) = k_0 (Rp - rd) ....(B)

We can re-arrange A and B and combine them.

$$\frac{f(20)}{Rp} = \frac{f(12.5)}{(Rp - rd)}$$

Divide through by f and re-arranging gives:

$$Rp - rd = \frac{12.5}{20} Rp$$

But we have previously calculated the penetration depth (Rp - rd) as 3.36×10^{-4} m (See 1).

Thus $3.36 \times 10^{-4} = \frac{12.5}{20} Rp$

$Rp = 5.38 \times 10^{-4}$ m.

Responses to Chapter 9 SAQs

9.1

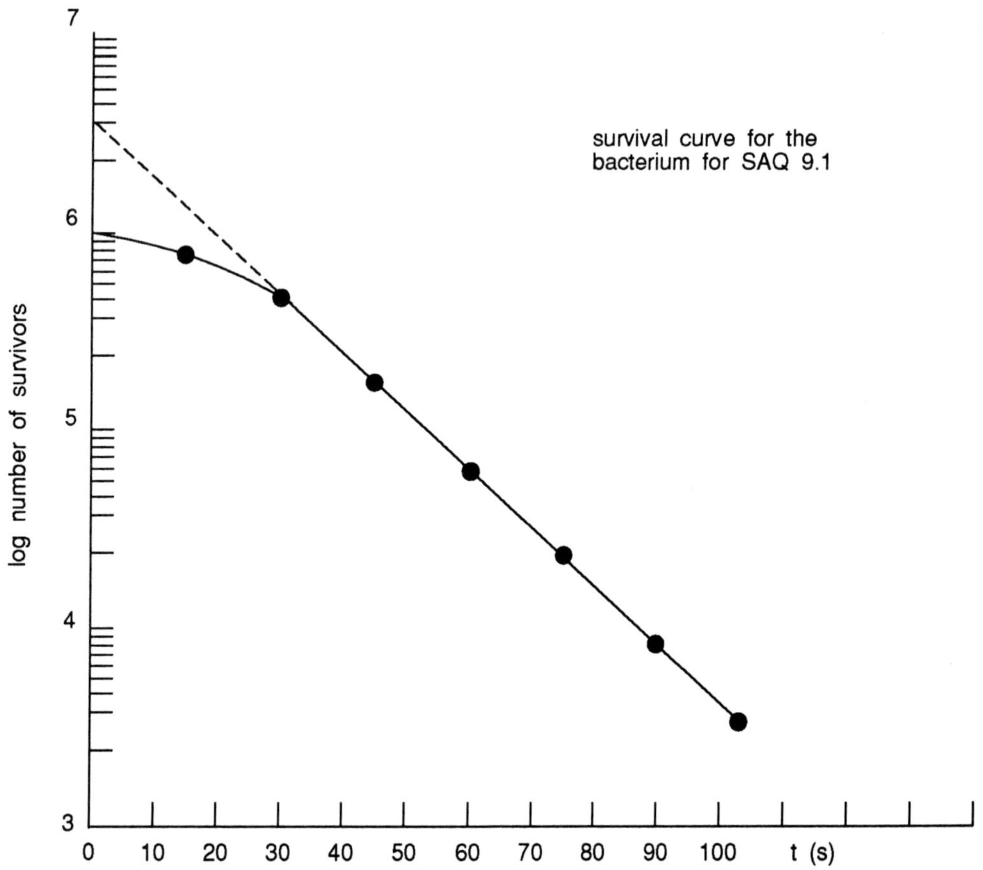

Since $N_t = 2.8 . 10^3$; $N_o = 1.0 . 10^6$; t = 105 sec

Thus $\ln 2.8 \times 10^3 - \ln 1.0 . 10^6 = - \delta . 105$

$7.937 - 13.815 = -\delta.105$

$\delta = 0.056 \text{ s}^{-1}$

It is probable that every colony forming (cfu) unit consists of three cells or requires three 'hits' to be inactivated. We can derive this from the ratio between the actual value of N_o and the extrapolated N_o.

9.2 $\delta = 0.056 \text{ s}^{-1}$ (from SAQ 9.1)
and D = 2.303/0.056 = 41.1 s = 0.69 min.

Note, however, although we have calculated D using δ if we use the data from the survival curve plotted in the response to SAQ 9.1 directly, we can obtained various values for D depending upon which part of the curve we use. Thus if we measure D for cell reduction from 10^6 to 10^5 we have a D-value of 50s. If, however, we use the reduction from 10^5 to 10^4, we would obtain a D value of about 33s.

9.3 z = 12.1

$T_1 = 121.1°C$; $T_2 = 115°C$; $D_1 = 0.69$; $D_2 = 2.2$;

So: $\frac{(121.1 + 273) - (115 + 273)}{z} = \log D_2 - D_1$ (see Equation 9.5) 2.2 -log 0.69

6.1 / z = 0.3424 - (-0.1612) and 6.1 = 0.5036 z. Therefore z = 12.1

9.4 We wish to reduce the count from 10^6 to 10^{-5}. Thus n = 11 (that is 11 decimal reductions); D = 0.69 min;

So: F = 11 . 0.69 = 7.59 min.

9.5 162 min

$$\text{Since } F_T^z = \frac{F_{121.1}^z}{10^{\left(\frac{T-121.1}{z}\right)}} \text{ (Equation 9.7)}$$

$F_T^z = F_{121.1}^z$ and $F_{121.1}^z = 7.59$ min; T = 105°C

$$\text{Thus } F_{105}^{12.1} = \frac{7.59}{10^{\left(\frac{105-121.1}{12.1}\right)}} = \frac{7.59}{10^{-1.33}} = 162 \text{ min}$$

9.6 For the lethality rates see Table 9.1. The reference temperature is 120°C.

So:

5 min	=	0.000	11 min	=	0.215
6 min	=	0.000	12 min	=	0.316
7 min	=	0.006	13 min	=	0.464
8 min	=	0.022	14 min	=	0.681
9 min	=	0.046	15 min	=	1.000
10 min	=	0.100	16-30 min	=	1.468 (x 15)
			31 min	=	0.022

Total lethality = 24.9 = 25. Since total lethality (S) = ln N_o/N_t (see Equation 9.10) then if we know the original number of cells (N_o) present, we can calculate how many will have survived (N_t) the process.

9.7 The main reason is that the food products themselves absorb the ionising radiation. Thus much of the radiation is not absorbed by the micro-organisms that may be present. This is especially true of those micro-organisms embedded in the product.

9.8 1) 1.46×10^{11} dynes m^{-2}

Using the relationship $r = 2\sigma/P$ (Equation 9.19) and given $r = 0.1$ µm and $\sigma = 73$ dyne cm

Then $0.1 \times 10^{-6} = \dfrac{2 \times 73 \times 100}{P}$ (note we have made all linear measurements in m)

Thus $P = \dfrac{1.4600}{0.1 \times 10^{-6}}$ dynes m^{-2}

$= 1.4600 \times 10^{11}$ dynes m^{-2}

2) 0.73×10^{11} dynes m^{-2}

We described the bubble point of kerosene as being exactly half the water bubble point.

3) All similar. The rapid use in the rate of bubble formation over a small pressure range indicates that the capillaries are all of rather similar radius.

4) The electron microscope measures all capillary spaces. The mercury intrusion method only measures those capillaries which are accessible from the two faces of the filter.

Thus the membrane labelled a) below should give the same result which ever method is used, b) would give a higher capillary volume using the electron microscope compared to the mercury technique.

a) 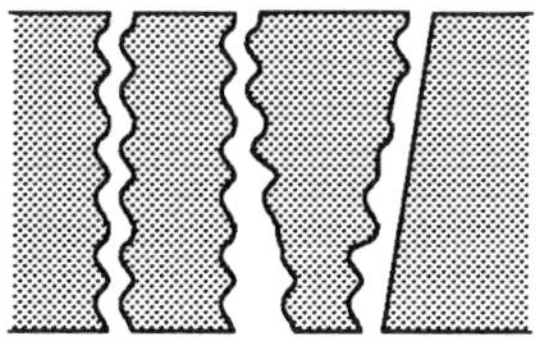

b) 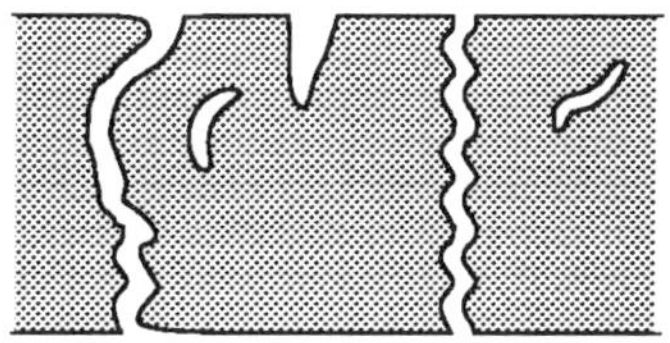

9.9 0.25 mm

We can use the relationship

$$L = L_{90} \log \frac{N_1}{N_2} \quad \text{(see Equation 9.29)}$$

Since $N_1 = 10^6\ m^{-3}$; $N_2 = 10^2\ m^{-3}$ $L = 1\ mm = 10^{-3}\ m$

then $10^{-3} = L_{90} \log \frac{10^6}{10^2} = L_{90} \cdot (6-2)$

Therefore $L_{90} = 0.25 \times 10^{-3}\ m = 0.25\ mm$

Suggestions for further reading

Chapter 2

Atkinson, B. and Mavituna, F. (1983), Gas-liquid mass transfer, In: Biochemical engineering and biotechnology handbook, MacMillan Publishers Ltd., Surrey, UK, pp. 727-801.

Bailey, J.E. and Ollis, D.F. (1977), Transport phenomena in microbial systems, In: Biochemical engineering fundamentals, McGraw-Hill Kogakusha Ltd., Tokyo, Japan, pp. 411-496.

Moo-Young, M. and Blanch, H. (1981), Design of biochemical reactors; mass transfer criteria for simple and complex systems, Adv. Biochem, Eng. 19 1-69.

Moser, A. (1988), Bioprocess Technology, Springer Verlag, New York, pp. 66-137.

Chapters 3-4

van 't Riet, K. and Tramper, J. (1991), Basic Bioreactor Design, Marcel Dekker Inc., New York.

Chapters 5-8

van 't Riet, K. and Tramper, J. (1991), Basic Bioreactor Design, Marcel Dekker Inc., New York.

Hooijmans, C.M. *et. al.*, (1990), Use of an oxygen microsensor for the determination of intrinsic kinetic parameters of an immobilised oxygen reducing enzyme. Biotechnology Bioengineering, 35, 1078.

Hooijmans, C.M. *et. al.*, (1990), Measurement of oxygen concentration gradients in gel immobilised Recombinant *E. coli*, Appl. Microbial Biotechnology, 33. 611-618.

Coulson, J.M. and Richardson, J.F. (1979), Chemical Engineering, Pergamon Press.

Vos, H. *et. al.*, (1990), Counterourrent fluidised bed reactor for immobilised biocatalysts, I modelling and simulation, Biotechnology Bioengineering, 36, 367-376.

Vos, H. *et. al.*, (1990), Effectiveness factor for spherical biofilm catalysts, Bioprocess. Eng., 5, 63-72.

Chapter 9

Books:

-Mulder, M. (1989), Basic Principles of Membrane Technology, Enschede.
-Russel, A.D., Hugo, W.B. and Ayliffe, G.A.J. (1982), Principles and Practice of Desinfection, Preservation and Sterilisation, Oxford.
-Wallhauser, K.H. (1984), Praxis der Sterilisaton Desinfektion Konservierung Keimidentifizierung Betriebshygiene, Stuttgart.

Articles:

-Bigelow, W.D. (1921), J. infect. Dis. 29, 528.
-Deindoerfer, F.H. & Humphry, A.E. (1959), Appl. Microbial 7, 256.
-Esty, J.R. & Meyer, K.F. (1922), J. infect. Dis. 31, 650.
-Patashnik, M. (1953), Food Technol. 7, 1.

Reviews:

-Brooks, C.H. and Russel, PD. (June 1986), Process Biochemistry, 77.
-Wallhaußer, K.H. (1987), Biotech-Forum 4, 90.

Appendix 1 - Modelling and effectiveness factors

For zeroth and first order heterogenous catalysis it is possible to get analytical solutions for the overall effectiveness factor (see Chapter 7). In biocatalysis Michaelis Menten kinetics is quite common. The substrate balance resulting from the coupled diffusion and (Michaelis Menten) reaction cannot be solved analytically and consequently no analytical expression for the effectiveness factor is available. A complete numerical solution can be obtained by numerically solving the substrate balance resulting in the substrate profile and the effectiveness factor. An alternative is to use approximate analytical expressions for the effectiveness factor. The latter method is based on the property that Michaelis Menten kinetics can be described as a combination of first and zeroth order kinetics.

A1.1 Numerical solution

If the overall kinetics of an immobilised biocatalyst obeying Michaelis Menten kinetics are determined, no analytical solution is available. The performance of the biocatalyst can be expressed by the overall effectiveness factor which is not a simple function of the Thiele modulus due to the non-linearity of the Michaelis Menten kinetics. If external mass transfer is considered the situation becomes even more complex. To solve this the overall effectiveness factor must be determined numerically as will be discussed below.

To calculate the overall effectiveness factor the flux at the interface must be known according to:

$$\eta_{ov} = \frac{\frac{3 D_e \left.\frac{ds}{dr}\right|_{Rp}}{Rp}}{Rp\ v(S_b)} \qquad \text{(E - A1.1)}$$

Consequently the concentration profile must be solved numerically. Once again the substrate balance for spherical particles:

$$\frac{d^2S}{dr^2} + \frac{2}{r}\frac{dS}{dr} = \frac{v(S)}{D_e} \qquad \text{(E - A1.2)}$$

For Michaelis Menten kinetics:

$$v(S) = V_{max}\frac{S}{K_M + S} \qquad \text{(E - A1.3)}$$

With the boundary conditions:

$$\frac{dS}{dr} = 0,\ r = 0 \text{ and } D_e\frac{dS}{dr} = k_L\,(S_b - S_i),\ r = Rp \qquad \text{(E - A1.4)}$$

The substrate concentration at the interface is not known and both the boundary conditions are first order differential equations (a two point boundary value problem). The second order differential Equation A1.2 is written as:

$$Z = \frac{dS}{dr}$$

$$\frac{dZ}{dr} + \frac{2}{r} Z = \frac{v\,(S)}{D_e} \qquad \text{(E - A1.5)}$$

Resulting in the boundary conditions:

$Z = 0, r = 0$ and $D_e\ Z = k_L\ (S_b\text{-}S_i), r = Rp$

(E - A1.6a)

Equations A1.4 and A1.5 are two first order differential equations which for example can be solved numerically by means of a Runge Kutta routine. The radius is divided into increments (integration steps), the substrate concentration and the derivative at each point are calculated by numerical integration of Equation A1.5 resulting in a substrate concentration profile. Suppose the integration starts at the centre of the particle. the first derivative of the substrate concentration with respect to the location is known from the boundary condition Z = 0 and can be used in the substrate balance. Integration of the substrate balance yields the substrate concentration and the derivative for the next point. But it can be seen that Equation A1.5 cannot be calculated at the centre of the particle because the value of the substrate concentration is unknown. This can be solved by choosing a starting value for the unknown substrate concentration at the particle centre, S_0. The substrate concentration and the derivative for the next points are sequentially calculated until the radius of the particle is reached. With the last integration step the value for the substrate concentration at the interface and the derivative at the interface are obtained. If the concentration profile is correct, the boundary condition at the interface must be valid. This can be checked by substituting the substrate concentration and the derivative in the boundary condition and comparing the left hand term of Equation 8.6 (see Chapter 8) with the right hand term. In other words, the estimate of the substrate concentration at the centre results after numerical calculation of the substrate concentration profile in a value for the flux at the interface which must match the flux described by the external mass transfer. If the flux does not match, the estimated concentration in the centre was not correct and a new estimate must be tried. This process of trial and error must go on until the boundary condition is fulfilled. In practice a certain tolerance level, for example 0.1% of the bulk substrate concentration is choosen for the fulfilment of the boundary condition. For the estimation of the substrate concentration at the centre a zero-point seeker, like bisection or regula-falsi, with a tolerance level can be used as the boundary condition can also be written as:

$$D_e\ Z - k_L\ (S_b - S_i) <= tol, r = Rp \qquad \text{(E - A1.6b)}$$

Using this numerical tool speeds up the convergence of the calculation. If the derivative at the interface is known the overall effectiveness factor can be calculated from Equation A1.1. In Figure A1.1 the procedure is shown schematically.

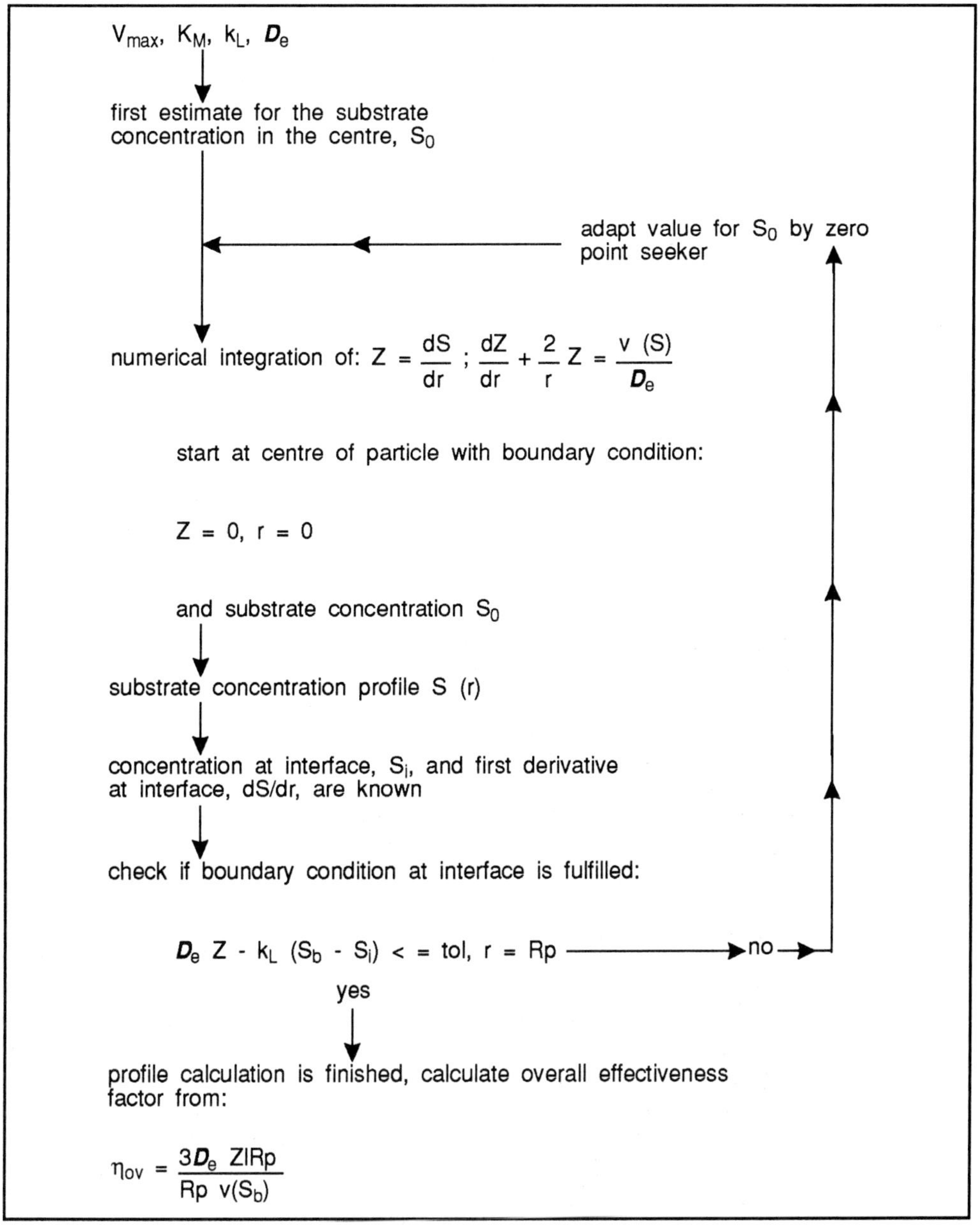

Figure A.1 Numerical calculation of the substrate concentration profile and the effectiveness factor for Michaelis Menten kinetics.

The numerical solution method for Michaelis Menten kinetics is outlined above, but the principles of the method can be applied for any form of substrate dependence kinetics (second order, substrate inhibition, etc). For kinetics which are also dependent on the product concentration (for example, product inhibition), the product balance must be solved simultaneously with the substrate balance. For Michaelis Menten kinetics as shown above the product concentration profile can be calculated directly from the substrate concentration profile as will be shown below.

Consider the reaction S -> n P, where n depicts the stoeichiometric relationship between substrate and product. The reaction rate of the substrate R_S and the reaction rate of the product R_P is equal to the flux at the surface of the particle which can be written as:

$$R_S = k_{L,S} (S_b - S_i) \text{ and } R_P = k_{L,P} (P_i - P_b) \qquad \text{(E - A1.7)}$$

In the case of a steady state behaviour no accumulation of product in the particle occurs:

$$R_P = n\, R_S \qquad \text{(E - A1.8)}$$

Resulting in:

$$n\, k_{L,S} (S_b - S_i) = k_{L,P} (P_i - P_b) \qquad \text{(E - A1.9)}$$

In the biocatalyst a similar relation can be derived. The two substrate balances for the substrate and the product are:

$$\frac{d^2S}{dr^2} + \frac{2}{r}\frac{dS}{dr} = \frac{v\,(S)}{D_{e_s}} \qquad \text{(E - A1.2)}$$

$$\frac{d^2P}{dr^2} + \frac{2}{r}\frac{dP}{dr} = \frac{v\,(P)}{D_{e_p}} \qquad \text{(E - A1.10)}$$

Due to the stoeichiometric relation:

$$v\,(P) = n\, v\,(S) \qquad \text{(E - A1.11)}$$

Combination of A1.2, A1.10 and A1.11 results in:

$$nD_{e_s}\left(\frac{d^2S}{dr^2} + \frac{2}{r}\frac{dS}{dr}\right) + D_{e_p}\left(\frac{d^2P}{dr^2} + \frac{2}{r}\frac{dP}{dr}\right) = 0 \qquad \text{(E - A1.12)}$$

Equation A1.12 can be solved with the boundary conditions:

$\frac{dS}{dr} = 0, r = 0$ no transport of substrate beyond the centre

$\frac{dP}{dr} = 0, r = 0$ no transport of product beyond the centre

$S = S_i$, $r = Rp$ surface substrate concentration

$P = P_i$, $r = Rp$ surface product concentration

The solution of Equation A1.12 is:

$$nD_{e_s} (S_i - S) = D_{e_p} (P - P_i) \qquad \text{(E - A1.13)}$$

By combination of Equation A1.13 with Equation A1.9:

$$n\, k_{L,S} (S_b - S_i) = k_{L,P} (P_i - P_b) \qquad \text{(E - A1.13a)}$$

the unknown product concentration at the surface of the particle can be eliminated resulting in the product concentration profile:

$$P = P_b + n \frac{k_{L,S}}{k_{L,P}} (S_b - S_i) + n \frac{D_{e_S}}{D_{e_P}} (S_i - S) \qquad \text{(E - A1.14)}$$

The substrate concentration profile was already obtained from the numerical solution of the substrate balance. With Equation A1.14 the product concentration profile can be calculated without solving the product balance. This can only be done if the substrate reaction rate is independent of the product concentration. If the substrate reaction rate is dependent on the product concentration Equation A1.12 cannot be solved.

A1.2 Approximate method

To obtain the overall effectiveness factor for an immobilised biocatalyst obeying Michaelis Menten kinetics the numerical solution of the substrate balance needs a lot of calculation. This led to the development of analytical methods which describe the overall effectiveness factor satisfactorily. The method described here uses an approximation which is based on the generalised Thiele modulus for Michaelis Menten kinetics, the analytical solution for first and zeroth order reactions and the knowledge that Michaelis Menten kinetics has first and zeroth properties.

A2.1 Generalised Thiele modulus for Michaelis Menten kinetics

In Chapter 8 the generalised Thiele modulus was presented for zeroth and first order kinetics. The advantage of the use of the generalised Thiele modulus was the almost complete independence from the overall effectiveness factor on geometry and reaction kinetics. The generalised Thiele modulus as described for first order and zeroth order kinetics is the result of the calculation of the generalised Thiele modulus for general irreversible reaction kinetics which is as follows:

$$\Phi = \sqrt{\left(\frac{V_s/A_s . v(S_s)}{2D_e . \int_0^{Ss} v(S)\, dS} \right)} \qquad \text{(E - A1.15)}$$

If for the reaction rate Michaelis Menten kinetics as described by Equation A1.3 is substituted the Generalised Thiele modulus for Michaelis Menten kinetics, Φ_{MM}, can be calculated yielding:

$$\Phi_{MM} = \frac{(Vp/Ap)^2 . V_{max}/2D_e\, (S_s/(S_s + K_M))^2}{S_s - K_M . 1n\, ((K_M + S_s)/K_M)} \qquad \text{(E - A1.16)}$$

If the value for the generalised Thiele modulus for Michaelis Menten kinetics can be calculated the overall effectiveness factor is, due to the definition of the generalised Thiele modulus, approximately the same as for first order and zeroth order kinetics with the same value for the generalised Thiele modulus.

Appendix 2

The major symbols used in the text (with examples of typical units)

Chapter 2

a = specific surface area (m^2 m^{-3})

A_s = total surface area of a liquid film (m^2)

C = concentration

C_{af} = concentration of antifoam (mol l^{-1}; kg m^{-3})

C_{co} = critical oxygen concentration (mol l^{-1}; kg m^{-3})

C_g = concentration of dissolved gas g (mol l^{-1}; kg m^{-3})

C_{gi} = concentration of dissolved gas at gas liquid interface (mol l^{-1}; kg m^{-3})

C_{gl} = concentration of dissolved gas in bulk liquid (mol l^{-1}; kg m^{-3})

C_{io} = initial oxygen concentration (mol l^{-1}; kg m^{-3})

C_{saa} = concentration of surface active agent (mol l^{-1}; kg m^{-3})

C^* = maximum (saturation) solubility in a solvent

C^*_s = maximum solubility in salt solution (mol l^{-1}; kg m^{-3})

C^*_w = maximum solubility in water (mol l^{-1}, kg m^{-3})

d_b = bubble (partial) diameter (m)

D_g = diffusion coefficient of g in liquid medium (n^2 s^{-1})

DOC = dissolved oxygen concentration (mol l^{-1}; kg m^{-3})

GTR = gas transfer rate (mol s^{-1} m^{-3}; kg s^{-1} m^{-3})

He_g = dimensionless Henry constant

H_g = Henry coefficient for g at a specific temperature

k_L = mass transfer coefficient (m s^{-1})

N = impeller speed (rad s^{-1})

OTR = oxygen transfer rate (mol s^{-1} m^{-3}; kg s^{-1} m^{-3}; m s^{-1})

P = power

p_g = partial pressure of g in gas phase (Pa)

p_t = total gas pressure (Pa)

R = gas constant (8.31 J mol^{-1} K^{-1})

r_b = bubble radius (m)

r_s = solute sphere radius (m)

s = surface renewal factor (s^{-1})

S_{gl} = glucose concentration (mol l^{-1}; kg m^{-3})

T = temperature (K or °C)

t = time (s)

t_e = exposure time (s)

v_b = particle rate (s^{-1})

V_g = volume of gas (m^3)

V_r = reaction volume (m^3)

V_s = volume of solvent

v_s = superficial gas velocity (m s^{-1})

X = biomass concentration (g l^{-1})

δ = thickness of a liquid film (m)

ε_g = gas hold up

η = viscosity (Pa.s)

λ = conductivity (Ω cm^{-1})

μ = specific growth rate (s^{-1}; h^{-1})

ϕ'' = mass flow per area (mol s^{-1} m^{-2}; kg s^{-1} m^{-2})

ϕ_g = gas flow (mol s^{-1})

$\phi_{oxygen,in}$ = mass flow of oxygen in (mol s^{-1})

$\phi_{oxygen,out}$ = mass flow of oxygen out (mol s^{-1})

Chapter 3

a = specific surface area (m^2 m^{-3})

d_b = bubble diameter (m)

D = stirrer diameter (m)

Fl = Flow number (aeration number)

Fr = Froude number (= N D/g)

g = gravitational constant (m s^{-2})

H = height of a vessel (m)

h_r = stirrer blade height (m)

k_L = mass transfer coefficient (m^2 s^{-1})

N = stirrer speed (rad s^{-1})

N_p = stirrer power number

P = power input (W)

p = pressure (Pa; bar)

P_A = power of a gas (W)

P_g = gassed power consumption

P_s = ungassed power consumption of a stirrer (W)

Re = Reynolds number (ND^2/h)

T = diameter of a vessel (m)

t = time (s)

t_c = circulation time (s)

t_m = mixing time (s)

V = volume (m^3)

V_g = volume of a gas (m^3)

V_l = volume of a liquid (m^3)

v_s = superficial gas velocity (m s^{-1})

v_{tip} = velocity of tip of stirrer blade (m s^{-1})

ε = volume fraction

η = viscosity (Pa.s)

λ = thermal conductivity

ρ = density (kg m^{-3})

ϕ_c = circulation capacity (m^3 s^{-1})

ϕ_g = volumetric gas flow rate (m^3 s^{-1})

ϕ_p = pumping capacity of a stirrer (m^3 s^{-1})

Chapter 4

a = specific surface area (m^2 m^{-3})

C_l = concentration in a bulk liquid

C^* = maximum (solubility) concentration

d = diameter of a bubble (m)

d_o = diameter of bubble at the sparger orifice

d_{vs} = Sauter mean diameter of a bubble (m)

D_g = effective gas dispersion coefficient (m^2 s^{-1})

D_l = effective dispersion coefficient (m^2 s^{-1})

d_o = diameter of a sparger orifice

g = gravitational constant (m s^{-2})

H = vessel height (m)

k_L = mass transfer coefficient (m s^{-1})

n = number of theoretical mixers

P = pores input (W)

Pe = Peclet number

T = vessel diameter (m)

t_c = circulation time (s)

v_{bs} = rise velocity of a bubble relative to a liquid

v_{lc} = liquid circulation velocity (m s^{-1})

v_o = gas velocity at the sparger orifice (m s^{-1})

v_s = gas superficial velocity (m s^{-1})

α = heat transfer coefficient

ε = gas hold up

η = viscosity of a liquid (Pas)

η_w = viscosity of water (Pas)

ρ = density (kg m^{-3})

σ = surface tension (subscript w = surface tension of water)

Chapter 6

C_A = concentration of component A (mol l^{-1}; kg m^{-3})

D = diffusion coefficient (m^2 s^{-1})

D_e = effective diffusion coefficient (m^2 s^{-1})

k_o = zeroth order rate constant (kg kg^{-1})

k_1 = first order rate constant (kg kg^{-1} s^{-1})

k_L = mass transfer coefficient (m s^{-1})

K_M = Michaelis constant (kg m^{-3}; mol l^{-1})

K_s = saturation constant (kg m^{-3}; mol l^{-1})

S = substrate concentration (mol l^{-1} kg m^{-3}) subscript b = bulk liquid, i = interphase, m = matrix

v = rate of reaction (rate of substrate utilisation) (kg m^{-3} s^{-1})

V_{max} = maximum reaction rate (kg m^{-3} s^{-1})

μ = specific growth rate (s^{-1})

$\phi_{M,A}$ = mass flow of component A (kg s^{-1})

$\phi''_{M,A}$ = mass flow of component A per unit area (kg m^{-2} s^{-1})

Chapter 7

A = area (m^2)

D_e = effective diffusion coefficient (m^2 s^{-1})

K = $\sqrt{K_i/D_e}$

K_M - Michaelis constant (kg m^{-3}; mol l^{-1})

k_n - nth order rate constant

K_s = substrate saturation constant (kg m^{-3}; mol l^{-1})

k_o = zeroth order rate constant (kg s^{-1})

k_1 = first order rate constant (kg kg^{-1} s^{-1})

rd = position in a biofilm at which there is substrate (m)

Rp = thickness of a biofilm (m)

[S] = substrate concentration (kg m^{-3}; mol l^{-1})

v = rate of reaction (substrate consumption) (kg m^{-3} s^{-1})

V_s = volume of a slab (m^3)

η = effectiveness factor. Subscripts i = internal, e = external, ov = overall , o = zeroth order kinetics, 1 = first order kinetics

μ = specific growth rate (s^{-1})

μ_{max} = maximum specific growth rate (s^{-1})

ϕ = Thiele modulus

Φ = generalised Thiele modulus

Chapter 8

D_e = effective diffusion coefficient (m^2 s^{-1})

h = column height (m)

K_M = Michaelis constant

k_o = zeroth order rate constant (kg kg^{-1})

K_s = substrate saturation constant

m = distribution coefficient

N = flux of gas into a column (m^3 m^{-2} s^{-1})

R = conversion rate

r = position in a biofilm (m)

R_p = thickness of a biofilm (m)

[S] = substrate concentration (kg m^{-3})

S_g = substrate concentration in a gas phase (kg m^{-3}; mol l^{-1})

S_i = input concentration of substrate (kg m^{-3})

S_l = substrate concentration in liquid phase (kg m^{-3}; mol l^{-1})

S_o = output concentration of substrate (kg m^{-3})

v_g = superficial gas velocity

V_{max} = maximum reaction rate

Y_{sp} = product yield (kg kg^{-1})

η = effectiveness factor

Chapter 9

A_m = area of a membrane (m^2)

D = decimal reproduction time (s)

E = activation energy (J mol^{-1})

k = hydraulic permeability

N = number of organisms. Subscript 0 = initial, t = time t

p = pressure

R = universal gas constant (J mol^{-1} K^{-1})

r = pore radius (m)

S = total lethality

t - time (s)

Z = temperature difference by which the D-valve is changed by a factor of D (°C; K)

δ = specific death rate (s^{-1})

ε = porosity

η = viscosity

ϕ_v = volume flux ($m^3\ m^{-2}\ s^{-1}$)

Appendix 3

Recommended values of physical constants

Physical constant	Symbol	Value
acceleration due to gravity	g	$9.81\ m\ s^{-2}$
Avogadro constant	N_A	$6.022.10^{23}\ mol^{-1}$
Boltzman constant	k	$1.380\ 10^{23}\ JK^{-1}$
Faraday constant	F	$9.649.10^{4}\ C\ mol^{-1}$
gas constant	R	$8.314\ JK^{-1}\ mol^{-1}$
molar volume of ideal gas of stp	V_m	$2.241.10^{-2}\ m^{3}\ mol^{-1}$
Planck constant	h	$6.628.10^{-34}\ Js$
velocity of light in a vacuum	c	$2.998\ 10^{8}\ m\ s^{-1}$

Appendix 4 - Dimensional Numbers

This appendix contains some frequently used dimensionless numbers, their definition and their meaning.

symbol	name	formula	meaning	used in
Ah	Arrhenius	$\frac{E}{RT}$	$\frac{\text{activation energy}}{\text{thermal energy}}$	chemical reactions
Bo	Bodenstein	$\frac{vL}{D}$	$\frac{\text{convection}}{\text{axial diffusion}}$	Bo = Pe diffusion in reactors
Bd	Bond	$\frac{\Delta\rho D^2 g}{\sigma}$	$\frac{\text{gravitational force}}{\text{surface tension}}$	Bd = Eo bubbles and drops
Da I	Damkohler I	$\frac{k_r D}{v}$	$\frac{\text{chemical reaction rate}}{\text{convection mass transfer rate}}$	chemical reactions
Da II	Damkohler II	$\frac{k_r D^2}{\mathcal{D}}$	$\frac{\text{chemical reaction rate}}{\text{diffusive mass transfer rate}}$	chemical reactions
Eo	Eotvos	$\frac{\Delta\rho D^2 g}{\sigma}$	$\frac{\text{gravitational force}}{\text{surface tension}}$	Eo = Bd bubbles and drops
f	Fanning fricition factor	$\frac{d\Delta p}{2\rho v^2 L}$	$\frac{\text{shear stress energy at the wall}}{\text{kinetic energy}}$	flow through tubes and channels
Fo	Fourier (mass)	$\frac{\mathcal{D}t}{d^2}$	$\frac{\text{process time}}{\text{diffusion time}}$	diffusion
Fo	Fourier (heat)	$\frac{\alpha t}{D^2}$	$\frac{\text{process time}}{\text{conduction time}}$	heat conduction
Fr	Froude	$\frac{v^2}{gH}$	$\frac{\text{inertia forces}}{\text{gravity forces}}$	
Fr	Froude (rotation)	$\frac{DN^2}{g}$	$\frac{\text{inertia forces}}{\text{gravity forces}}$	mixing with free surfaces
Gz	Graetz	$\frac{\alpha L}{D^2 v}$	$\frac{\text{conductive heat transfer}}{\text{convective heat transfer}}$	heat transfer to flowing media
Gr	Grashof	$\frac{D^3 g}{v^2}\frac{\Delta\rho}{\rho}$	$\frac{\text{buoyancy forces}}{\text{viscous forces}}$	free convection
Ha	Hatta	$\frac{\sqrt{\mathcal{D} k_r}}{k}$	$\frac{\text{mass transfer with chemical reaction}}{\text{mass transfer without chemical reaction}}$	mass transfer with chemical reaction

Le	Lewis	$\frac{\alpha}{D}$	$\frac{\text{thermal boundary layer thickness}}{\text{mass transfer boundary layer thickness}}$	$Pe = \frac{Sc}{Pr}$ combined heat and mass transfer
Nu	Nusselt	$\frac{hD}{\lambda}$	$\frac{\text{total heat transfer}}{\text{conductive heat transfer}}$	heat transfer
Pe	Peclet (heat)	$\frac{vD}{\alpha}$	$\frac{\text{convective heat transfer}}{\text{conductive heat transfer}}$	heat transfer in flowing media
Pe	Peclet (mass)	$\frac{vd}{D}$	$\frac{\text{convective mass transfer}}{\text{diffusive mass transfer}}$	mass transfer in flowing media
Po	Power number	$\frac{P}{\rho N^3 D^5}$	$\frac{\text{power added}}{\text{power transferred to kinetic energy}}$	stirred vessels and pumps
Pr	Prandtl	$\frac{v}{\alpha}$	$\frac{\text{hydrodynamic boundary layer thickness}}{\text{thermal boundary layer thickness}}$	heat transfer in flowing media
Re	Reynolds	$\frac{\rho v D}{\eta}$	$\frac{\text{inertia forces}}{\text{viscous forces}}$	flow
Sc	Schmidt	$\frac{v}{D}$	$\frac{\text{hydrodynamic boundary layer thickness}}{\text{mass transfer boundary layer thickness}}$	mass transfer in flowing media
Sh	Sherwood	$\frac{kD}{D}$	$\frac{\text{total mass transfer}}{\text{diffusive mass transfer}}$	mass transfer
T	Thiele modulus	$D\sqrt{\frac{k_r}{D}}$	$\frac{\text{chemical reaction rate}}{\text{diffusive mass transfer}}$	$T = \sqrt{DaII}$ chemical reactions
We	Weber	$\frac{\rho v^2 d}{\sigma}$	$\frac{\text{inertia forces}}{\text{surface tension forces}}$	We = Eo . Fr bubbles and drops

Index

A

B

C

U

V

W

X

Y

Z